Lecture Notes in Biomathematics

Managing Editor: S. Levin

55

Modelling of Patterns in Space and Time

Proceedings of a Workshop held by the Sonderforschungsbereich 123 at Heidelberg July 4–8, 1983

Edited by W. Jäger and J. D. Murray

Springer-Verlag
Berlin Heidelberg New York Tokyo 1984

Editors

Willi Jäger
Institut für Angewandte Mathematik
Im Neuenheimer Feld 294, 6900 Heidelberg, Federal Republic of Germany

James D. Murray
Mathematics Institute, Centre for Mathematical Biology
University of Oxford
24–29 St. Giles', Oxford OX1 3LB, Great Britain

AMS-MOS Subject Classification (1980): 34 C xx, 34 D xx, 35xx, 60 T xx,
73 P xx, 76 E xx, 76 Z xx, 92 A 05, 92 A 09, 92 A 10, 92 A 15, 92 A 17, 92 A 40

ISBN-13: 978-3-540-13892-1 e-ISBN-13: 978-3-642-45589-6
DOI: 10.1007/978-3-642-45589-6

PREFACE

This volume contains a selection of papers presented at the workshop "Modelling of Patterns in Space and Time", organized by the Sonderforschungsbereich 123, "Stochastische Mathematische Modelle", in Heidelberg, July 4-8, 1983. The main aim of this workshop was to bring together physicists, chemists, biologists and mathematicians for an exchange of ideas and results in modelling patterns. Since the mathematical problems arising depend only partially on the particular field of applications the interdisciplinary cooperation proved very useful. The workshop mainly treated phenomena showing spatial structures. The special areas covered were morphogenesis, growth in cell cultures, competition systems, structured populations, chemotaxis, chemical precipitation, space-time oscillations in chemical reactors, patterns in flames and fluids and mathematical methods.

The discussions between experimentalists and theoreticians were especially interesting and effective. The editors hope that these proceedings reflect at least partially the atmosphere of this workshop.

For the convenience of the reader, the papers are ordered alphabetically according to authors. However, the table of contents can easily be grouped into the main topics of the workshop.

For practical reasons it was not possible to reproduce in colour the beautiful pictures of patterns shown at the workshop. Since a larger number of half-tone pictures could be included in this volume, the loss of information has, however, been kept to a minimum.

The workshop has already stimulated cooperation between its participants and this volume is intended to spread this effect.

We would like to thank all participants that contributed to the success of the workshop and also the authors of these proceedings who have not only summarized their results but also initiated new research. The assistance of P. Große in the preparation of this volume is also gratefully appreciated. Last but not least, we acknowledge the support of the Deutsche Forschungsgemeinschaft in sponsoring the workshop and thus making this volume possible.

Heidelberg, July 1984 The editors

LIST OF AUTHORS

ALT, W.; Sonderforschungsbereich 123, Universität Heidelberg, Im Neuenheimer Feld 293, D-6900 Heidelberg

AVNIR, D.; Department of Organic Chemistry, Hebrew University of Jerusalem, Jerusalem 91904, Israel

BOON, J.-P.; Faculté des Sciences, C.P. 231, Université Libre de Bruxelles, B-1050 Bruxelles, Belgium

BROWN, R.A.; Department of Chemical Engineering and Materials Processing Center, Massachusetts Institute of Technology, Cambridge, MA 02139, USA

BURGESS, A.E.; Department of Chemistry, Glasgow College of Technology, Cowcaddens Road, Glasgow G4 OBA, U.K.

BUSSE, F.H.; Department of Earth and Space Sciences and Institute of Geophysics and Planetary Physics, University of California, Los Angeles, CA 90024, USA

CHILDRESS, St.; Courant Institute of Mathematical Sciences, New York University, New York, N.Y. 10012, USA

COOMBS, J.P.; Department of Microbiology, University College, Newport Road, Cardiff DF2 1TA, Wales

ERNEUX, Th.; Department of Engineering Sciences and Applied Mathematics, Northwestern University, Evanston, Illinois 60201, USA

FIFE, P.C.; Department of Mathematics, University of Arizona, Tucson, Arizona 85721, USA

GOMATAM, J.; Department of Mathematics, Glasgow College of Technology, Cowcaddens Road, Glasgow G4 OBA, U.K.

GYLLENBERG, M.; Helsinki University of Technology, Institute of Mechanics, SF-02150 Espoo 15, Finland

HARRIS, A.K.; Department of Biology, Wilson Hall (046A), University of North Carolina, Chapel Hill, North Carolina 27514, USA

HERPIGNY, B.; Faculté des Sciences, C.P. 231, Université Libre de Bruxelles, B-1050 Bruxelles, Belgium

HOPPENSTEADT, F.C.; Department of Mathematics, University of Utah, Salt Lake City, Utah 84112, USA

HUDSON, J.L.; Department of Chemical Engineering, University of Virginia, Charlottesville, VA 22901, USA

JAFFE, S.; Department of Microbiology, University College, Newport Road, Cardiff CF2 1TA, Wales

JÄGER, W.; Institut für Angewandte Mathematik, Universität Heidelberg, Im Neuenheimer Feld 294, D-6900 Heidelberg

KAGAN, M.C.; Department of Organic Chemistry, Hebrew University of Jerusalem, Jerusalem 91904, Israel

KEENER, J.P.; Department of Mathematics, University of Utah, Salt Lake City, Utah 84112, USA

KEMMNER, W.; Institut für Zoologie II, Universität Heidelberg, Im Neuenheimer Feld 230, D-6900 Heidelberg

KESHET, Y.; Department of Applied Mathematics, Weizmann Institute of Science, 76100 Rehovot, Israel

LAUFFENBURGER, D.A.; Department of Chemical Engineering, University of Pennsylvania, Philadelphia, PA 19104, USA

MAREK, M.; Prague Institute of Chemical Technology, 16628 Prague 6, ČSSR

MATKOWSKY, B.J.; Department of Engineering Sciences and Applied Mathematics, Northwestern University, Evanston, Illinois 60201, USA

MEINHARDT, H.; Max-Planck-Institut für Virusforschung, D-7400 Tübingen

MEISELS, E.; Department of Organic Chemistry, Hebrew University of Jerusalem, Jerusalem 91904, Israel

MÜLLER, S.C.; Max-Planck-Institut für Ernährungsphysiologie, Rheinlanddamm 201, D-4600 Dortmund

MURRAY, J.D.; Centre for Mathematical Biology, Mathematical Institute, University of Oxford, Oxford OX1 3LB, England

NICOLAENKO, D.; Los Alamos National Lab., Los Alamos, USA

NISHIURA, Y.; Kyoto Sangyo University, Kyoto 603, Japan

ODELL, G.M.; Department of Mathematics, Rensellaer Polytechnic Institute, Troy, N.Y. 12181, USA

OSTER, G.F.; Department of Biophysics, University of California, Berkeley, CA 94720, USA

PELEG, S.; Department of Computer Science, Hebrew University of Jerusalem, Jerusalem 91904, Israel

PLESSER, Th.; Max-Planck-Institut für Ernährungsphysiologie, Rheinlanddamm 201, D-4600 Dortmund

PÖPPE, Ch.; Sonderforschungsbereich 123, Universität Heidelberg, Im Neuenheimer Feld 293, D-6900 Heidelberg

REISS, E.L.; Department of Engineering Sciences and Applied Mathematics, Northwestern University, Evanston, Illinois 60201, USA

RÖSSLER, O.E.; Institute for Physical and Theoretical Chemistry, University of Tübingen, D-7400 Tübingen

ROSEN, R.; Department of Physiology and Biophysics, Dalhousie University, Halifax, N.S., Canada B3H 4H7

ROTHE, F.; Lehrstuhl für Biomathematik, Universität Tübingen, Auf der Morgenstelle 28, D-7400 Tübingen

SCHAAF, R.; Sonderforschungsbereich 123, Universität Heidelberg, Im Neuenheimer Feld 293, D-6900 Heidelberg

SEGEL, L.A.; Department of Applied Mathematics, Weizmann Institute of Science, 76100 Rehovot, Israel

SHI, J.S.; Department of Mathematics, University of Michigan, Ann Arbor, Michigan 48109, USA

SMOLLER, J.A.; Department of Mathematics, University of Michigan, Ann Arbor, Michigan 48109, USA

TAUTU, P.; Department of Mathematical Models, Institute of Documentation, Information and Statistics, German Cancer Research Center, D-6900 Heidelberg

UNGAR, L.A.; Department of Chemical Engineering and Materials Processing Center, Massachusetts Institute of Technology, Cambridge, MA 02139, USA

VENZL, G.; Institut für Theoretische Physik, Physik-Department der Technischen Universität München, D-8046 Garching

WELSH, B.J.; Department of Mathematics, Glasgow College of Technology, Cowcaddens Road, Glasgow G4 OBA, U.K.

WIMPENNY, J.; Department of Microbiology, University College, Newport Road, Cardiff CF2 1TA, Wales

TABLE OF CONTENTS

CONTRACTION PATTERNS IN A VISCOUS POLYMER SYSTEM

Wolfgang Alt
Sonderforschungsbereich 123
Universität Heidelberg
Im Neuenheimer Feld 293
D-6900 Heidelberg

The explanation of cell locomotion is one of the current biological research topics. During the last decade it became apparent that the active forces which deform the shape of a motile cell (amoeboid cell, fibroblast or leukocyte, for example) and which finally lead to its deplacement are provided by contractile filaments, being present within the cell plasma, in particular near the cell membrane.

Amoeboid locomotion, see figure 1a, has been studied extensively. The mathematical model of G. Odell [7] describes the interaction and transport of filaments in an endoplasmatic flow governed by sol-gel-transitions within various pseudopods of a moving cell. Further extensions of this model include the viscoelastic properties of actomyosin gels and the control by diffusing substances (like Ca^{++}), see the recent work of G. Oster, G. Odell [9] in this volume.

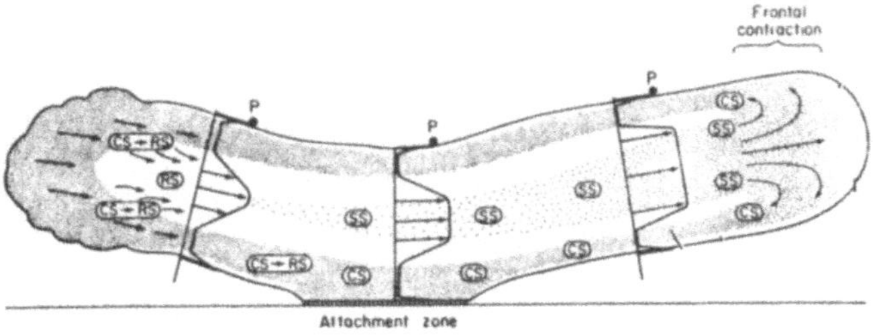

Figure 1a. Amoeboid locomotion, cp. [1].

Figure 1b. Leukocyte locomotion, cp. [10] .

The locomotion of fibroblasts or leukocytes, for example, seems to be based on similar, but slightly different mechanisms, see figure 1b. Periodical protrusion of flat membrane extensions (lamellipods), their attachment to some underlying substrate and their partial withdrawal opposite to the direction of cell displacement (ruffling) suggest that actin filaments, distributed inside the (hyalo-)plasma of the lammellipodium, and their interactions play the most important role, see [8].

In order to understand the possible function of the contractile actomyosin polymer system itself, simple mathematical models should be investigated, which are able to reproduce the following observed phenomena:

(a) formation of (at least transiently) stable contraction centers,

(b) oscillatory competition between different contraction centers.

In a first attempt M. Dembo, F. Harlow and the author [4] proposed a model for a highly viscous "fluid" of actin bundles (or filaments) which attract each other via binding to myosin polymers. The actin bundles are formed (nucleated and polymerized) from a large reservoir of actin monomers, but when density increases they are depolymerized and disassembled. Assuming that mutual attraction, shearing forces (viscosity) and friction dominate other forces (inertia, or elasticity as in [9]) we get the following simplified balance equations for the mean concentration $u(t,x)$ and the mean velocity $v(t,x)$ of polymer bundles, for more details see [2] :

Mass balance:

$$\partial_t u + \nabla_x \cdot (uv) = f(u) \tag{1}$$

where the kinetic growth function

$$f(u) = N_+(u) - u \cdot N_-(u)$$

contains a density dependent nucleation term $N_+(u)$ and a disassembly rate $N_-(u)$. Typically f has exactly one zero, without restriction at u=1, and f decreases for $u \geq 1$ with

$$f'(1) = -\eta < 0 .$$

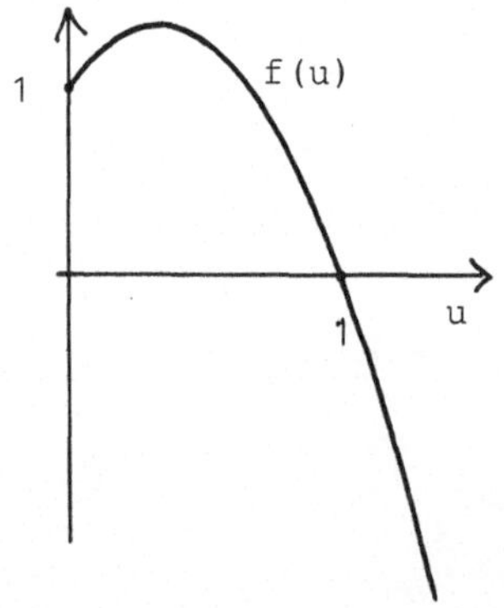

Figure 2. Kinetic growth function f as in (19) with $\rho=3$, $\eta=4$.

Force balance (High viscosity limit)

(2) $$\nabla_x \cdot \{\mu(u)\nabla_x v\} + \nabla_x \psi(u) - \varphi(u)v = 0 .$$

The coefficients in this linear, inhomogeneous elliptic equation for v can be density dependent, typically we assume

Viscosity: $\mu(u) = u^m$, $m \geq 0$

Attractivity: $\psi(u) = \frac{\psi}{p} u^p$, $p > 0$

Friction: $\varphi(u) = \phi \cdot u$.

Neglecting any forces from a possible membrane surrounding the considered domain Ω (induced by interaction with polymers or by hydrostatic pressure) we can suppose the

Boundary condition

(3) $$v = 0 \quad \text{on } \partial\Omega$$

or some periodicity condition, depending on the geometrical configuration.

In the one-dimensional case $\Omega = [b_-, b_+]$ with no friction ($\varphi \equiv 0$) the elliptic problem (2),(3) for the velocity v can be solved:

(4) $$\boxed{\partial_x v = \frac{1}{\mu(u)}\Big(K - \psi(u)\Big) \quad , \quad v(\cdot, b_\pm) = 0}$$

so that the characteristic direction in the hyperbolic equation (1) is a functional of the density $u(t,\cdot)$:

(5) $$v(\cdot, x) = \int_{b_-}^{x} \frac{1}{\mu(u)}\Big(K - \psi(u)\Big) .$$

Here the integration constant $K = K(t)$ is

(6) $$K = \int_\Omega \frac{\psi(u)}{\mu(u)} \Big/ \int_\Omega \frac{1}{\mu(u)} .$$

Using (4) the hyperbolic equation (1) can be rewritten as

(7) $$\boxed{\partial_t u + v \cdot \partial_x u = \frac{u}{\mu(u)}\Big(\psi(u) - K\Big) + f(u) =: F_K(u)} .$$

In the following we will investigate the steady state solutions of system (4), (7) by analytic means. Afterwards their stability properties and the occurrence of oscillatory solutions are illustrated by numerical examples.

Steady Contraction Patterns (1-dimensional)

The unique constant steady state $u_* \equiv 1$, $v_* \equiv 0$ with $K_* = \frac{\psi}{p}$ is easily seen to be locally stable iff

(8) $$\psi - \eta = F'_{K_*}(1) < 0 .$$

Thus, instability occurs if the attractivity factor ψ exceeds the depolymerization factor η.

Nontrivial steady states are described by the corresponding degenerate ODE-system (4), (7). Looking for monotone (increasing) positive solutions $u = u(x)$ on $\Omega = [b_-, b_+]$ we are lead to an ODE for $v = v(u)$, $u_- < u < u_+$, $u_\pm = u(b_\pm)$, which can be solved explicitly by

(9) $$v = c \cdot v_K(u) = c \cdot \exp \int_1^u \frac{K - \psi(z)}{\mu(z) \cdot F_K(z)} dz$$

with an arbitrary constant $c > 0$. In order to fulfill the boundary conditions for v we must have

(10) $$F_K(u_\pm) = 0 \quad , \quad 0 < u_+ < u_- \quad ,$$

and

(11) $$F_K(u) > 0 \quad \text{for} \quad u_+ < u < u_- \quad ,$$

where the constant K has to be chosen with

(12) $$\psi(u_-) < K < \psi(u_+) \quad .$$

Equation (7) then gives

(13) $$\frac{du}{dx} = \frac{F_K(u)}{c \cdot v_K(u)} \quad ,$$

so that the constant c is proportional to the length of the interval

(14) $$L = b_+ - b_- = c \int_{u_-}^{u_+} \frac{v_K(u)}{F_k(u)}$$

whereas the free constant K is determined by the integrability condition for (1), namely

(15) $$0 = \frac{1}{c} \int_\Omega f(u)dx = \int_{u_-}^{u_+} \frac{v_K(u)\cdot f(u)}{F_K(u)} du \quad .$$

From (9) and (13) we deduce the asymptotic behavior:

(16) $$v_K(u) \sim |u_\pm - u|^{\alpha_\pm} \qquad \text{for } u \to u_\pm$$

and

(17) $$|u_\pm - u(x)| \sim |b_\pm - x|^{1/\alpha_\pm} \qquad \text{for } x \to b_\pm \; ,$$

where

(18) $$\alpha_\pm = \frac{K - \psi(u_\pm)}{\mu(u_\pm)\cdot F_K'(u_\pm)} > 0 \; .$$

Example: Let us consider the special case of a quadratic growth function (see figure 2)

(19) $$f(u) = \eta(1-u) - \rho(1-u)^2$$

with $0 \leqq \rho < \eta$, a quadratic attractivity ($p = 2$) and a linear viscosity ($m = 1$). Then

(20) $$F_K(u) = \frac{\psi}{2} u^2 - K + f(u) = a(u-u_-)(u_+ - u)$$

with

$$a = \rho - \frac{\psi}{2} \; , \quad b = 2\rho - \eta, \; c = K - \eta + \rho$$

and

(21) $$u_\pm = \frac{1}{2a}(b \pm \sqrt{b^2 - 4ac}) \; .$$

Conditions (10) and (11) are satisfied if a,b and c are positive. Together with condition (12) this means that the free parameter K can vary in the range

(22) $$0 < \eta - \rho < K < \frac{\psi}{2}$$

and the other parameters have to obey the inequalities

(23) $$\boxed{2(\eta - \rho) < \psi \leqq \eta < 2\rho}$$

or, in the "unstable" case,

(24) $$\boxed{\rho < \eta < \psi < 2\rho} \, .$$

Since F_K in (20) is quadratic, we obtain an explicit integral in (9)

(25) $$v_K(u) = (u - u_-)^{\alpha_-}\cdot(u_+ - u)^{\alpha_+}\cdot u^{-\frac{K}{au_+u_-}}$$

so that the integrability condition (15) is equivalent to

$$0 = \phi(K) = \int_{u_-}^{u_+} (u-u_-)^{\alpha_- -1} \cdot (u_+ - u)^{\alpha_+ -1} \cdot u^{-\frac{K}{au_+u_-}} \cdot f(u)\,du \;. \tag{26}$$

This is a <u>nonlinear equation for K</u> since $u_\pm$ in (21) and

$$\alpha_\pm = \pm \frac{\frac{\psi}{2}u_\pm^2 - K}{a(u_+-u_-)}$$

depend on K. By computing $\phi(K)$ for the boundary values of the range of K in (22) we can get a sufficient criterion for the existence of a solution of (26):

<u>LEMMA 1</u>: Under condition (23) or (24) we have

$$\lim_{K \nearrow \psi/2} \phi(K) \begin{cases} > 0 & \text{if } \psi < \eta \\ < 0 & \text{if } \psi > \eta \end{cases}$$

and

$$\lim_{K \searrow (\eta-\rho)} \phi(K) = \phi_*$$

with

$$\phi_* = \int_0^{\frac{b}{a}} u^{\frac{a}{b^2}(\eta-\rho)-1} \left(\frac{b}{a} - u\right)^{\alpha_+ -1} f(u)\,du \;. \tag{27}$$

Recall that $a = \rho - \psi/2$ and $b = 2\rho - \eta$. Here $\alpha_+ = (\frac{\psi}{2}\, b^2/a^2 - \eta + \rho)/b$.

Thus we conclude the

<u>PROPOSITION</u>. *In the example with* $p = 2$, $m = 1$ *and* f *as in (19) there exists a strictly <u>increasing steady state</u>* u *with positive* v *on* Ω *if the parameter values* ψ, η *and* ρ *with (23), (24) satisfy*

$$\phi_* \cdot (\psi - \eta) > 0. \tag{28}$$

In analogy there exists a decreasing steady state, see figure 3.

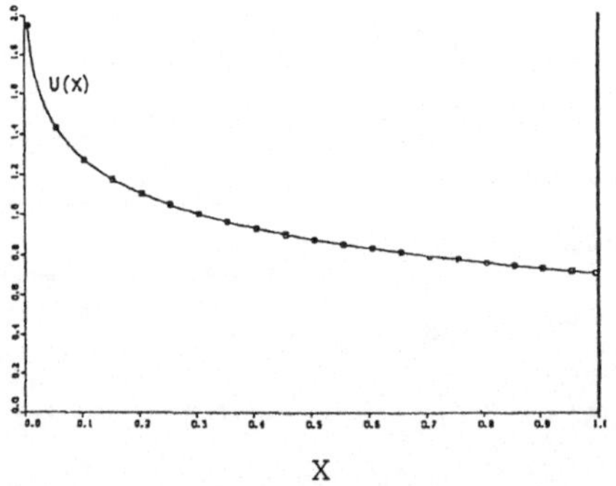

<u>Figure 3</u>. Monotone decreasing steady contraction pattern obtained as asymptotic state for a time dependent solution of (4),(7). $\rho=3$, $\eta=4$, $\psi=5$.

This steady contraction pattern corresponding to a zero of Φ in (26) exists on any interval of arbitrary length, see (14). Such a scaling invariance property, which generally holds for system (1), (2) with vanishing friction $\varphi \equiv 0$, might be important for the biological function of the actomyosin system. It is not found in pattern formations described by autonomous reaction-diffusion-models, cp. [5], [9].

Bifurcation analysis

Consider the general situation with nonlinearities f, μ and ψ as before ($\varphi \equiv 0$). We want to analyse the behavior of system (4), (7) near the parameter value $\psi = \eta$, where the constant steady state $u_* \equiv 1$, $v_* = 0$ changes stability, see (8). For

$$\varepsilon = \psi - \eta$$

introduce the new variables

$$(29) \qquad \begin{cases} u = 1 + \varepsilon z \\ v = \varepsilon w \\ K = K_* - \varepsilon\lambda_o - \varepsilon^2\lambda \end{cases}$$

and the slow time transformation

$$\tau = |\varepsilon| t .$$

Then after an initial layer correction

$$(30) \qquad z_o(t,x) = \frac{z(o,x)}{1 + \eta\overline{z(o,\bullet)}t}$$

with corresponding $w_o(t,x)$ and $\lambda_o(t) = -\eta\overline{z_o(t,\bullet)}$, system (4), (7) in zeroth order approximation is equivalent to

$$(31) \qquad \begin{cases} \pm\partial_\tau Z + W\partial_x Z = (1 + \frac{\lambda_o}{\varepsilon})Z - aZ^2 + \lambda \\ \partial_x W = -\eta Z \\ W(\cdot, b_\pm) = 0 \end{cases}$$

Here ($\pm$) refers to the signum of $\varepsilon = \psi - \eta$, and

$$(32) \qquad a = -(\frac{p-1}{2} + m - 1)\eta - \frac{1}{2}f''(1) .$$

Because of (31) we must have $\overline{Z} = 0$, so that necessarily

$$\lambda(\tau) = (a+\eta)\int_{b_-}^{b_+} Z^2(\tau,x)\,dx \;.$$

For each finite T_o the approximations u,v,K in (29) with $z(t,\cdot) = z_o(t,\cdot) + Z(\frac{\tau}{|\varepsilon|},\cdot)$ and $w(t,\cdot) = w_o(t,\cdot) + W(\frac{\tau}{|\varepsilon|},\cdot)$ are valid modulo $O(\varepsilon^2)$ for all times t with

$$\begin{cases} 0 \leqq t \leqq \dfrac{T_o}{\varepsilon} & \text{if } \varepsilon > 0 \\ 0 \leqq t < \infty & \text{if } \varepsilon < 0 \;. \end{cases}$$

The steady states of the approximate system (31) with $\lambda_o = 0$ can be analysed in the same way as in the example above, since the right-hand side is a quadratic polynomial in Z:

$$P_\lambda(Z) = a(Z - Z_-)(Z_+ - Z)$$

with two zeros $Z_- < 0 < Z_+$, namely

$$Z_\pm = \frac{1}{2a}(1 \pm \omega) \;, \quad \omega = \sqrt{1+4a\lambda}.$$

In analogy to the conditions on F_K we get the following solutions, provided a in (32) is positive:

$$W = C\,(Z - Z_-)^{\alpha_-}(Z_+ - Z)^{\alpha_+} \tag{33}$$

with

$$\alpha_\pm = \frac{\eta}{2a}\left(1 \pm \frac{1}{\omega}\right) > 0 \;. \tag{34}$$

The constant C is determined by the length $L = b_+ - b_-$ according to (14) whereas the analogous integrability condition (15) is now replaced by $\overline{Z} = 0$, namely

$$0 = \Psi(\lambda) = \int_{Z_-}^{Z_+} Z\cdot(Z - Z_-)^{\alpha_- - 1}\,(Z_+ - Z)^{\alpha_+ - 1}\,dZ \tag{35}$$

This is a nonlinear equation for λ in the range $0 < \lambda < \infty$. An analysis similar to lemma 1 in the example above leads to

LEMMA 2. Under the condition $a > 0$ we have

$$\lim_{\lambda \searrow 0} \Psi(\lambda) > 0$$

and

$$\lim_{\lambda\to\infty} \lambda^{\frac{1}{2}(1-\frac{\eta}{a})} \Psi(\lambda) = a^{-\frac{\eta}{2a}} \cdot \Psi_o$$

with

$$(36) \qquad \Psi_o = \int_0^1 (1 + \frac{\eta}{2a} r \cdot \ln \frac{1-r}{1+r})(1 - r^2)^{\frac{\eta}{2a} - 1} dr .$$

By the mean value principle applied to (35) we conclude the

THEOREM: (Bifurcation criterion)

For general nonlinearities as in (1), (2) with vanishing friction ($\varphi \equiv 0$), $a > 0$ *in (32) and*

$$(37) \qquad \Psi_o < 0 ,$$

there exists at least one smooth, local branch of nontrivial monotone (increasing) steady contraction states parametrized over $\varepsilon = \psi - \eta$

$$(38) \qquad \begin{aligned} u_\varepsilon &= 1 + \varepsilon Z + O(\varepsilon^2) \\ v_\varepsilon &= \varepsilon W + O(\varepsilon^2) \\ K_\varepsilon &= K_* - \varepsilon^2 \lambda + O(\varepsilon^3) \end{aligned}$$

Here (Z,W,λ) *is a stationary solution of (31) with* $\lambda_0 = 0$ *corresponding to a zero of (35).* W *and* Z *have the analogous asymptotic properties as in (16), (17) with* $\alpha_\pm$ *defined in (34).*

Figure 4. Bifurcation diagram for steady state solutions (—). The additional branch of periodic solutions (ooo) is suggested by simulations, see (B) below.

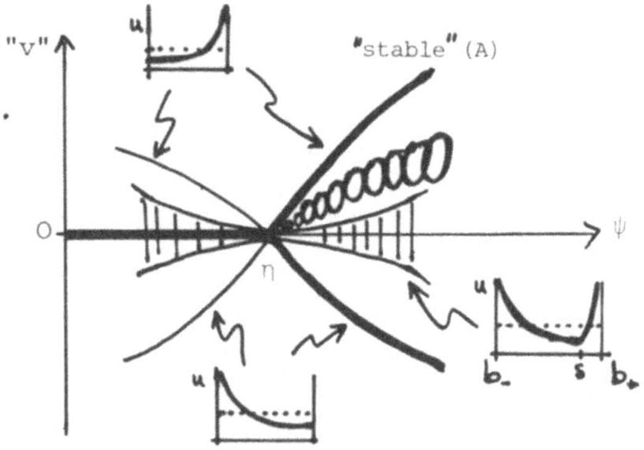

The resulting bifurcation diagram (figure 4) contains not only the two branches of monotone in- or decreasing solutions with one contraction center at the boundary, but also a continuum of bimodal solutions with two contraction centers. They can easily be constructed by composing a monotone decreasing solution on $[b_-,s]$ and an increasing solution on $[s,b_+]$ belonging to the same constant K_ε.

Besides the fact that positivity, monotonicity and regularity of the density distributions are preserved for finite times, see [2], [4], not much is known analytically about the asymptotic behavior of system (4), (7) and the local stability of the steady contraction patterns described above. In the situation of the example we always get an upper solution $u(t,\cdot) \equiv U(t)$ satisfying

$$\dot{U} = \frac{\psi}{2}U^2 + f(U)$$

and converging to u_+^o in (21) with $K = 0$, provided $\psi < 2\rho$, which is the necessary condition for the existence of steady contraction patterns, see (22) and (23). However, this a-priori boundedness of all solutions does not imply any convergence result since the hyperbolic system (4), (7) has no compactness properties.

Nevertheless, all numerical simulations of system (4), (7) show, under the conditions above, convergence either to steady contraction patterns or to periodic solutions. In particular we can formulate the following "empirical" rules:

(A) Monotone (e.g. decreasing) contraction patterns, which are formed as final steady states from monotone initial data, are stable under perturbations creating a slope $u_o'(b_+) \leq 0$ at the low-density boundary b_+. The perturbations are attracted and finally absorbed by the contraction center at b_- (figure 5).

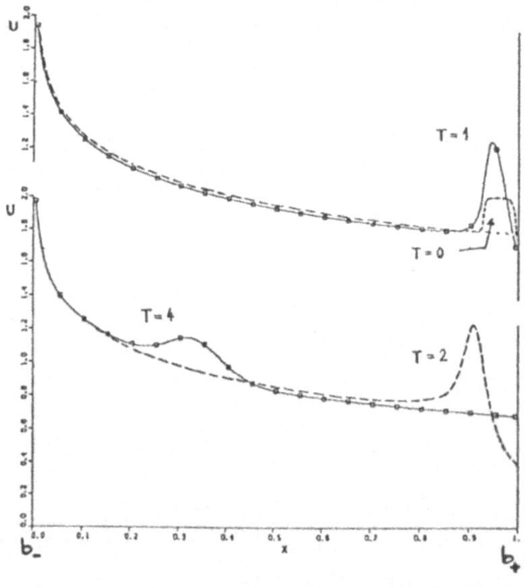

Figure 5. Local perturbation (T=0) of the steady contraction pattern in figure 4, growing to a transient secondary contraction center, which is pulled to the left by the primary center.

(B) Non-monotone initial data with at least two maxima usually lead to bimodal density distributions with two competing contraction centers at the boundaries. Their masses and the velocity between them finally undergo regular oscillations , [4: Fig. 12b]. Perturbations with one hump usually are attracted and absorbed by one of the two contraction centers (figure 6).

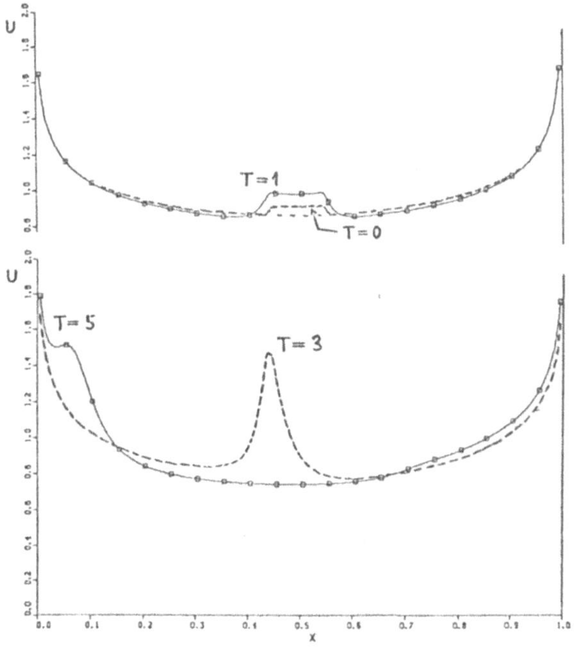

Figure 6. Local non-symmetric perturbation (T=0) of a bimodal pattern with two (slightly oscillatory) competing contraction centers. It first grows to a quite strong autonomous center, before it is pulled by and unified with the left marginal contraction center.

The numerical simulations use a semidiscretization method and have been produced by M. Dembo (1983, Los Alamos Scientific Laboratory), whom I want to thank for initiation and continuing cooperation in this project.

As models for pattern formation of epithelial or mesenchymal cell sheets, similar systems have been proposed, see the contribution of J. Murray [6] in this volume. Very closely related to equations (4),(7) is the system analysed by S. Childress and J.K. Percus [3: equations (25), (27) and (31)].

REFERENCES

1. ALLEN, R.D. and TAYLOR, D.L.: The molecular basis of amoeboid movement. In:"Molecules and Cell Movement". Eds. S. Inoué, R.E. Stephens. Raven Press, New York 1975

2. ALT, W. and DEMBO, M.: A contraction-disassembly model for intracellular actin gels. In: Proc. Equadiff. Conf. Würzburg, Aug. 1982. To appear in Springer Lect. Notes

3. CHILDRESS, S. and PERCUS, J.K.: Modeling of cell and tissue movements in the developing embryo. Lect. Math. in the Life Sciences, 14, S. 59-88 (1981)

4. DEMBO, M., HARLOW, F. and ALT, W.: The biophysics of cell surface motility. In: "Cell Surface Phenomena". Eds. Delisi, Wiegel, Perelson. Marcel Dekker, to appear 1983

5. MEINHARDT, H.: Models of biological pattern formation. Acad. Press 1982

6. MURRAY, J.: Mechanical models of morphogenesis: Mesenchymal patterns. (This volume)

7. ODELL, G.M. and FRISCH, H.L.: A continuum theory of the mechanics of amoeboid pseudopodium extension. J. Theor. Biol. 50, 59-86 (1975)

8. ONISHI, R.: Cinematographic analysis of ruffling movement of L cells. In:"Cell Motility, Book B". Cold Spring Harbor Conf. on Gel Prolif. Vol. 3. Eds. Goldman, Pollard and Rosenbaum. S. 251-261, 1976

9. OSTER, G.F. and ODELL, G.M.: A mechanochemical model for plasmodial oscillations in physarum. (This volume)

10. STOSSEL, T.P.: The mechanism of leukocyte locomotion. In: "Leucocyte Chemotaxis". Eds. Gallin, Quie. S. 143-160. Raven Press, New York 1978

FORMATION AND EVOLUTION OF SPATIAL STRUCTURES IN CHEMOTACTIC BACTERIAL SYSTEMS

Jean-Pierre BOON and Brigitte HERPIGNY
Faculté des Sciences, C.P. 231
Université Libre de Bruxelles
B -1050 Bruxelles, Belgium

1. INTRODUCTION

From the "conjecture that any motile organism is tactic to at least one stimulus" (Segel, 1983) one may infer that the function of motility - and of altered motility - is the optimization of ecological environment. In particular, chemotaxis induces organisms to move to "better" regions. For instance the dynamics of a motile bacterium will be modified in a chemically inhomogeneous medium, such that the cell undergoes a net displacement towards regions richer in attractant (see e.g. Lauffenburger, 1983) ; then biased cell motion, in sufficiently dense populations, may result into cooperative phenomena producing spatial and temporal structures in chemotactic bacterial systems (for a review see e.g. Adler, 1978 ; see also Adler, 1966, and Koshland, 1977). An example is given in Fig. 1 which shows bacteria E. coli forming a migrating band in response to the glucose gradient created in the medium by the bacteria themselves.

Structures developed by chemotactic bacterial populations have been known for quite some time. Various experimental studies have been concerned with traveling bands (Adler, 1966), migrating rings (Adler, 1966) and non-uniform distributions (Dahlquist, Lovely, and Koshland, 1972) ; more recently laser light scattering spectroscopy has been used to investigate the detailed dynamics of bacterial motion (for a review see Chen and Hallett, 1982 ; see also Boon, 1983). Correspondingly, theoretical work has developed and mathematical models have been presented for the analysis of experimental observation (Keller and Segel, 1971 ; Nossal, 1972 ; Lovely and Dahlquist, 1975 ; for reviews see Boon, 1975 ; Segel, 1978 ; Nossal, 1980 ; Lapidus and Levandowsky, 1981).

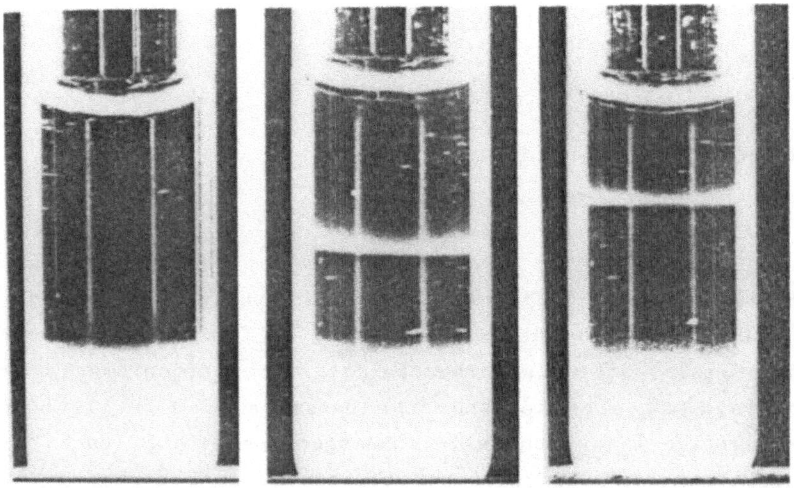

(a) (b) (c)

FIGURE 1. Migration band of E. coli. (a) initial time : a suspension of E. coli is layered in the lower part of a cell containing EDTA with glucose (10^{-4} M) as attractant ; (b) after 105 minutes and (c) after three hours : by consuming the glucose, bacteria create an attractant gradient in response to which the cells migrate towards the region richer in glucose and form a traveling band . Height of cell is 3 cm (Herpigny, 1983).

The Keller and Segel model (1971) provides the basic set of equations for the development of theoretical analyses ; such analyses have been rather successful for the interpretation of experimental situations where a bacterial population responds chemotactically to one single attractant or when the system is modeled as such. Predictions become less obvious when the complexity of the system increases, for instance if more than one attractant is present or if competition occurs between species. Leaving aside the competition-type problem (see e.g. Lauffenburger and Calcagno, 1982) we consider the first class of systems where the key points in the analysis are the multiplicity of the mechanisms involved and the modeling of the response function. In each case, the evolution of the system is triggered by an initial spatial inhomogeneity, whether in the distribution of one of the substrates or in the bacterial density distribution. Once the system has evolved, it will develop a spatial structure as a steady state phenomenon in open or infinite systems whereas in closed systems such structures will appear as transients. Of the latter class let us mention a simple example. An initially uniformly distributed suspension of S. typhymurium responds chemotactically to a step gradient of serine by moving up gradient, thereby forming a bacterial distribution profile with a peak centered around the location of the steepest change in attractant (Dahlquist, Lovely, and Koshland, 1972). In a similar experiment inspired by the one just described, a suspension of E. coli is subject to a glucose concentration gradient in an oxy-

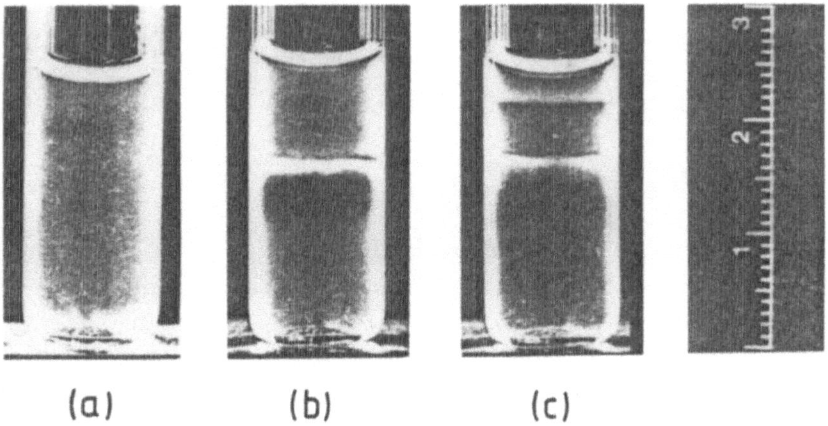

FIGURE 2. Band formation in E. coli suspension. (a) initial time : bacteria are uniformly distributed in an oxygen satured (2×10^{-4} M) medium with a glucose gradient ($0 - 10^{-3}$ M) at mid heigh of the cell ; (b) after 10 minutes : a bacterial band forms in the upper zone where glucose is initially absent ; (c) after 37 minutes : two bands have appeared in the upper half of the cell. Full scale on the right is 3 cm. (Boon and Herpigny, 1982).

gen satured medium ; the system develops a rather complex and somewhat unexpected structure and reaches a (quasi-steady) state where two bacterial bands separated by a depleted zone have formed in the region where glucose is initially absent, as shown in Fig. 2. This constitutes one of the cases that will be analyzed on the basis of the theory presented in section 2. Applications will be discussed in section 3.

2. THEORY

Our theoretical analysis starts from the two basic equations initially put forward by Keller and Segel for the modeling of cellularaggregation in amoebae (1970) and of traveling bands of chemotactic bacteria (1971). The theory presented here was developed by Boon and Herpigny (1982) as a generalization of the Keller and Segel model, based on the phenomenological equations for the conservation of bacterial density

$$\partial_t b + \nabla . \underline{J} = G \qquad (1)$$

and for the evolution of substrate (attractant) concentration

$$\partial_t c = D \nabla^2 c - k b \qquad (2)$$

Here $b \equiv b(\underline{r}, t)$ is the bacterial density ; G, the bacterial growth term ; and $\underline{J} \equiv \underline{J}(\underline{r}, t)$, the sum of the current due to motility, $\underline{J}_\mu = -\mu \nabla b$, with μ the motility coefficient, and of the chemotactic current $\underline{J}_c = \underline{v}_c b$, with $\underline{v}_c$, the chemotactic velocity. In the second equation $c \equiv c(\underline{r}, t)$ is the attractant concentration ; on the r.h.s. the first term is the diffusion term, with D, the diffusion coefficient, and the second term describes substrate consumption, with k, the consumption rate factor. When more than one substrate is present, additional equations of type (2) are required for the description of the system. Because of the functional form of the chemotactic term (with $\underline{v}_c = f(c)$) and of the consumption term (with $k = f(c)$), these equations, subject to the appropriate initial and boundary conditions, become highly non-linear and may have different branches of solutions indicative of various types of system behavior (For nonlinear aspects of the Keller and Segel model, see also Childress and Percus, 1983).

The generalized model presented here concerns a system with one bacterial population and two attractants in linear geometry ; 3-d extension to several species and more than two attractants (and/or repellents) should be straightforward, although mathematically more involved. When expressed in terms of reduced quantities the new set of equations reads (Boon and Herpigny, 1982)

$$\partial_\tau B = \partial_x^2 B - \partial_x (B\, \partial_x F_c) \quad ; \quad \partial_x F_c = V_c (C_i , C_j) \tag{3}$$

$$\partial_\tau C_i = \alpha_i\, \partial_x^2 C_i - \gamma_i (f_i\, g_j + \beta_{ij}\, f'_i\, f_j)\, B \tag{4}$$

where the subscripts i, j = 1, 2 refer to the two attractants. Before going into a detailed analysis of the various functions entering the equations, we rewrite the above set in block diagram form so as to clarify the physical meaning of the various terms. Each equation is based on a conservation equation with a diffusion term and a systematic flux term ; their interpretation follows from the various mechanisms governing the evolution of the system. For the sake of illustration we consider a bacterial system with two attractants, oxygen (subscript 1) and glucose (subscript 2), as described in section 1.

$\dot{B}$ = [MOTILITY FLUX] + [CHEMOTACTIC FLUX] (3a)

$\dot{C}_1$ = [OXYGEN DIFFUSION] + [O_2 CONSUMPTION FOR GL. METABOLISM + ENDOGENEOUS RESPIRATION] (4a)

$\dot{C}_2$ = [GLUCOSE DIFFUSION] + [AEROBIC METABOLISM + FERMENTATION] (4b)

The scaled variables are $x = z v_0/\mu$ for the spatial coordinate, $\tau = t v_0^2/\mu$ for the time variable, where v_0 measures the amplitude of the chemotactic response (see Appendix). $B = b(x,\tau)/b_0$ is the reduced bacterial density and $C_1 = c_1(x,\tau)/c_{1,0}$ and $C_2 = c_2(x,\tau)/c_{2,0}$ are the reduced oxygen and glucose concentrations respectively ; the subscript 0 denotes the maximum initial time value. The parameter

$\alpha_i = D_i/\mu$ measures the reciprocal efficiency of motility versus diffusion. The scaled chemotactic velocity V_c, a function of the concentrations C_1 and C_2, is called the response function, defined as the gradient of the chemotactic potential F_c (see Appendix). F_c should be modelled such that the response function be in agreement with experimental observation (Mesibov, Ordal, and Adler, 1973) showing that in the low concentration domain the chemotactic response is not triggered $(F_c \simeq 0)$ and in the upper concentration zone saturation inhibits chemotaxis $(\partial_x F_c \simeq 0)$. These requirements are satisfied by the model function introduced by Boon and Herpigny (1982) ; with the additivity assumption for non-competitive attractants, one has

$$F_c = \sum_i F_i \quad ; \quad F_i = (q_i C_i)^2 (1 + (q_i C_i)^2)^{-1} \qquad (5)$$

with $q_i = c_{i,o} / c_i^s$, where c_i^s is a "threshold" concentration that follows from the sigmoid form of the chemotactic function and whose value can be evaluated from experimental data (Herpigny, 1983).

Substrate consumption can occur according to two mechanisms : simultaneous consumption of the two substrates (here aerobic metabolism of glucose) and consumption of one of the substrates in the absence of the other (here fermentation and endogeneous respiration). These mechanisms are described by the last two terms on the rhs of Eq. 4. Consumption of glucose in the presence of oxygen, and reciprocally, is expressed by $f_1' f_2$, and $f_2' f_1$, where the f's are hyperbolic functions according to Michaelis-Menten consumption mechanism (see e.g. Lin and Segel, 1974)

$$f_i = (n_i C_i)(1 + n_i C_i)^{-1} \qquad (6)$$

with $n_i = c_{i,o} / c_i^*$, such that for low concentrations ($c_i \ll c_i^*$), the regime is linear, and at high concentrations $(c_i \gg c_i^*)$ saturation occurs. In a region where j (or i) is absent, because it was never present or because the medium has been locally depleted, i (or j) is consumed alone. The latter mechanism is taken into account by the term $f_i g_j$, where g_j is an inhibition function of the type

$$g_j = (1 + (p_j C_j)^\eta)^{-1} \quad ; \quad \eta \geq 4 \qquad (7)$$

with $p_j = c_{j,0} / c_j^-$; when the concentration of j drops below the value c_j^-, g_j switches on f_i to trigger endogeneous respiration (i = 1, j = 2) or fermentation (i = 2, j = 1).

The mechanisms described above are governed by the chemical kinetics constants with scaling $\beta_{ij} = k_{ij} / k_i^\infty$ and $\gamma_i = k_i^\infty b_0 \mu c_{i,0}^{-1} v_0^{-2}$, where k_{ij} is the rate constant for the consumption of i in the presence of j and k_i^∞, the rate constant for the consumption of i alone. Note that the motility coefficient μ depends on the mechanical and physiological conditions ; in particular, since there are local changes in the medium, $\mu = \mu(\nu, c_i, c_j)$, with ν the kinematic viscosity. However in the applications discussed in section 3, μ can be considered as a constant, as it was shown by dynamical light scattering experiments that no substantial variation of μ occurs in the concentration ranges of interest. (Herpigny, 1983)

Eqs. (3) and (4) with the functions (5), (6) and (7) constitute the basic set of the general model. This set of equations must be complemented by initial and boundary conditions for B, C_1, and C_2 relevant to the specific problem considered. In the next section, we present applications, starting with the experiment described in Section 1.

3. APPLICATIONS

Application of the theoretical model requires evaluation of the parameters entering the equations and specification of initial conditions.The computation of the experimental parameters is discussed elsewhere (Boon and Herpigny, 1982 ; Herpigny, 1983). Three specifically different situations will be considered here and classified according to the initial state of the system (see Table 1)

Experiments were performed with E. coli suspensions in a vertical cylindrical cell (diam. 1 cm, height 5 cm) and the spatial profile of the bacterial density distribution was measured as a function of time by integrated light scattering intensity from a laser beam swept vertically through the cell (Herpigny, 1983).

TABLE 1 : Initial conditions			Long time
Bacterial density	Oxygen concentration	Glucose concentration	Observed phenomenon
A : uniform	uniform	step gradient at mid-cell	Band formation
B : inoculum	uniform	uniform	Band migration
C : uniform	gradient at meniscus	_	Accumulation

In the first experimental situation (denoted A in Table 1), bacteria are uniformly distributed in an oxygen satured medium with a glucose concentration gradient at mid-height of the cell (Fig.2a). Because, as soon as the sample is prepared, diffusion tends to round off the step gradient, the initial glucose gradient is modeled, in the computation, by the Fermi function, $C_2(x,\tau=0) = (1+\exp ax)^{-1}$, with $ax = z/\bar{z}$, where $\bar{z}$ measures the spatial range of the initial gradient. Experimental and theoretical results are shown in Fig. 3 for three typical states of the system, as time evolves. (Figs. 3.a, b, and c correspond to the photographs of Fig.2.a, b, and c respectively) Good agreement is found between the experimental data and the theoretical profile of the bacterial density distribution.

Now in order to explain the system evolution, further analysis is necessary. The keypoint in the analysis is the response function V_c given by Eqs. (A.4) and (5) ; $V_c(x)$ is shown in Fig. 4 along with the bacterial density profile B(x) and the corresponding evolution of the oxygen and glucose concentration profiles ($C_1(x)$ and $C_2(x)$ respectively). The instantaneous chemotactic response to the glucose gradient is reflected by a negative peak in V_c (Fig. 4, $\tau = 0.1$) : the x-axis being oriented from left to right, the area where V_c is negative is a zone where bacteria move chemotactically from right to left (bacteria swimming downwards in the actual experiment) towards the region richer in glucose. Since aerobic metabolism is most favorable, oxygen will be consumed simultaneously with glucose and so an

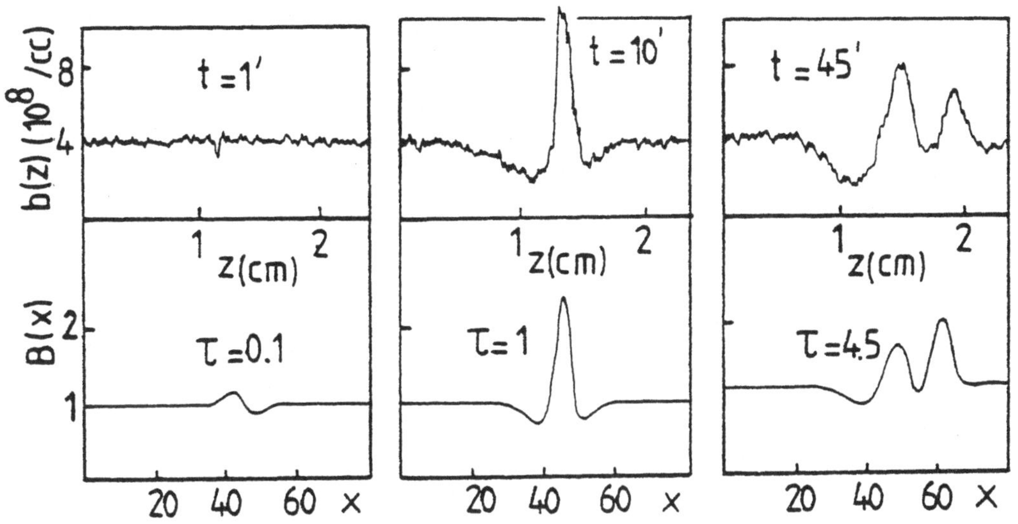

FIGURE 3. Bacterial density profiles for the experiments shown in Fig. 2. Upper traces are experimental (the spatial coordinate is along the vertical axis of the cell (z in cm) and times are in minutes) ; lower curves are theoretical (x and τ are the reduced space and time variables respectively with scaling : x = 33 for z = 1 cm and τ = 0.1 for t = 1 min.)

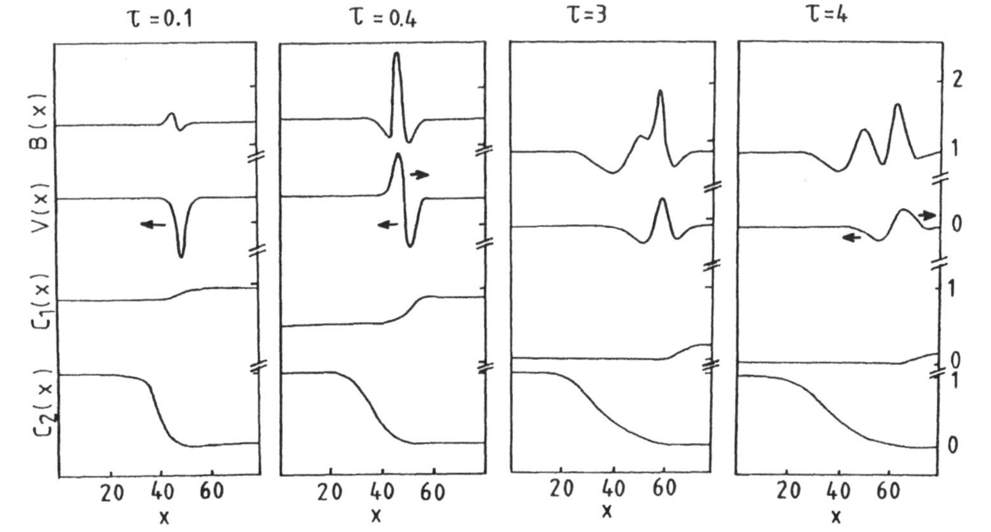

FIGURE 4. Theoretical model computation for the experiment shown in Fig. 2. B(x) : bacterial density distribution ; V(x) : chemotactic response function ; $C_1(x)$: oxygen concentration ; $C_2(x)$: glucose concentration (x and τ , see caption of Fig. 3). Arrows indicate direction of bacterial motion.

oxygen concentration gradient will set up (Fig. 4, τ = 0.4). In response to the latter gradient a zone with a positive peak in V_c appears indicating bacteria moving left to right (upwards in the experiment). As a result bacteria tend to accumulate around the zone of sign change in V_c and the bacterial distribution shows a central peak surrounded by two depleted zones. Now, since here oxygen is the limiting factor in aerobic metabolism, oxygen disappears faster than glucose in the lower half of the cell where fermentation will take place. However because the gradient $\partial_x C_2$ has become very smooth, the chemotactic effect towards glucose is very weak and no bacterial band forms towards the bottom of the cell. By enhanced consumption in the middle region of the cell, the two gradients have now separated and specific attraction towards oxygen alone has started : a second band forms in response to the oxygen gradient (Fig. 4, τ = 3). Oxygen is consumed by endogeneous respiration and the second band migrates from left to right (upwards in the experiment)(Fig. 4, τ =4). Note the inversion of the oscillation in $V_c(x)$ (compare Fig. 4, τ = 0.3, and Fig. 4, τ = 4). From this stage on, all effects tend to smooth out gradually and the system reaches eventually a steady state where spatial distributions have become uniform. Notice that here the evolution of the system is analyzed from the mechanisms that govern the successive transient states. Indeed in the present (and next) application of the model all interesting features appear as transients whereas the steady state is unimportant.

In the second experiment (referred to as B in Table 1), a bacterial suspension is layered in the lower part of a cell with medium containing oxygen and glucose. Soon after the experiment has started bacteria form a band that separates from the inoculum (see Fig. 5, t = 1 min.) ; about one hour later, a second band has appeared (Fig. 5., t = 55 min), then smoothes out, while the first band grows and migrates (Fig. 5, t = 140 min.). The results of the theoretical model are given in the lower boxes of Fig. 5. Bacteria first consume oxygen and glucose simultaneously, creating an oxygen gradient and a glucose gradient, in response to which they form a traveling band (Fig. 5, τ=.75 ; note the peak in the response function indicating bacterial motion from left to right) Aerobic metabolism is controled primarily by the oxygen concentration if the second attractant is in excess (here glucose). As a result there is a residual glucose gradient inducing fermentation and the formation of a second band (Fig. 5, τ = 5 ; note that the response function is now double peaked). When time

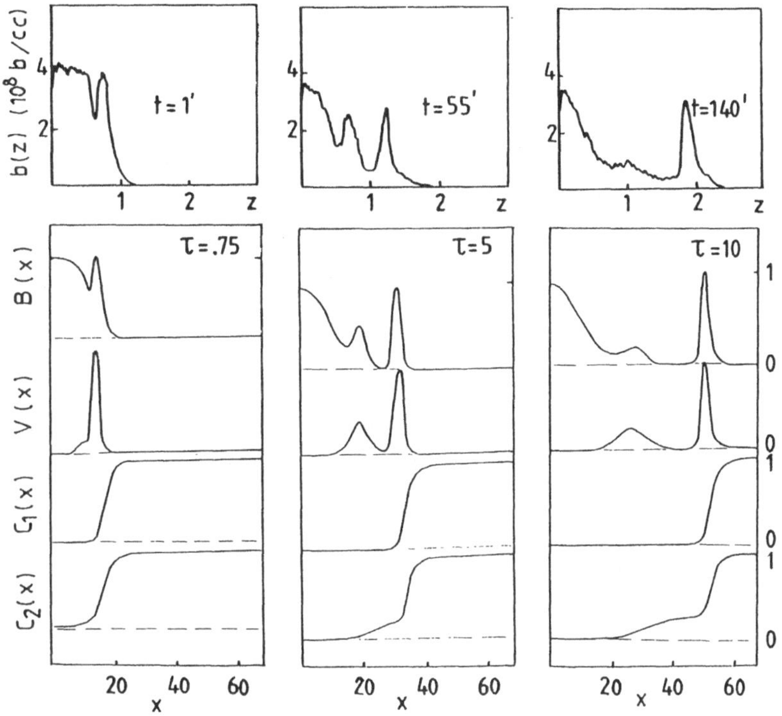

FIGURE 5. Band migration in response to two attractants. For symbols see caption of Figs. 3 and 4.

evolves the first band continues to grow while migrating whereas the second band broadens and decreases in intensity as the second peak in the response function becomes weaker due to spreading of the residual glucose gradient (Fig. 5 ; τ = 10). Eventually the system reaches a uniformly distributed steady state when the two substrates have been consumed completely. Migration velocities are 1.45 $\mu m.sec^{-1}$ and 0.54 $\mu m.sec^{-1}$ for the first and second band respectively ; the corresponding velocities as computed from the theoretical model are 1.57 $\mu m.sec^{-1}$ and 0.71 $\mu m.sec^{-1}$ respectively (Herpigny, 1983).

The third application (denoted C in Table 1) differs from those discussed above in that in this case the system (provided it is sufficiently large) reaches a non-uniform stationary state. The experimental conditions are the following : bacteria are uniformly suspended in a cell containing an aqueous medium with no attractant and the cell is open to air so that oxygen diffuses into the medium through the upper surface. As a result an oxygen gradient sets up to which the bacteria will respond chemotactically ; the resulting effect of the chemotactic response wil now depend on competition between substrate diffusion and substrate consumption. The mathematical formulation of the problem reduces to a set of two coupled equations

$$\partial_\tau B = \partial_x^2 B + \partial_x (B \partial_x F_c(C)) \tag{8}$$

$$\partial_\tau C = \alpha \partial_x^2 C - \gamma f(C) B \tag{9}$$

with boundary conditions

$$C(x=0, \forall \tau) = 1 \; ; \; \partial_x C|_{x=L} = \partial_x B|_{x=L} = \partial_x B|_{x=0} = 0 \tag{10}$$

where $x=0$ is the position of the meniscus and $x=L$ $(L>>1)$ the coordinate of the bottom of the cell. The competition effect is measured by the ratio $\gamma/\alpha \div k/D$; since $\alpha\,(=D/\mu)$ is a given fixed quantity, the control parameter is $\gamma = \gamma_0 b_0$, where γ_0 is a constant and b_0, the initial bacterial density, is a variable quantity.

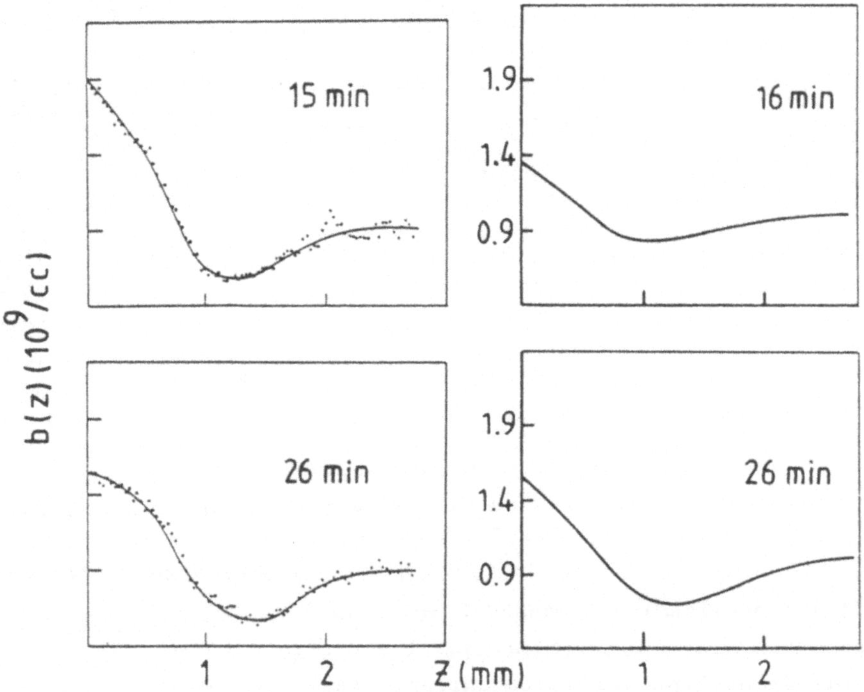

FIGURE 6. Formation of bacterial band in reponse to oxygen diffusion. Comparison of experimental results (left ; Wang and Chen, 1982) with theoretical model (right ; Herpigny, 1983).

In an experiment performed by Wang and Chen, with $b_0 = 10^9/cc$ (Wang, 1982), they observe accumulation of bacteria below the meniscus as shown in Fig. 6.a. The corresponding results as obtained from the theoretical model with γ = 50 (the value of γ for b_0 = 10^9) are given in Fig. 6.b. The results of a more detailed analysis are displayed in Fig. 7 (the graphs in the upper boxes are for γ = 50). Oxygen diffuses into the cell from the top surface thereby creating an attractant gradient (see $C(x)$ in Fig. 7) which in turn induces a peak in the response function $V_C(x)$. As a result bacteria are attracted to the top of the cell and by consuming oxygen enhance the gradient thereby increasing the response ; consequently more bacteria move toward the meniscus by depleting the nearby lower zone (see the evolution of $B(x)$ in Fig. 7, upper boxes).

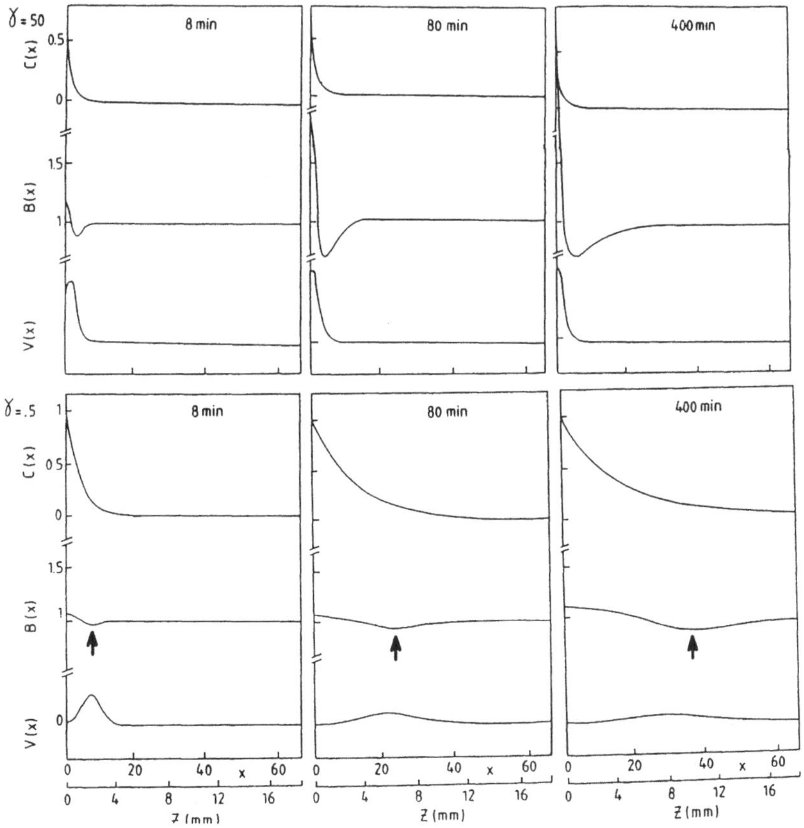

FIGURE 7. Bacterial response to oxygen diffusion for two values of consumption versus diffusion ratio : γ = 50 (upper boxes) and γ = 0.5 (lower boxes). C(x) = oxygen concentration ; B(x) = bacterial density distribution (arrows indicate the position of the minimum in B(x)) ; $V_C(x)$ = response function. Spatial coordinate is oriented such that top of cell is on the left (x = 0).

When the value of the ratio of consumption versus diffusion is reversed, ($\gamma = 0.5$) diffusion outgrows consumption, and bacteria accumulate weakly near the meniscus but the depletion zone moves away from the surface as a consequence of the displacement of the maximum in the response function (see Fig. 7, lower boxes). Whether these results for different values of the initial parameter are indicative of branching of solutions can be analyzed from the stationary state solutions of Eqs. (8) and (9). The system is solved in the spatial coordinate and it is found that x behaves like $(\alpha/\gamma)^{1/2}$, that is oxygen penetration into the medium is proportional to $\gamma^{-1/2}$. Therefore, all other conditions being equal, the position of the peak in the response function will be displaced away from the meniscus according to the power law $\gamma^{-1/2}$. As a result the minimum in $B(x)$ (position of the dip in Fig. 7) is a linear function of $\gamma^{-1/2}$ and the system exhibits no bifurcation.

It may be conjectured that the presence of a second attractant would bias the competition problem and could possibly lead to different branches of solutions. Similarly competition between bacterial species would also modify the system behavior. Such and further applications can be analyzed on the basis of the theory, which in this respect may also serve the purpose to suggesting new experiments.

ACKNOWLEDGEMENTS

This work was supported by the "Fonds National de la Recherche Scientifique" (F.N.R.S., Belgium).

APPENDIX

Nossal (1972) presented a justification of a specific form of the chemotactic current on the basis of a modified Langevin equation. Following a similar approach, we give here a more general derivation based on the non-Markovian Langevin equation for the drift velocity $\underline{v}_c$

$$\dot{\underline{v}}_c(\underline{r},t) = -\int_0^t dt'\, K(t-t')\, \underline{v}_c(\underline{r},t') \quad + \quad \underline{F}(\underline{r},t) \qquad \text{(A.1)}$$

where $\underline{F}$ is the chemotactic force. Eq. (A.1) is solved formally to yield

$$\underline{v}_c(\underline{r},t) = M(t)\,\underline{v}_c(\underline{r},0) \; + \int_0^t dt'\, M(t-t')\, \nabla V(\underline{r},t') \qquad \text{(A.2)}$$

where M(t) is a memory function related to the kernel K(t) through their Laplace transforms as $M(s) = (s + K(s))^{-1}$, and $\underline{F}$ has been expressed in terms of the chemotactic potential $V(\underline{r},t)$. In linear geometry, the spatial vector $\underline{r}$ is replaced by z, the vertical direction along which the chemotactic potential acts. Introducing the scaled variables $x = z\, v_o/\mu$ and $\tau = t\, v_o^2/\mu$ where v_o denotes the strength of the chemotactic potential, we obtain

$$v_c(x,\tau) = M(\tau)\, v_c(x,0) \; + \; v_o \int_0^\tau d\tau'\, M(\tau-\tau')\, \partial_x F_c(x,\tau') \qquad \text{(A.3)}$$

where F_c is the dimensionless chemotactic function. The characteristic time for chemotactic effects is $t_c = \mu/v_o^2$ and the characteristic kinematic time over which memory effects take place is $t_m = m/\zeta$, where m is the bacterial mass density and ζ the friction coefficient related to viscous drag inherent to bacterial motion in the fluid medium. When t_c and t_m are of the same order of magnitude, (A.3) must be considered ; an ansatz must then be introduced to model the memory function $M(\tau)$ (e.g. $M(\tau) = \exp(-\tau tc/tm)$; Nossal, 1972). On the other hand, if t_m is much shorter than t_c (cases treated in the present paper yield $t_m \sim 10^{-6}$ sec and $t_c \sim 10^2$ sec), memory effects decay very fast on the chemotactic time scale ; then one can set $M(\tau) = \delta(\tau)$ and (A.3) reduces to

$$V_c \equiv v_c(x,\tau)/v_o \; = \; \partial_c F_c(C)\, \partial_x C(x,\tau) \qquad \text{(A.4)}$$

(A.4) establishes the form of the response function V_c in terms of the chemotactic function F_c, for instantaneous chemotactic response.

REFERENCES.

ADLER, J. (1966) Science 153, 708.

ADLER, J. (1978) in The Harvey Lectures (Academic Press, New York) Series 72, 195.

BOON, J.P. (1975) in Membranes,Dissipative Structures and Evolution, G. Nicolis and R. Lefever, eds. (John Wiley Interscience, New York) 169.

BOON, J.P. (1983) in The Application of Laser Light Scattering to the Study of Biological Motion, J.C. Earnshaw and M.W. Steer, eds. (Plenum Press, New York).

BOON, J.P. and HERPIGNY, B. (1982) "Bacterial chemotaxis and band formation : Response to the simultaneous effects of two attractants" (preprint).

CHEN, S.M. and HALLETT, F.R. (1982) Quart. Revs. Biolphys. 15, 131.

CHILDRESS, S. and PERCUS, J.K. (1983) in the present volume.

DAHLQUIST, F.W., LOVELY, P.S. and KOSHLAND, D.E.Jr. (1972) Nature New Biol. 15, 131.

HERPIGNY, B. (1983) Ph. D. Thesis, University of Brussels, Belgium.

KELLER,E.F. and SEGEL, L.A. (1970) J. Theor. Biol. 26, 399.

KELLER,E.F. and SEGEL, L.A. (1971) J. Theor. Biol. 30, 235.

KOSHLAND, D.E.Jr. (1977) in Advances in Neurochemistry, B.W. Agranoff and M.H. Aprison, eds. (Plenum Press, New York) vol. 2, 277.

LAPIDUS, I.R. and LEVANDOWSKY, M. (1981) in Biochemisty and Physiology of Protozoa, (Academic Press, New York) Vol. 4, 235.

LAUFFENBURGER, D.A. (1983) in the present volume.

LAUFFENBURGER, D.A. and B. CALCAGNO, P. (1983) Biotechnol. and Bioeng. (to be published).

LIN, C.C. and SEGEL, L.A. (1974), Mathematics applied to Deterministic Problems in the Natural Sciences, (Mc Millan Publ. Co, New York) chapter 10.

LOVELY, P.S. and DAHLQUIST, F.W. (1975) J. Theor. Biol. 50, 477.

MESIBOV, R., ORDAL, G.W. and ADLER, J. (1973) J. Gen. Physiol. 62, 203.

NOSSAL,R. (1972) Math. Biosc. 13, 397.

NOSSAL,R. (1972) Exptl. Cell. Res. 75,138.

NOSSAL,R. (1980) in Biological Growth and Spread, W. Jäger, M. Rost, and P. Tautu, eds (Springer-Verlag, Berlin) 410.

SEGEL, L.A. (1978) in Studies in Mathematical Biology, Vol. 15, Part 1, S.A. Levin, ed. (The Mathematical Association of America,

Washington) 156.
SEGEL, L.A. (1983) "Taxes in cellular ecology" (preprint).
WANG, P.C. (1982) Ph. D. Thesis, MIT (1982).

PATTERN FORMATION IN DIRECTIONAL SOLIDIFICATION: THE NONLINEAR EVOLUTION OF CELLULAR MELT/SOLID INTERFACES

R. A. Brown and L. H. Ungar
Department of Chemical Engineering and
Materials Processing Center
Massachusetts Institute of Technology
Cambridge, MA 02139

Abstract

The formulation of moderate amplitude two-dimensional cellular interfaces arising in directional solidification of a binary alloy is probed by finite element analysis coupled with computer-implemented perturbation methods for tracing families of interface shapes. Results for a "one-sided" solidification model show stable finite amplitude cells irrespective of whether the initial evolution from the planar state is subcritical or supercritical with respect to changes in the applied temperature gradient. At large amplitude a discontinuous change in the stable interface morphology causes a halving of the spatial wavelength. The occurrence of this transition is linked to the existence of second-order critical points for bifurcation from the planar form. Grain angles which mark the boundaries of individual crystals are imperfections to the bifurcations from the planar front and cause the transition to cellular forms to occur smoothly over a range of temperature gradient.

1. Introduction

The interface separating a solid from its solidifying melt is not morphologically simple but can assume a variety of forms ranging from a mildly sinusoidal shape, to a large amplitude three-dimensional cellular surface and finally to a complicated tree-like dendritic structure; see Woodruff (1973) for examples of each of these morphologies. These intricate patterns point to the rich mathematical structure that must be described by models of microscopic solidification. The identification of these transitions is one of the outstanding problems of pattern formation arising in applied physics. In this short paper, we review recent calculations of large amplitude cellular interfaces with the goal of elucidating some of the transitions in melt/solid interface morphology.

The basic physics describing the transition to nonplanar interface shapes in directional solidification of alloys was described qualitatively by Rutter and Chalmers (1953). The rigorous linear stability theory for the evolution of small amplitude sinusoidal disturbances to a flat interface was developed by Mullins and Sekerka (1963, 1964) who obtained an exact relationship between the critical temperature gradient $\tilde{G}$ at the onset of instability, the growth speed V, the spatial wavelength λ of the disturbance and thermophysical parameters of the alloy. Disturbances spanning a continuous range of wavelengths $\tilde{\lambda}_1(\tilde{G})<\tilde{\lambda}<\tilde{\lambda}_2(\tilde{G})$ were found to be unstable for temperature gradients less than a minimum value $\tilde{G} = \tilde{G}_c$. At $\tilde{G} = \tilde{G}_c$ the interface was unstable only to the disturbance with the single wavelength $\tilde{\lambda} = \tilde{\lambda}_c$. Only a few studies have attempted to trace interface morphology beyond the loss of stability of the planar form (Wollkind and Segal, 1970; Langer, 1980; Mathur and Dee, 1983; McFadden and Coriell, 1983; Sriranganathan et al., 1983). With the exception of the work of McFadden and Coriell (1983) each of these studies is an asymptotic analysis valid only for melt/solid interfaces that are slightly perturbed from a plane. Wollkind and Segal (1970) used an expansion correct to second order in the deformation amplitude to characterize the evolution of weakly nonlinear cellular interfaces. Two types of behavior were discovered and are summarized in Fig. 1 as sub- and super-critical bifurcations from the planar interface with ε representing a measure of the amplitude of the deformed interface. For low growth rates, cellular interfaces were found only for values of $\tilde{G}$ greater than $\tilde{G}_c$ and were unstable to small amplitude perturbations, as is typically the case for subcritical branching. At high growth rates the deformed interfaces evolved toward lower values of G and were stable, corresponding to supercritical bifurcation. These asymptotic results gave no indication of the fate of the interface morphology when the initial cellular interface was unstable.

When the perfectly planar melt/solid interface exists, transitions between interfaces with different symmetry occur as bifurcations and are easily detected with perturbation techniques as demonstrated in §3. Grain boundaries along a melt/solid interface disrupt the planar shape and have been demonstrated experimentally to serve as focal points for the onset of the cellular instability with decreasing temperature gradient (Schaefer and Glicksman 1970; Ayers and Schaefer 1973). An attempt to explain this result by including small grain angles in a linear stability analysis (Coriell and Sekerka 1973) failed. As is discussed in another paper (Ungar and Brown 1983b),

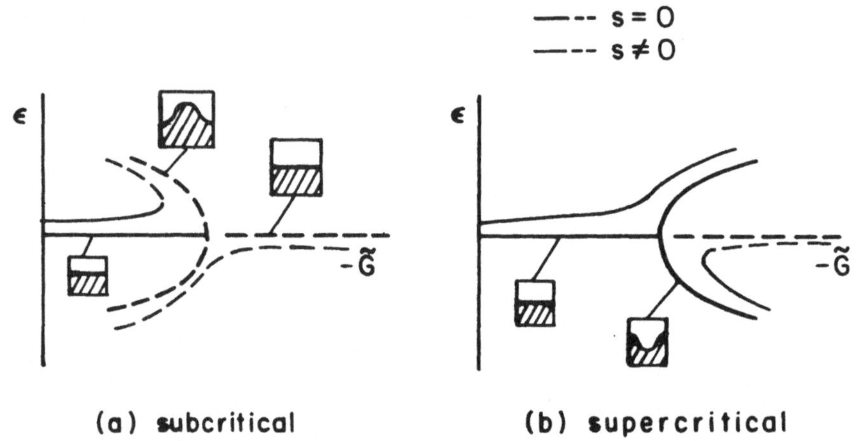

Figure 1. Families of small amplitude cellular interfaces characterized as either subcritical or supercritical bifurcation in terms of G. Curves for nonzero s represent imperfection caused by grain boundary.

grain boundaries are imperfections to the bifurcation structure and fission the intersection of the planar and cellular families into non-intersecting branches. Even small grain angles have a singular effect on the structure of interface families which is not accounted for in the regular expansion of Coriell and Sekerka, but that is systematically incorporated in an asymptotic expansion by Ungar and Brown (1983b). The transition from an almost planar to an undulating morphology then occurs smoothly over a range of temperature gradients as represented by the curves for $s \neq 0$ in Fig. 1. Finite element calculations which demonstrate this transition are included in §4.

2. One-Sided Solidification Model and Finite-Element Analysis

In unidirectional solidification a solid is grown at speed V from its melt in an applied constant temperature gradient that is strictly perpendicular to the growth direction far from the melt/solid interface. We represent the melt/solid interface and the field variables in a rectangular cartesian coordinate system attached to the melt/solid interface; this configuration is shown in Fig. 2a. In the one-sided solidification model the thermal conductivities of melt and crystal are taken to be the same and latent heat release is ignored. With these simplifications the imposed temperature gradient is constant in both phases and is not altered by changes in interface shape. The model is "one-sided" in the sense that the shape of the interface is set solely by the deformation of the solute composition field in the melt and its interaction with the imposed temperature gradient $\tilde{G}$.

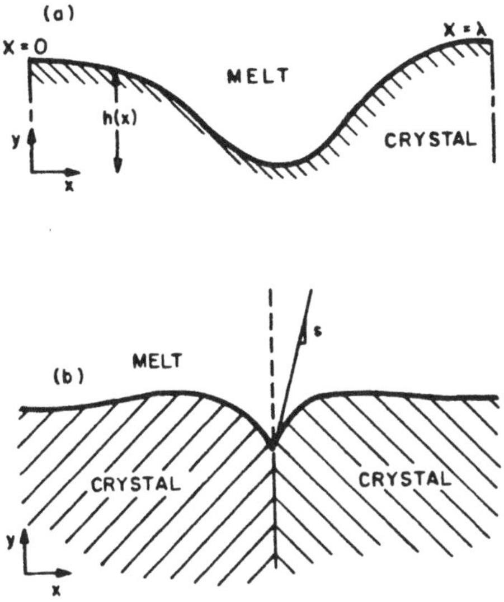

Figure 2. Schematics of melt/solid interfaces (a) with and (b) without grain boundary. The wavelength of the cells is denoted as λ.

The steady free-boundary problem which governs the one-sided model is written in dimensionless form with lengths, temperatures and concentrations scaled with a characteristic wavelength λ^*, the melting temperature for the binary alloy at a planar interface T_m^o and the bulk concentration of the melt $\tilde{c}_\infty$, respectively. The shape of the melt/solid interface is taken as a single-valued function $y \equiv h(x)$. The equations are in the melt ($0 \le x \le \lambda \equiv \tilde{\lambda}/\lambda^*$, $h(x) \le y \le \infty$)

$$\nabla^2 c + P \frac{\partial c}{\partial y} = 0 \ , \tag{1}$$

at the melt/solid interface ($y = h(x)$, $0 \le x \le \lambda$)

$$Gh - mc - \Gamma \frac{\partial^2 h}{\partial x^2} \left\{ 1 + \left(\frac{\partial h}{\partial x}\right)^2 \right\}^{-3/2} = 0 \ , \tag{2}$$

$$\frac{\partial c}{\partial y} - \frac{\partial h}{\partial x}\frac{\partial c}{\partial x} - (k-1)\, P\, c = 0 \ , \tag{3}$$

and along the other boundaries of the melt,

$$c(x,\infty) = 0, \ (\partial c/\partial x)_{x=0} = (\partial c/\partial x)_{x=\lambda} = 0 \ . \tag{4}$$

In these equations, $P \equiv V\lambda^*/\mathcal{D}$ is the dimensionless growth rate or Peclet number, $\Gamma \equiv \tilde{\Gamma}/\lambda^*$ is the capillary constant, $G \equiv \tilde{G}\lambda/T_m^o$ is the dimensionless temperature gradient, k is the partition coefficient for the dilute component between melt and solid, and $m \equiv \tilde{m}\,\tilde{c}_\infty/T_m$ is the dimensionless slope of the liquidus line. The boundary conditions (4) force the interface to have wavelength λ. In addition we set

$$(\partial h/\partial x)_{x=0} = s, \ (\partial h/\partial x)_{x=\lambda/2} = 0 \quad , \tag{5}$$

where $s > 0$ is the slope of a grain boundary appearing at $x = 0$. When s is nonzero the interface has a discontinuous indentation near this point, as shown in Fig. 2b.

When s is zero the interface is perfectly smooth and the planar shape ($h(x) = 0$, $0 \le x \le \lambda$) is a solution for all values of G and P. We have analyzed the evolution of small amplitude cellular interfaces from the planar state by a classical bifurcation analysis carried out in terms of the amplitude of the interface deformation (Ungar and Brown 1983a). This study revealed that families of cellular interfaces $G = G(\varepsilon)$ with shape approximated by

$$h(x;\varepsilon) = \varepsilon \cos(\omega_n x) \ , \tag{6}$$

evolved from the simple bifurcation points

$$G_c^{(n)} = \frac{m(k-1)P(\omega_n - P)}{k(\omega_n + (k-1)P)} - \Gamma\omega_n^2 \ , \tag{7}$$

where $\omega_n \equiv 2n\pi\lambda^{-1}$. All these bifurcations were found to be one-sided, i.e., $(dG/d\varepsilon)_{\varepsilon=0} = 0$, and the direction of the evolution was given by $G^{(2)} \equiv (d^2G/d\varepsilon^2)_{\varepsilon=0}$, which depended on the growth rate and other physical properties. Results are shown in Fig. 3 for the thermophysical properties listed in Table I, which are similar to the lead-antinomy (Pb-Sb) system described by Morris and Winegard (1967). At this growth rate and concentration the cellular forms with the highest critical temperature gradient $G_c^{(1)}$ were almost precisely on the border between sub- and super-critical evolution, so it was impossible to make assertions about the stability of the evolving family. The second family branched subcritically at $G = G_c^{(2)}$ and was unstable to two modes of deformation.

Table I. Thermophysical Properties and Dimensionless Groups Representative of PbSb.

Property	Symbol	Value
Segregation Coefficient	k	0.4
Bulk Concentration of Sb	$\tilde{c}_\infty$	0.02 wt. %
Slope of Liquidus	$\tilde{m}$	-5K/wt. %
Diffusivity	$\mathcal{D}$	$2x10^{-5}$ cm^2/sec
Reference Melting Temperature	$\tilde{T}_m$	600 K
Capillary Length	$\tilde{\Gamma}$	$8.2x10^{-9}$ cm
Reference Length Scale	λ^*	$1x10^{-2}$ cm
Growth Rate	V	$1.6x10^{-3}$ cm/sec
Dimensionless Slope of Liquidus	m	$-1.67x10^{-4}$
Capillary Constant	Γ	8.2×10^{-7}
Peclet Number	P	0.8

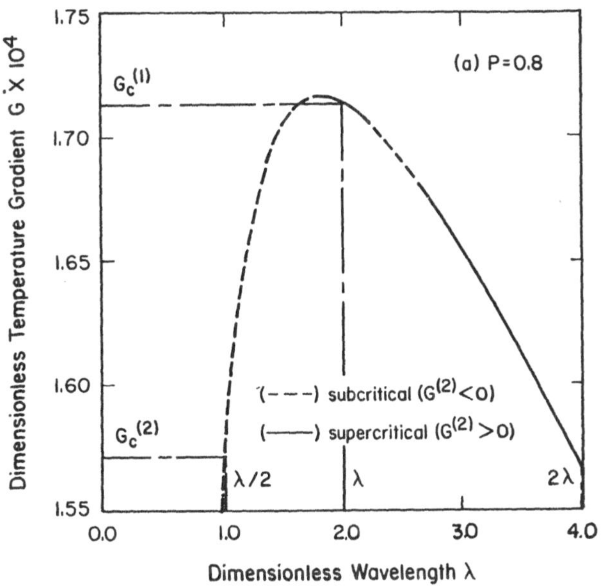

Figure 3. Critical values of temperature gradient $G_c^{(1)}$ predicted as a function of wavelength for the Pb-Sb system with P = 0.8 . Whether the first family of cellular forms evolves sub- or super-critically is denoted by the solid or dashed curve.

The finite element method developed by Ettouney and Brown (1983) has been modified to calculate large amplitude cellular interfaces given by the one-sided model. As described by Ungar and Brown (1983a), the free-boundary problem defined by Eqs. (1-5) is transformed to a domain with fixed and finite boundaries. In this representation, the unknown interface shape h(x) appears throughout the equation set. Both the concentration field and interface shape are represented by expansions of quadratic finite element basis functions and the equations are reduced to a nonlinear algebraic set by Galerkin's method. This set is solved by Newton's method as implemented by Ettouney and Brown (1983) combined with computer-implemented perturbation techniques for tracing solution families as a parameter is varied, for detecting solution bifurcation, for jumping between solutions and determining their stability. Each of these algorithms is analogous to classical asymptotic techniques in analysis, but is implemented numerically in terms of solutions to a large set of algebraic equations. The details of these methods are described elsewhere (Keller 1977; Yamaguchi et al. 1983).

3. Large Amplitude Cellular Interfaces

The families of cellular interfaces emanating from the critical temperature gradients ($G_c^{(1)}$, $G_c^{(2)}$) for P = 0.8 and λ = 2.0 are represented in Fig. 4 by the maximum deflection Δ computed as

$$\Delta \equiv \max_{0 \le x \le \lambda} \{h(x)\} - \min_{0 \le x \le \lambda} \{h(x)\} . \tag{8}$$

This wavelength was approximately fifteen percent above the critical value $\lambda_c \equiv \tilde{\lambda}_c/\lambda^*$ and was very close to the transition between subcritical and supercritical bifurcation. The two families branching from $G = G_c^{(1)} \simeq 1.71 \times 10^{-4}$ had cellular interfaces with wavelength $\lambda = 2.0$ and evolved almost vertically from the planar form. The shapes in these two families were identical up to a reflection about the line x = 0 and were classified as the 1U and 1D branches; the 1 signified that the fundamental wavelength of the cells was λ and the U and D denoted whether the solid extended into (U) or away from (D) the melt at x = 0.

Both the 1U and 1D families evolved subcritically (to higher values of G) as the deflection increased until they reversed direction at a limit point $G \equiv G_\ell^{(1)} \simeq 1.745 \times 10^{-4}$ and continued to lower values of G. These interfaces regained stability at this limit point and remained stable up to a second limit point $G_\ell^{(2)} \simeq 1.681 \times 10^{-4}$. Sample melt/solid interfaces in the 1U family are shown in Fig. 5a for half the wavelength. From nearly sinusoidal forms for small Δ, the interface devel-

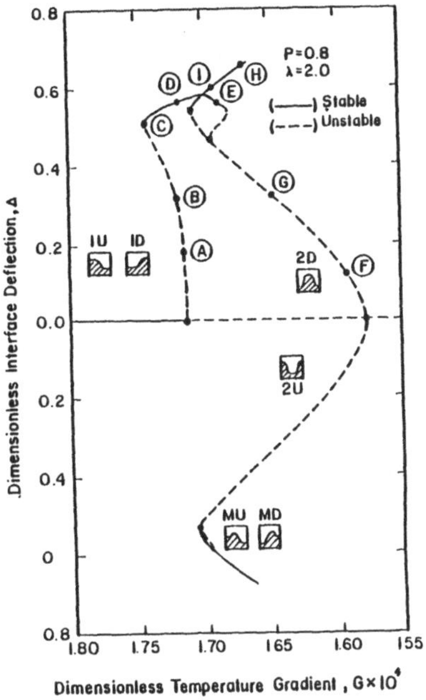

Figure 4. Families of cellular interfaces for P = 0.8 and λ = 2.0 as represented by the interface deflection Δ. Letters refer to sample interface shapes in Fig. 5.

oped a deep and narrow groove separating a large, almost planar plateau for shapes with $\Delta \approx 0.5$. More highly deformed interfaces developed a depression in the plateau with decreasing G.

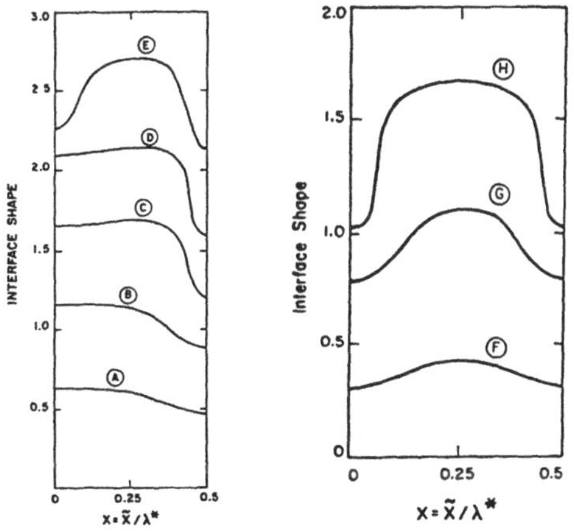

Figure 5. Sample interface shpaes in the first and second families of cellular forms. Letters refer to points shown in Fig. 4.

The two families bifurcating at $G = G_c^{(2)} \simeq 1.577 \times 10^{-4}$ were named 2U and 2D because each had interfaces with wavelengths of $\lambda/2$ so two complete cells existed in the spacing $0 \leq x \leq 2$. The 2U shapes protruded into the melt at $x = 0$ and the 2D ones were identical except for a lateral shift of $x = \pm 0.5$. The nonlinear structure of these two families differed because introducing a shape perturbation which ruptures the plane of symmetry about $x = 0$ in the 2U and 2D forms leads to distinct interface shapes. We have differentiated these families by plotting them separately in Fig. 4.

As shown in Fig. 4, the 2U and 2D families bifurcated subcritically (increasing G), but reversed direction at a limit point $G \equiv G_\ell^{(3)} \simeq 1.71 \times 10^{-4}$. Interface shapes in the 2D family are shown in Fig. 5 and have deep grooves like the 1U and 1D shapes, but at half the wavelength. The 1U and 1D shapes merged with the 2D forms at a simple bifurcation point, ($G_s^{(1)} \simeq 1.694 \times 10^{-4}$), just before the limit point in the 2D family. Here the depression in the interfaces in the 1U and 1D families became identical with the primary groove and halved the wavelength. The 2U family evolved identically to the 2D branch, except that no connection between the 2U and either the 1U or 1D families was found. A secondary bifurcation point was located in the 2U family, as shown in Fig. 4, and marked the beginning of two new families with shapes resembling a combination of 2U with 1U and 1D forms.

Beyond the limit point both the 2U and 2D forms were stable to disturbances with wavelength λ. For values of G within ten percent of $G_c^{(1)}$, these results predict an abrupt transition between interfaces with wavelength λ and $\lambda/2$. This transition would be detected at temperature gradients near $G_\ell^{(2)}$.

The spacing between the first two critical values ($G_c^{(1)}$, $G_c^{(2)}$) depends on the value of wavelength λ. The evolution of the nonlinear morphological structure with changes in λ clarifies the connectivity between shape families demonstrated in Fig. 4 for $\lambda \simeq \lambda_c$. It is easily shown by geometrical reasoning that for some wavelength $\lambda_d > \lambda_c$ the critical points $G_c^{(1)}$ and $G_c^{(2)}$ coincide and for $\lambda > \lambda_d$ the family emanating from the first critical value (now $G_c^{(2)}$) has interfaces with wavelength $\lambda/2$. The wavelength $\lambda \equiv \lambda_d$ is a multiple critical point with respect to the planar family and has been found to signal the occurrence of secondary bifurcations in the vicinity of this critical parameter set (λ_d, $G = G_c^{(1)} = G_c^{(2)}$) in many applications (Keener 1976; Hall and Walton 1979; Iooss and Joseph 1980).

The shape families bifurcating from $G_c^{(1)}$ and $G_c^{(2)}$ are represented in Fig. 6 for $P = 0.8$ and wavelengths at and on either side of an estimated value for the double point $\lambda = \lambda_d = 2.64$. For $\lambda = 2.60$ the ordering of the modes was the same as for $\lambda = 2.0$ except that the secondary bifurcation points along the 2U and 2D families had moved close to the planar form. All four bifurcation points ($G_c^{(1)}$, $G_c^{(2)}$, $G_s^{(1)}$, $G_s^{(2)}$) coalesced as λ was increased to λ_d. At $\lambda \equiv \lambda_d$ the 1U and 1D form disappeared entirely and only the 2U, 2D, MU and MD families evolved from the point $G = G_c^{(1)} = G_c^{(2)}$. The critical values for the (2U,2D) and (1U,1D) bifurcations exchanged order for $\lambda > \lambda_d$ and the MU and MD had only 1U and 1D components at the bifurcation from the plane. Interfaces in both families had components with wavelength $\lambda/2$ for finite amplitude.

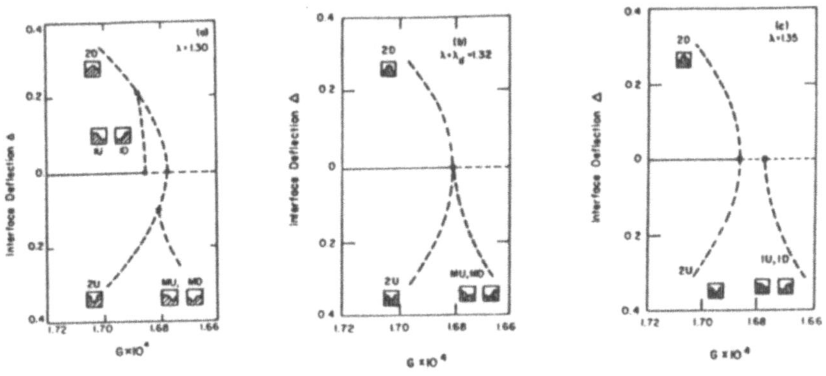

Figure 6. Families of cellular interfaces as λ is varied through the double critical point λ_c for P = 0.8.

4. Imperfection Caused by Grain Boundary

Introducing a grain boundary defined by the angle s in eq. 5 caused interfaces deformation for all values of temperature gradient. The primary family of interface shapes calculated for the properties in Table 1 are plotted in Fig. 1 for grain angles between 0° (s = 0) and 89° (s = 5). For these parameters, the planar forms for $G > G_c^{(1)}$ and the cellular forms bifurcating from $G = G_c^{(1)}$ are ruptured for nonzero grain angles into a continuous family of interface shapes. For the lowest values of s, the shape family has a limit point where it reverses direction to higher values of G. The large amplitude interfaces regain stability at a second limit point, just as is the case for the cellular interfaces without the grain boundary (s = 0). Sample interface shapes computed for s = 1.0 (45°) are shown in Fig. 8.

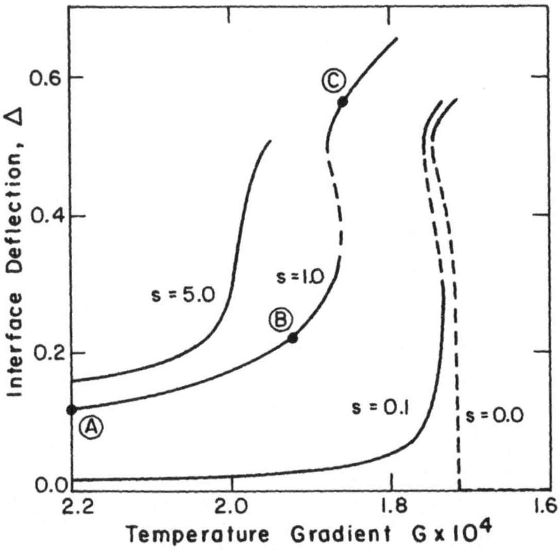

Figure 7. Families of cellular interfaces for P = 0.8 and λ = 2.0 as a function of the grain angle s present at x = 0 .

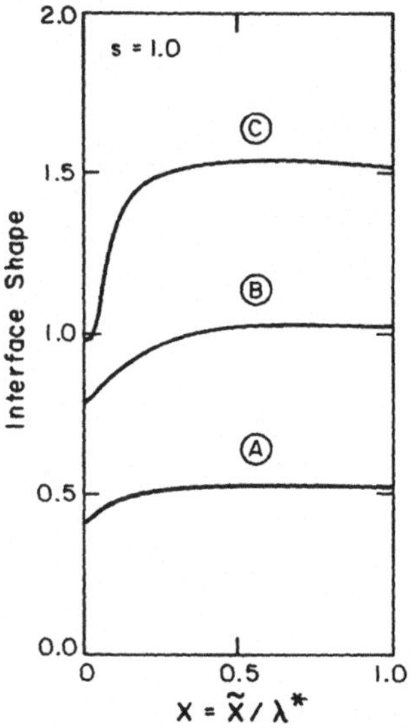

Figure 8. Sample interface shapes for s = 1.0 . Letters refer to points in Fig. 7.

At large grain angles the subcritical branch disappears altogether and the calculations predict a smooth transition to large interface deflections.

5. Summary

The complicated portrait of the steady, two-dimensional interface shapes possible for the limited range of temperature gradient span in the calculations presented here gives an indication of the rich mathematical structure that may be needed to describe transitions in melt/solid interface morphology. Barring instabilities to three-dimensional forms, the results suggest that large amplitude cellular interfaces are possible and that the fundamental wavelength of these forms may be halved abruptly at a critical value of the temperature gradient which is lower than the value for the onset of morphological instability.

These stable, large-amplitude cellular forms existed even though the first cellular forms evolved subcritically and were unstable. This last result points to the danger of attempting to infer the large amplitude behavior of a nonlinear system from results, like those displayed in Fig. 3, which are obtained by amplitude expansions for the weakly nonlinear system.

The results shown here cover a range of temperature gradient only 15% lower than the critical value $G_c^{(1)}$ and are only a first step toward explaining the transitions between cellular, pre-dendritic and dendritic interface morphologies. Continuing the calculations to higher amplitude cellular forms will require generalizing the representation of the melt/solid interface to account for interfaces with shapes that are multivalued with respect to the transverse direction x. Then the arms along the cells which correspond to the onset of pre-dendritic morphologies will appear as time-periodic traveling waves and may be detected by methods for analyzing Hopf bifurcations.

6. References

Ayers, J.D. and Schaefer, R.J., in Solidification and Casting of Metals, Metals Society, London (1979).

Coriell, S.R. and Sekerka, R.F., J. Crystal Growth 19, 285, (1973).

Ettouney, H.M. and Brown, R.A., J. Comput. Physics 49, 118 (1983).

Hall, P. and Walton, I.C., J. Fluid Mech. 90, 377 (1979).

Iooss, G. and Joseph, D.D., Elementary Stability and Bifurcation Theory, Springer-Verlag, New York, (1980).

Keener, J.P., Stud. Appl. Math. 55, 187(1976).

Keller, H.B., In Applications of Bifurcation Theory (P.H. Rabinowitz, ed.) Academic Press, New York (1977).

Langer, J.S., Rev. Mod. Phys. 52, 1 (1980).

Mathus, G. and Dee, D., Physical Review B submitted (1983).

McFadden, G.B. and Coriell, S.R., Physica D, in press (1983).

Morris, L.R. and Winegard, W.J., J. Crystal Growth 5, 361 (1969).

Mullins, W.W. and Sekerka, R.F., J. Appl. Phys. 34, 323 (1963).

Rutter, J.W. and Chalmers, B., Can. J. Phys. 31, 5 (1953).

Schaefer, R.J. and Glicksman, M.E., Met. Trans. 1, 1973 (1970).

Srianzanathan, R., Wollkind, D.J. and Oulton, D.B., J. Crystal Growth 62, 265 (1983).

Ungar, L.H. and Brown, R.A., Physical Review B., in press (1984).

Ungar, L.H. and Brown, R.A., Physical Review B., submitted (1984).

Wollkind, D.J. and Segal, L.A. Philos. Trans. R. Soc. Lond. A268, 351(1970).

Woodruff, D.P., The Solid-Liquid Interface, Cambridge University Press, London, (1973).

Yamaguchi, Y., Chang, C.J. and Brown, R.A., Philos. Trans. R. Soc. Lond. in press (1984).

EVOLUTION OF 3-D CHEMICAL WAVES IN THE BZ REACTION MEDIUM

A E Burgess, B J Welsh and J Gomatam

1. INTRODUCTION

It was our first intention to bring to the Pattern Workshop samples of the BZ reagent solutions and demonstrate at firsthand our experimental design for direct observation of evolving 3-D chemical waves. However carriage restrictions imposed on the transport of chemical reagents together with the vagaries experienced with live demonstrations determined that the experiments should instead be video-recorded at our Glasgow laboratory. Nonetheless, we wish to share at Heidelberg the sense of our excitement when this relatively simple experimental operation allows the unhindered development of such delicate 3-D wave structures to be observed in situ.

2. EXPERIMENTAL DESIGN

Details for the preparation of the reaction mixture are given elsewhere [1] but a typical solution used for the experiments shown here requires 10 cm^3 of each of the following reagents to be added together :

cerous nitrate	5 mM
potassium bromate	0.35 M
sulphuric acid	1.5 M
malonic acid	1.2 M

This mixture is stirred continuously in an open cylinder (100 mm high, 27 mm diam) and sufficient time is allowed for the reaction to pass through its early stages. At about 5 minutes (which allows ample time for the initial surge of active oxidant to be quenched thereby preventing bleaching of dyestuff) 0.6 cm^3 of the indicator dye ferroin 25 mM is added. It is known that oxygen catalyses the ceric ion reduction by malonic acid [2] and that induced oxidation by dissolved molecular oxygen is a common occurrence when oxidants such as bromate participate in slow reactions [3]. The BZ reaction mixture is sensitive to oxygen transferred from the air [4,5] and so steady stirring for a further 35 minutes eliminates any phase gradients and the reaction progresses to a stage where bulk oscillation is becoming regular. The solution can now be poured smoothly into a prepared set of test tubes. Semi-micro tubes 75 mm long and 10 mm outer diameter are particularly appropriate. This size comfortably holds about 3 cm^3 of solution and the diameter is most satisfactory in two important respects :

it is just about right for clear cross-sectional viewing of the developing wave patterns and is well below the diameter where distortion of patterns by convection currents becomes significant.

Each tube is sealed with a close-fitting rubber stopper which is quite deliberately inserted with firm thumb pressure. Sealing the tube is important since it appears that atmospheric oxygen encourages the formation of a wide variety of colliding wave forms. A more detailed account of this phenomenon will be published elsewhere. Equally the pressure under the stopper serves to diminish and in most instances almost prevents the formation of gas bubbles. A sudden spate of gas bubbles may still occur in the odd tube and a trail of destruction to any patterns follows in its wake. In other tubes an occasional small bubble may appear and rise through the solution but the patterns punctured along its way have a remarkable capacity to repair such limited type of damage, an event which is easily captured by video-recording. We suggest that the pressure condition created by the stoppers increases sufficiently the solubility of the gas (Henry's law [6]) and inhibits bubble formation. By wetting the glassware as part of the preparation procedure we find that the bubble problem is further diminished. So the tubes are thoroughly soaked in deionized water. Then excess of water on the inner surface is shaken off before the reaction mixture is slowly poured in. Nucleation sites and air entrapment are both minimised in this way.

3. PATTERNS IN EVOLUTION

The filled tubes must now be maintained in a still environment and free of temperature fluctuations from draughts and lighting equipment. A 3-tube video camera is set in front of these tubes against a background view of diffuse light. After only a short interval 3-D patterns start to evolve [Fig. 1]. Most of the structures formed may be complicated initially and not easily identified but often the development of some dominant wave form merely requires longer time. Since it is unlikely for every tube to form undistorted or recognizable structures a battery of 10 or 12 will give an adequate choice for any one session of study.

With the relatively low concentration of indicator present and a gradual degradation in dye intensity the visual colour contrast of the waves is restricted to pastel blue and pale orange. However the video camera enables some enhancement of the colour contrast to be achieved. Also by selecting the black and white mode further enhancement of definition of wave form can be perceived.

4. CONCLUSIONS

An edited sequence of patterns in evolution was presented at the Workshop. Although the video-recording cannot be matched for completeness and continuity, the photographs in the figures give a representative sample of the structures obtained. Examples are shown of the two kinds of patterns most frequently seen, simple scrolls and sphere-like structures [Fig. 2-5] typical of the forms which have been described [4,7-10]. Surprisingly the tubes with well-established wave forms are insensitive to small mechanical perturbations. The opportunity to perturb any of the tubes by tapping or tipping them was illustrated in the video-recording.

This experimental procedure has recently been used to test the effect of different shapes of container including a prismatic vessel [Fig. 3] on the patterns formed. No apparent effect was observed and it is concluded that the pattern adopted is independent of the vessel shape.

This design of experiment makes possible the direct observation of evolving 3-D wave structures (up to 5 h with the given conditions) and complements the earlier procedures for determining resultant wave forms[8,9,11-13].

Acknowledgements : We are indebted to G Cooper for his help with the video-recording, the British Council for some financial assistance and the Glasgow College of Technology for its continuing research support.

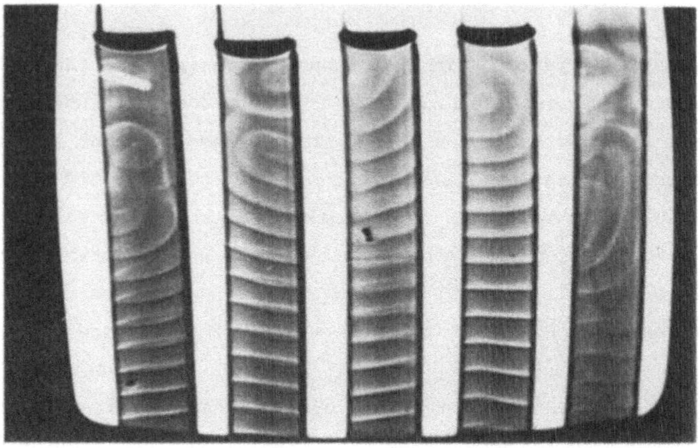

Fig 1. BZ reaction mixture in sealed tubes (10 mm diam). Early patterns with some dominant waves forming as seen in cross-sectional view about 30 minutes after pouring. Photographed from video-recording.

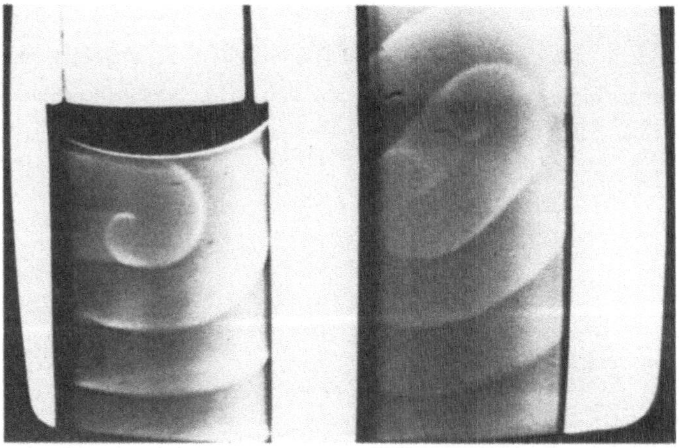

Fig 2. Simple scroll waves viewed in close-up about 1 hour after pouring. Notice no evidence of distortion from convection and that gas bubbles too are absent. The meniscus is seen above the scroll in the left tube and two small impurities may be observed in the right. Photographed from video-recording.

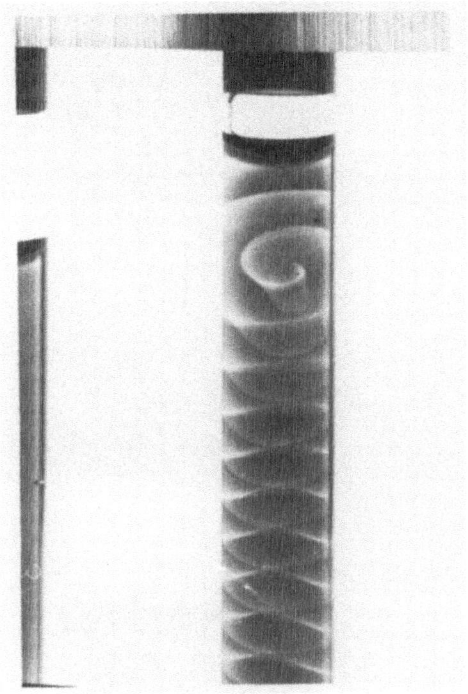

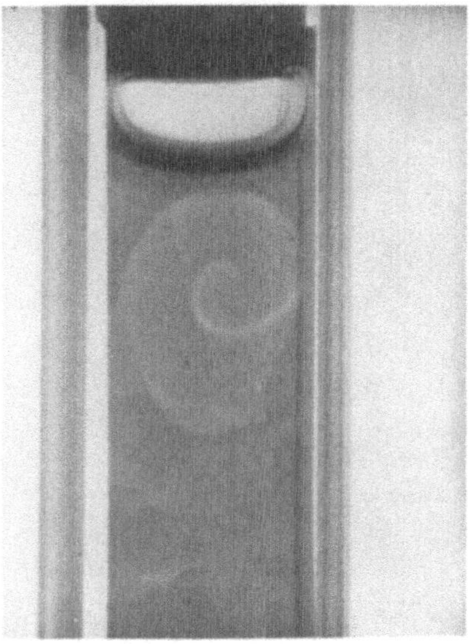

Fig 3. Scroll waves photographed directly with a clear view on the left of the developing waves being annihilated upon contact with the tube wall. The course of the pattern can be traced and appears unbounded by the wall. The scroll on the right is evolving in a square-sided glass cuvette (1 cm across) in a similar way.

The design of experiment is most important. Variation in concentration does not affect the occurrence of scroll and related waves provided the change is not too drastic.

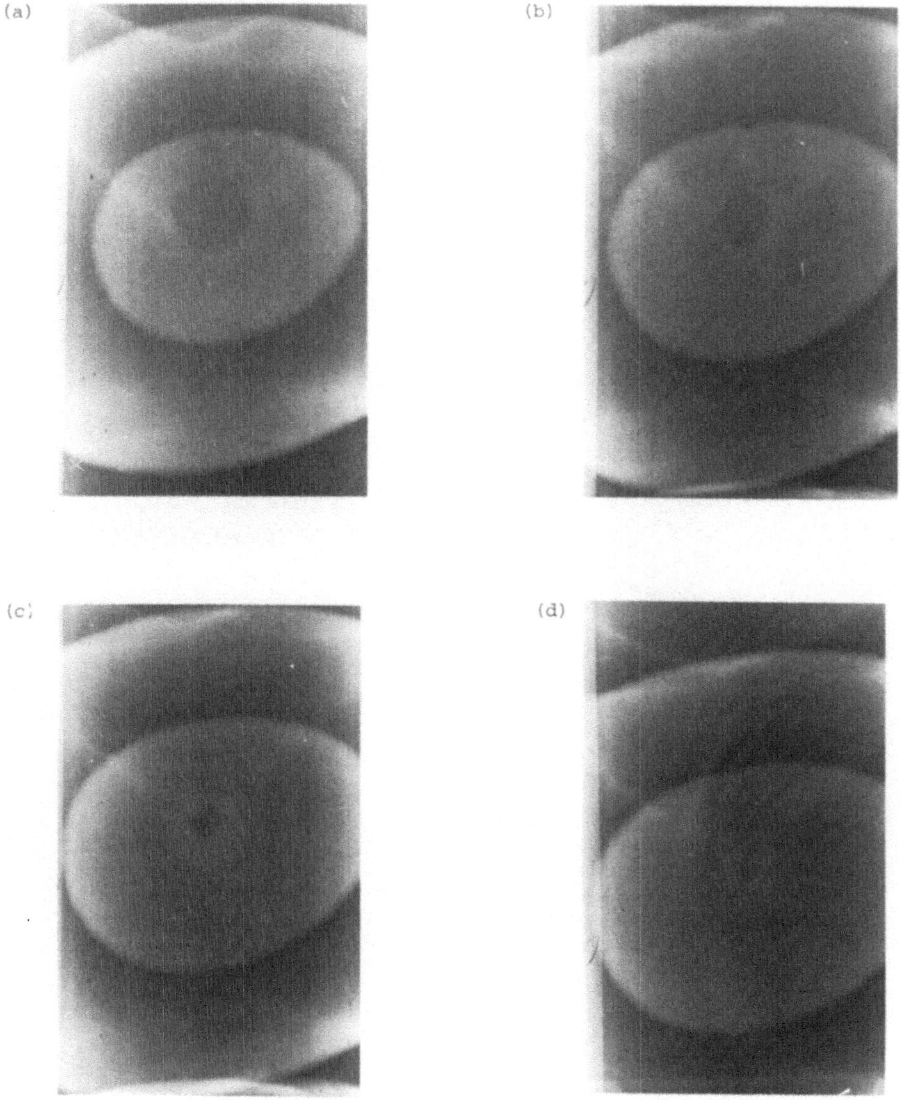

Fig 4. (a - d) Evolution of a sphere-like structure photographed at regular intervals over one period of evolution of about 2 minutes. Ambient room temperature maintained near 22°C.

(a) (b)

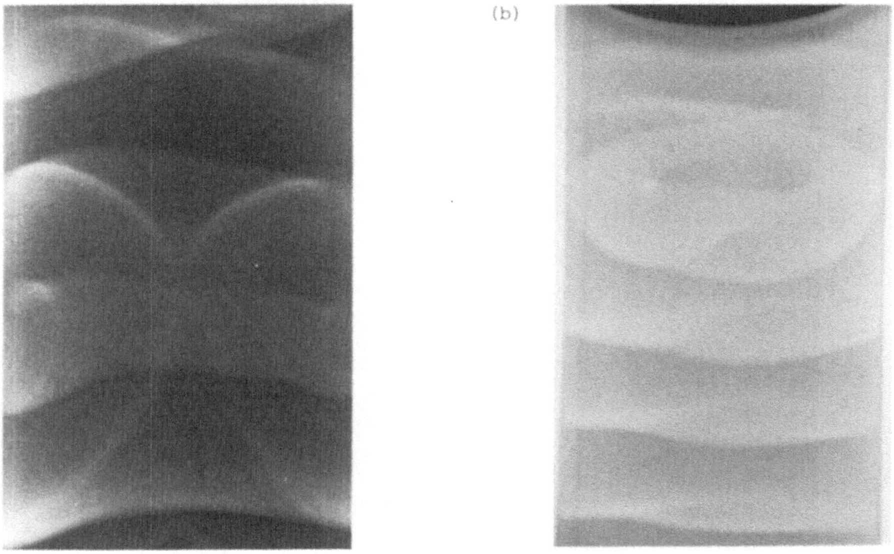

Fig 5.(a) Scroll wave sheared by the wall of the tube as it evolves.

(b) Toroidal scroll wave with an indent. In some perspectives it may appear to be a twisted toroidal scroll surface.

(a) (b)

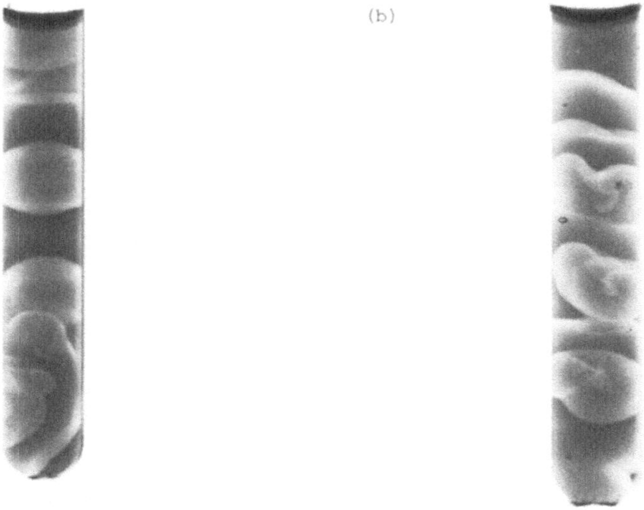

Fig 6.(a,b) Two tubes showing some unusual structures following perturbation of initial structures by lightly tapping or tipping the tubes.

REFERENCES

1. B J Welsh, J Gomatam, A E Burgess : Nature 304, 611 (1983)

2. W Geiseler, K Bar-Eli : In Modelling of Chemical Reaction Systems ed by K H Ebert, P Deuflhard, W Jäger (Springer, Berlin, Heidelberg, New York 1981)

3. E Bishop : In Indicators p545 ed by E Bishop (Pergamon Press, Oxford 1972)

4. A T Winfree : In Geometry of Biological Time (Springer, Berlin, Heidelberg, New York 1980)

5. G Patony, Z Noszticzius : React. Kinet. Catal. Lett. 17, 187 (1981)

6. S Glasstone : In Textbook of Physical Chemistry p696 (2nd edition, MacMillan, London 1960) or similar Physical Chemistry textbook

7. A T Winfree : Science 181, 937 (1973)

8. A T Winfree : JCS Faraday Trans. 9, 38 (1974)

9. A T Winfree , S H Strogatz : Singular Filaments Organise Chemical Waves in 3-Dimensions; 2 Twisted Waves (to appear in Physica D)

10. J Gomatam: J Phys. A 15, 1463 (1982)

11. H G Busse : J Phys. Chem. 73, 750 (1969)

12. M Herschkowitz-Kaufman : C R Acad. Sc, Paris t270C 1049 (1970)

13. A Pacault, C Vidal : J chim. phys. 79, 691 (1982)

Contributed by:

Burgess, Arthur E
Department of Chemistry
Glasgow College of Technology
Cowcaddens Road
Glasgow G4 OBA
UK

Co-authors:

Welsh, Brian J and
Gomatam, Jagannathan
Department of Mathematics
Glasgow College of Technology
Cowcaddens Road
Glasgow G4 OBA
UK

PATTERNS OF BIFURCATIONS IN PLANE LAYERS AND SPHERICAL SHELLS

BY F.H. BUSSE

Department of Earth and Space Sciences and
Institute of Geophysics and Planetary Physics
University of California, Los Angeles
Los Angeles, California 90024

1. INTRODUCTION

Among the numerous processes of pattern generation in biological, hydrodynamical and other physical systems the formations of patterns by bifurcations that break the symmetry of a basic state with respect to two or more dimensions are especially intriguing. These processes exhibit universal geometric properties on the one hand and a high sensitivity with respect to external parameters on the other hand. The contrast between the general geometric aspects of the generated patterns and their strong dependence on the parameters of a particular process make them a very attractive subject of theoretical and experimental research.

In this paper some aspects of the pattern formation in bifurcations from basic states that are homogeneous and isotropic with respect to planar and spherical surfaces will be reviewed. This subject is far from being well understood; but in the weakly nonlinear limit of the bifurcation process the beginning of a general theory have emerged. As will be discussed in section 2, the problem of pattern formation for small amplitudes of the bifurcating solution can be formulated without reference to the physical conditions under which bifurcation occurs. The mathematical description of the system and its parameters enter the problem only through the numerical value of certain coefficients which determine the stability of competing patterns and select the physically realized one. For an important subclass of problems it is possible to derive a pattern evolution equation which guarantees the existence of at least one stable steady solution as will be shown in section 3.

While the discussion of sections 2 and 3 is focussed on the planar case, a similar general approach can be applied to the problem of bifurcations from a spherically symmetric basic state. The theory is much richer in this case and has not yet been completed. A brief outline of the results will be given in section 4. Thermal convection in a plane layer heated from below or in a spherical shell heated from within represents a special example of the theory described in this review and will be referred to in several instances. Some aspects of the problem of secondary bifurcations will be discussed in the concluding section.

2. PATTERN FORMATION IN THE PLANAR CASE

To simplify the following discussion we shall assume that the solutions $\underset{\sim}{X}$ which bifurcate from a basic state described by $\underset{\sim}{X} = 0$ are steady. Analogous considerations will apply to the case of standing oscillations. The basic state is homogeneous and isotropic with respect to directions normal to the unit vector $\underset{\sim}{\lambda}$, and $\underset{\sim}{X}$ satisfies an equation of the type

$$(\underset{\approx}{W} + R\underset{\approx}{U}) \cdot \underset{\sim}{X} = \underset{\sim}{Q}(\underset{\sim}{X}, \underset{\sim}{X}) \tag{2.1}$$

where $\underset{\approx}{W}$ and $\underset{\approx}{U}$ are linear operators and R is the control parameter. $\underset{\sim}{Q}$ is a nonlinear operator which for simplicity will be assumed to be quadratic in $\underset{\sim}{X}$.

An example of the operator equation (2.1) are the equations governing steady convection in a horizontal layer heated from below (see, for example, Busse (1978)),

$$\nabla^2 \Delta_2 \Phi - \Delta_2 \Theta = \underset{\sim}{\delta} \cdot (\underset{\sim}{\delta}\Phi \cdot \nabla \underset{\sim}{\delta}\Phi) P^{-1} \tag{2.2a}$$

$$\nabla^2 \Theta - R \Delta_2 \Phi = \underset{\sim}{\delta}\Phi \cdot \nabla \Theta \tag{2.2b}$$

where R is the Rayleigh number, P is the Prandtl number, $\underset{\sim}{\lambda}$ points in the vertical direction, and the operator Δ_2 is defined by

$$\Delta_2 \equiv \nabla^2 - (\underset{\sim}{\lambda} \cdot \nabla)^2$$

Equation (2.2a) can be derived from the Navier-Stokes equations by introducing the general representation

$$\underset{\sim}{v} = \nabla \times (\nabla \times \underset{\sim}{\lambda}\Phi) + \nabla \times \underset{\sim}{\lambda}\psi \equiv \underset{\sim}{\delta}\Phi + \nabla \times \underset{\sim}{\lambda}\psi \tag{2.3}$$

for the solenoidal velocity field v. Because $\underset{\sim}{v}$ does not exhibit a vertical component of vorticity to the order in which the problem will be considered here (Schlüter et al., 1965), the second term in the representation (2.3) can be neglected and only Φ enters equations (2.2). The vector $\underset{\sim}{X}$ thus consists of the two components Φ and Θ in the case of equations (2.2). Typical boundary conditions which must be added to equations (2.2) are given by

$$\Phi = \underset{\sim}{\lambda} \cdot \nabla \Phi = \Theta \quad \text{at } \underset{\sim}{\lambda} \cdot \underset{\approx}{x} = \pm\frac{1}{2} \tag{2.4}$$

in the case of rigid top and bottom boundaries with high heat conductivity.

Because of the isotropy of the basic state the linear eigenvalue problem

$$(\underset{\approx}{W} + R_o \underset{\approx}{U}) \cdot \underset{\sim}{X}_o = 0 \tag{2.5}$$

exhibits infinitely degenerate eigenvalues R_o. From the physical point of view only the lowest eigenvalue R_o is of interest to which the attention will be restricted in the following. Using a Cartesian system of coordinates (x,y,z) with z-coordinate in the direction of $\underset{\sim}{\lambda}$ we can write the eigenvectors in the form

$$\underset{\sim}{X}_o = \underset{\sim}{f}(z)\, w(x,y) \equiv \underset{\sim}{f}(z) \sum_{n=-N}^{N} c_n \exp\{i \underset{\sim}{k}_n \cdot \underset{\sim}{x}\} \tag{2.6}$$

with $\underset{\sim}{k}_n \cdot \underset{\sim}{\lambda} = 0$, $\underset{\sim}{k}_{-n} = -\underset{\sim}{k}_n$, $|\underset{\sim}{k}_n| = \alpha$.

Because of the isotropy of the basic state R_o depends only on the absolute value of the vectors $\underset{\sim}{k}_n$. The value α at which R_o reaches a minimum is called the critical value α_c. Since $\underset{\approx}{X}_o$ is real and its amplitude is undetermined it is appropriate to impose the conditions

$$c_{-n} = c_n^+ , \quad \sum_{n=-N}^{N} c_n c_n^+ = 1$$

where the superscript $^+$ indicates the complex conjugate. The summation limit N does not need to be finite. The function space of the functions $w(x,y)$ consists of all almost periodic functions satisfying

$$\Delta_2 w(x,y) + \alpha^2 w(x,y) = 0 \tag{2.7}$$

Starting with the general solution (2.6) of the linear problem, we solve the nonlinear problem (2.1) by assuming an expansion in powers of the amplitude parameter ε,

$$\begin{aligned} X &= (\underset{\sim}{X}_o + \varepsilon \underset{\sim}{X}_1 + \varepsilon^2 \underset{\sim}{X}_2 + \ldots) \\ R &= R_o + \varepsilon R_1 + \varepsilon^2 R_2 + \ldots \end{aligned} \tag{2.8}$$

Several definitions of ε are possible, among which the definition

$$\varepsilon = \langle \underset{\sim}{X}_o^* \cdot \underset{\approx}{U} \cdot \underset{\sim}{X}_o \rangle \tag{2.9}$$

is particularly convenient. Here the angular brackets indicate a suitable average over the domain of the problem. In the case of the convection problem (2.2), for example, the average over the fluid layer is used. $\underset{\sim}{X}_o^*$ denotes the adjoint solution to solution (2.6), i.e. $\underset{\sim}{X}_o^*$ satisfies the adjoint equation to equation (2.5) with the same x,y-dependence as $\underset{\sim}{X}_o$,

$$\underset{\sim}{X}_o^* = \underset{\sim}{f}^*(z) \sum_{n=-N}^{N} c_n \exp\{i \underset{\sim}{k}_n \cdot \underset{\sim}{x}\} \equiv \sum_{n=-N}^{N} c_n \underset{\sim}{X}_{on}^* \tag{2.10}$$

The terms of order ε^2 in equation (2.1) give rise to the inhomogeneous linear problem

$$(\underset{\approx}{W} + R_o \underset{\approx}{U}) \cdot \underset{\sim}{X}_1 = -R_1 \underset{\approx}{U} \cdot \underset{\sim}{X}_o + \underset{\sim}{Q}(\underset{\sim}{X}_o, \underset{\sim}{X}_o) \tag{2.11}$$

The Fredholm alternative requires that the right hand side must be orthogonal to all solutions of the adjoint homogeneous problem. In accordance with definition (2.10) a complete set of these solutions can be written in the form

$$\underset{\sim}{X}_{om}^* \equiv \underset{\sim}{f}^*(z) \exp\{i \underset{\sim}{k}_m \cdot \underset{\sim}{x}\}, \quad -\infty \le m \le \infty \tag{2.12}$$

A non-trivial contribution on the right hand side of the solvability condition

$$R_1 c_{-m} = \langle \underset{\sim}{X}_{om}^* \cdot \underset{\sim}{Q}(\underset{\sim}{X}_o, \underset{\sim}{X}_o) \rangle , \quad -N \le m \le N \tag{2.13}$$

is obtained only if the condition

$$\underset{\sim}{k}_m + \underset{\sim}{k}_n + \underset{\sim}{k}_\ell = 0 \tag{2.14}$$

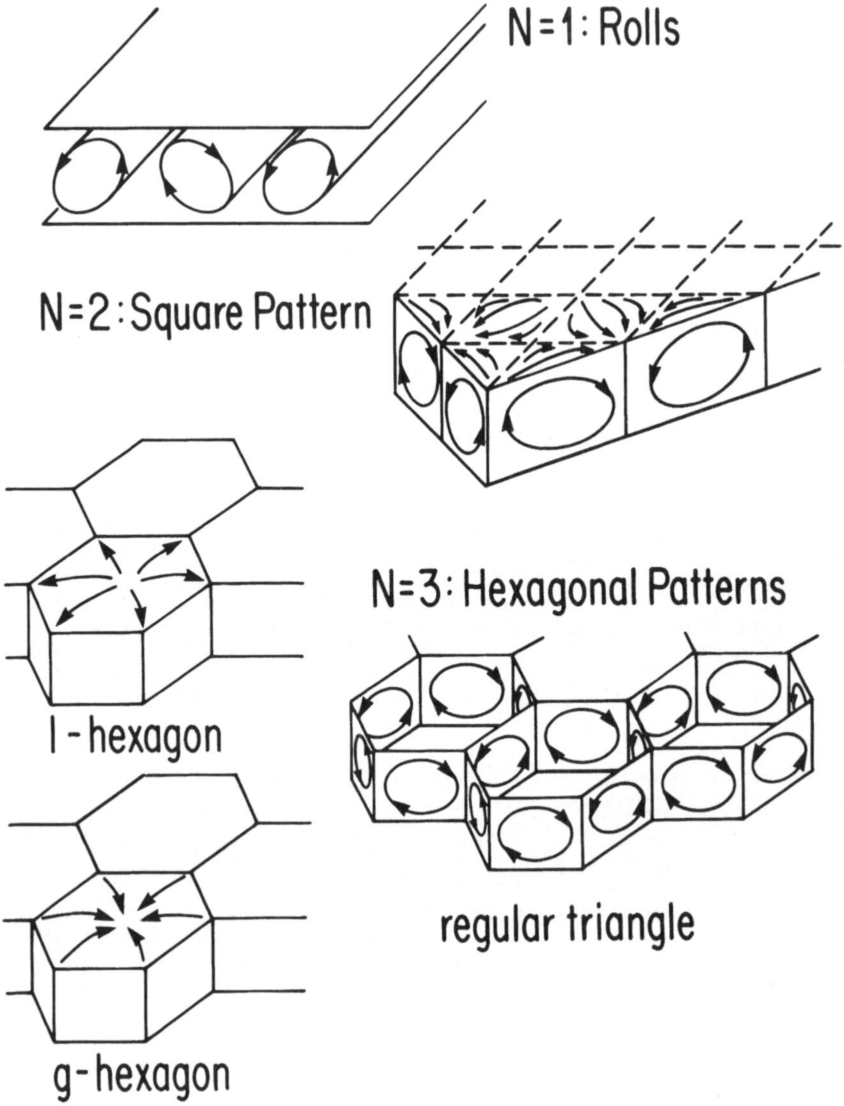

Figure 1. Patterns of convection in a layer heated from below. The regular triangle pattern is shown only as an example of a more general class of patterns fitting the hexagonal lattice. There are no physical conditions known under which the regular triangle pattern can occur as a convection pattern.

is met which implies an angle of 120° between the three $\underset{\sim}{k}$-vectors. For many applications the value of R_1 obtained from equations (2.13) either vanishes or is small in which case it is appropriate to add the non-vanishing terms to the equations derived from the solvability condition of the order ε^3. The main result in the order ε^2 is that $\underset{\approx}{X}_1$ can be written in the form

$$\underset{\sim}{X}_1 = \sum_{\ell,p} \underset{\sim}{F}(z,\underset{\sim}{k}_\ell \cdot \underset{\sim}{k}_p) c_\ell c_p \exp\{i(\underset{\sim}{k}_\ell + \underset{\sim}{k}_p)\cdot \underset{\sim}{x}\} \tag{2.15}$$

The equation of the order ε^3 gives rise to the solvability condition

$$R_2 c_{-m} = \langle \underset{\sim}{X}^*_{om} \cdot [\underset{\sim}{Q}(\underset{\sim}{X}_o, \underset{\sim}{X}_1) + \underset{\sim}{Q}(\underset{\sim}{X}_1, \underset{\sim}{X}_o)] \rangle \tag{2.16}$$

Evaluation of the right hand side yields non-vanishing contributions wherever the four $\underset{\sim}{k}$-vectors involved in each term add to zero,

$$\underset{\sim}{k}_m + \underset{\sim}{k}_n + \underset{\sim}{k}_\ell + \underset{\sim}{k}_p = 0. \tag{2.17}$$

Accordingly a system of equations of the following form is obtained for the as yet unknown coefficients c_m,

$$\begin{aligned}(R - R_o)\, c_{-m} &= (\varepsilon R_1 + \varepsilon^2 R_2 + \ldots)c_{-m} \\ &= \varepsilon\beta \sum_{\ell,p=-n}^{N} c_n c_m \quad (\underset{\sim}{k}_m + \underset{\sim}{k}_\ell + \underset{\sim}{k}_p) \\ &+ \quad c_{-m} \sum_{n=1}^{N} c_n c_m^+ \, A(\underset{\sim}{k}_n \cdot \underset{\sim}{k}_m,\ \underset{\sim}{k}_n \times \underset{\sim}{k}_m \cdot \underset{\sim}{\lambda}) + \ldots\end{aligned} \tag{2.18}$$

As mentioned earlier the terms of the order ε are regarded as small, i.e. $\beta \ll A(\alpha^2, 0)$, and terms of the order ε^3 or higher are not given explicitly since they are negligible in general. Because of the constraint (2.17) the function A can depend only on the two scalar invariants of two $\underset{\sim}{k}$-vectors. Unless the problem involves a sense of rotation about the axis indicated by $\underset{\sim}{\lambda}$, the dependence of A on $\underset{\sim}{k}_n \times \underset{\sim}{k}_m \cdot \underset{\sim}{\lambda}$ vanishes.

Among the solutions of equations (2.18) those corresponding to a regular distribution of $\underset{\sim}{k}$-vectors are of special interest. These regular solutions are given by

$$|c_1|^2 = \ldots = |c_N|^2 = N/2 \tag{2.19}$$

and include all commonly observed patterns in the weakly nonlinear regime of bifurcation from basic states with planar isotropy. Actually, only the pattern corresponding to N = 1, N = 2, and N = 3 are periodic in the plane and thus are expected to be preferred in physical realizations of the bifurcation problem. Examples of these patterns are shown in figure 1 in the case of convection. It is noteworthy that the freedom of the phases of the coefficients c_n is equilvalent to the translations of a single solution in the x,y-plane in the cases N = 1, N =2, while in the case N = 3 a one-parameter class of different solutions is permitted by (2.19) of which only the most symmetric examples are shown in figure 1. In the case of finite β equations (2.18) require a real product

$c_1 c_2 c_3$ which excludes all but the ℓ- and g-hexagon solutions (Busse, 1967a). But Buzano and Golubitsky (1983) and Golubitsky et al. (1983) have shown that the regular triangle solution could in principle describe a stable form of convection in the case $\beta = 0$. Using group theoretical methods these authors give a complete description of patterns fitting the hexagonal lattice and also include the effects of terms of higher order than those discussed in this review.

3. AN EVOLUTION EQUATION

In order to discuss the time dependence of the evolving patterns and the stability of steady patterns a derivative with respect to time must be added to the basic equation (2.1),

$$(\underset{\approx}{W} + R\underset{\approx}{U}) \cdot \underset{\sim}{X} = \underset{\approx}{V} \cdot \frac{\partial}{\partial t} \underset{\sim}{X} + \underset{\sim}{Q}(\underset{\sim}{X}, \underset{\sim}{X}) \qquad (3.1)$$

Since we are interested only in the weak time dependence by which solutions approach their steady states and not in the general discussion of the initial value problem, it is sufficient to restrict the attention to the slow time scale, $t \sim \varepsilon^{-2}$. Accordingly the expressions derived in the preceding section remain valid except that the coefficients c_n are now time dependent and a time derivative must be added in equations (2.18). In the case when A is a function only of $\underset{\sim}{k}_n \cdot \underset{\sim}{k}_m$ the resulting equations can be written in the form

$$M \, d C_m^+/dt = - \frac{\partial}{\partial C_m} F(C_{-N}, \ldots, C_N) \qquad (3.2)$$

where the definitions

$$F(C_{-N}, \ldots, C_N) \equiv \frac{-1}{2}(R - R_o) \sum_{n=-N}^{N} |C_n|^2 + \frac{1}{3}\beta \sum_{\ell,m,p} C_\ell C_m C_p$$

$$\delta(\underset{\sim}{k}_\ell + \underset{\sim}{k}_m + \underset{\sim}{k}_p) + \frac{1}{4} \sum_{n,m} A(\underset{\sim}{k}_n \cdot \underset{\sim}{k}_m) |C_n|^2 \, |C_m|^2$$

$$C_n(t) \equiv \varepsilon \, c_n(t)$$

have been used. Equations (3.2) guarantee the existence of at least one stable steady state characterized by a minimum of the function F (Busse, 1967a). There may be several stable solutions corresponding to several local minima. In the case of a convection layer rotating about a vertical axis the condition for the existence of the function F is not met and the situation that no stable steady state exists does indeed occur (Küppers and Lortz, 1969).

4. PATTERN FORMATION IN THE SPHERICAL CASE

In principle the mathematical problem of pattern formation in bifurcations from spherically symmetric basic states is simpler than in the planar case. Because of the finite domain of the sphere the degeneracy of the eigenvalue is finite and a larger variety of mathematical tools is available to treat the problem. The actual analysis is quite complex, however, since each discrete wavenumber ℓ denoting the degree of the spherical harmonics describing the bifurcating solution requires separate considerations. The problem of pattern formation in the spherical case is thus much richer and has only partially been solved.

Because of the general formulation of the pattern formation problem adopted in this review, only a few changes must be made in order to apply the discussion of section 2 to the spherical case. The unit vector λ must be replaced by the unit vector in the radial direction and the Cartesian system of coordinates must be replaced by a spherical system of coordinates (r, θ, ϕ). In place of expression (2.6) we obtain for the eigenvectors of the linear eigenvalue problem

$$\underset{\sim}{X}_o = \underset{\sim}{f}(r)\, w(\theta,\phi) \equiv \underset{\sim}{f}(r) \sum_{m=-\ell}^{\ell} c_m Y_\ell^m (\theta,\phi) \qquad (4.1a)$$

with

$$c_{-m} = c_m^+ , \quad \sum_{m=-\ell}^{\ell} |c_m|^2 = 1 \qquad (4.1b)$$

where the functions Y_ℓ^m represent an orthonormalized set of spherical harmonics,

$$Y_\ell^m(\theta,\phi) \equiv \exp\{im\phi\} P_\ell^m(\cos\theta)\ [(2\ell+1)(2-\delta_{mo})(\ell-m)!/(\ell+m)!]^{1/2} \qquad (4.2)$$

The extra factor in expression (4.2) has been introduced to insure that the average of $|Y_\ell^m|^2$ over the unit sphere becomes unity.

The eigenvalue R_o assumes discrete values as a function of the integer parameter ℓ in contrast to the continuous dependence on the wavenumber α in the planar case. Of physical interest is the value which minimizes R_o. The $(2\ell + 1)$-fold degeneracy of the eigenvector (4.1a) has two different origins. First there is the orientational degeneracy reflecting the manifold of solutions that can be generated from a given solution by shifting it on the spherical surfaces or reflecting it with respect to the center of the sphere. The second degeneracy is the pattern degeneracy by which we denote the coexistence of different solutions of the form (4.1a) which cannot be transformed into each other. In contrast to the vector notation employed in expression (2.6) the coordinate dependent notation in the spherical case does not permit an easy separation of the two kinds of degeneracy.

The orientational degeneracy corresponds to two degrees of freedom of transformation on the sphere. No pattern degeneracy thus exists in the case $\ell = 1$ where only two coefficients c_n can be chosen independently. But for $\ell \geq 2$ a non-trivial problem of pattern degeneracy does exist and the solvability conditions in the higher orders in ε of the nonlinear problem must be analyzed to determine the possible steady patterns for a given value of ℓ. Since this problem has been discussed recently from a general point of view (Busse and Riahi, 1983) only the main results will be quoted here:

(i) Because of the property

$$Y_\ell^m (\theta,\phi) = (-1)^\ell\, Y_\ell^m(\pi-\theta, \phi+\pi) \qquad (4.3)$$

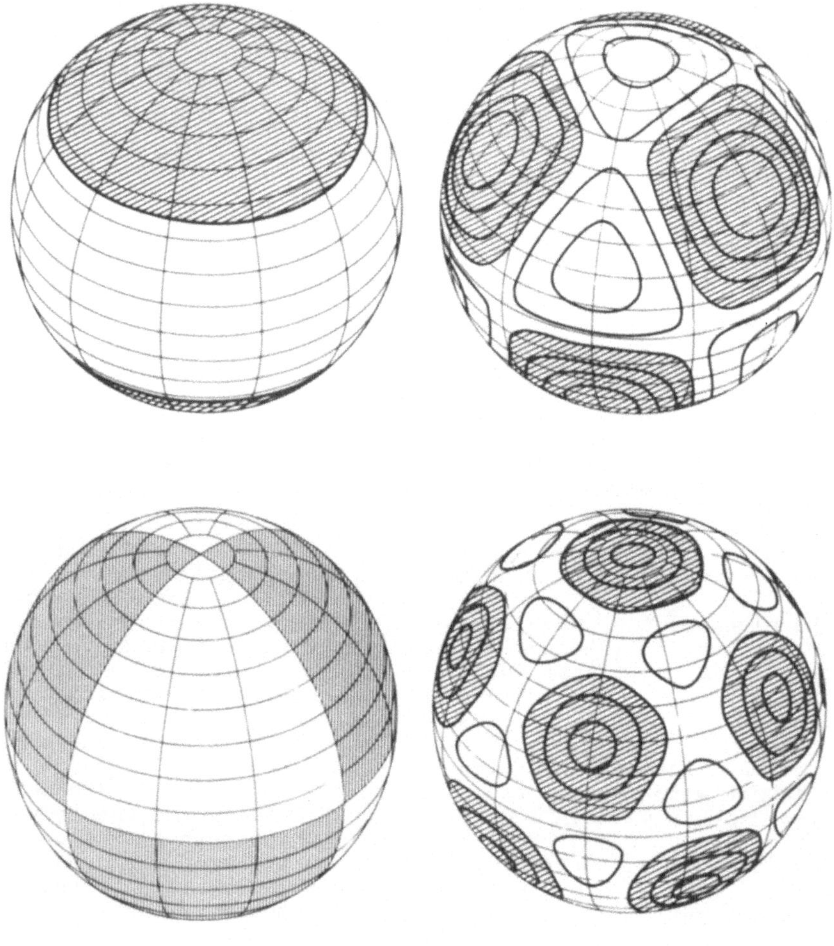

Figure 2. Preferred patterns of solutions bifurcating from a spherically symmetric state. The cases $\ell = 2, 3$ (left) and $\ell = 4, 6$ (right) are shown. It is possible that instead of the solution with tetrahedronal symmetry the solution described by $Y_3^3(\theta,\phi)$ may be preferred in special cases.

of spherical harmonics, the average over the sphere of triple products of spherical harmonics with odd degree ℓ vanish. Accordingly the coefficient R_1 determined by equations of the form (2.13) must vanish for odd ℓ. By the same reasoning it can be shown that all coefficients R_n with odd subscripts n vanish in the case of odd ℓ.

(ii) Since R_n with odd subscript n does not vanish in general for even ℓ, there always exist two physically different solutions for a given pattern. The two solutions differ by the sign of the coefficients c_n and have the same relationship to each other as the ℓ- and g-hexagon solutions.

(iii) Since R_2 is typically positive in bifurcation problems characterized by a minimum of the control parameter R for the physical onset of bifurcation, subcritical bifurcation can only occur for a negative value of R_1, i.e. in the case of even ℓ. This gives a slight edge to even ℓ modes in the competition with odd ℓ modes. In addition a negative εR_1 causes an onset of hysteresis of the kind discussed in the case of hexagonal convection (Busse, 1967a).

Among the more specific results that have been obtained in cases of low integers ℓ we mention the remarkable property that the axisymmetric pattern which always corresponds to a possible steady solution is stable only in the cases $\ell = 1$ and $\ell = 2$. The preferred patterns which are exhibited by stable steady solutions in the neighborhood of the critical value R_o of the control parameter are shown in figure 2. It should be mentioned that the solution for $\ell = 3$ exhibiting tetrahedronal symmetry competes with a solution exhibiting a purely sectorial structure which may also be stable depending on details of the problem.

5. CONCLUDING REMARKS

The general treatment of the weakly nonlinear problem of pattern formation has been particularly successful in the case of thermal convection where it was first developed. The transition of convection in hexagonal cells to convection rolls observed at low Rayleigh numbers (Silveston, 1958) can readily be understood on the basis of a theory of the type discussed in this review (Segel and Stuart, 1962; Busse, 1962, 1967a). At high Rayleigh numbers transitions to new patterns of convection occur which cannot be described by solutions in the form (2.6). The transition from steady rolls to steady bimodal convection is characterized by the appearance of a second wavenumber indicating the instability of the thermal boundary layer generated by large amplitude convection (Busse, 1967b). Other nonlinear processes are responsible for further transitions to oscillatory bimodal convection and spoke pattern convection which have been observed experimentally (Busse and Whitehead, 1974). Obviously the assumption of a bifurcation from a homogeneous and isotropic basic state is no longer satisfied, even in the approximate sense, in the case of these higher transitions and each case thus requires a separate theoretical treatment.

In biological problems of pattern formation the condition of isotropy of a basic state is not often realized. But the example of the skeleton pattern shown in figure 3 demonstrates that the general theory reviewed in this paper appears to be applicable in this particular case. The coverage of the sphere with hexagons and an occasional pentagon is just what must be expected in the limit of high values of ℓ when the problem of pattern formation of the sphere approaches the corresponding problem in the plane.

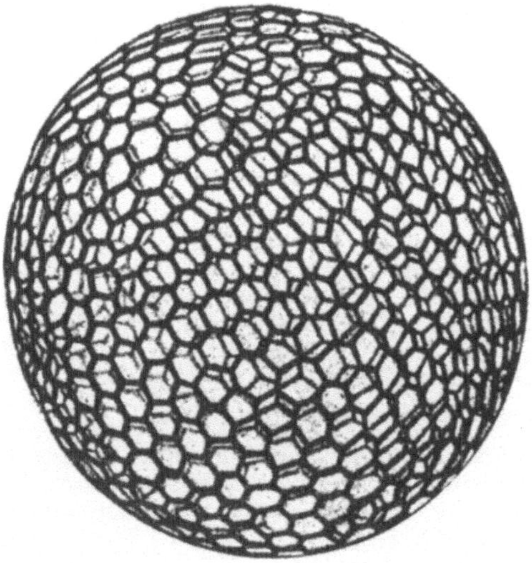

Figure 3. Skeleton of the radiolaria Aulonia hexagona, from Thompson (1942).

REFERENCES

Busse, F.H., The stability of finite amplitude cellular convection and its relation to an extremum principle, J. Fluid Mech. 30, 625-649, 1967a.

Busse, F.H., On the stability of two-dimensional convection in a layer heated from below, J. Math. Phys. 46, 140-150, 1967b.

Busse, F.H., Nonlinear properties of convection, Rep. Progress in Physics 41, 1929-1967, 1978.

Busse, F.H., and N. Riahi, Patterns of convection in spherical shells II, J. Fluid Mech. 123, 283-302, 1982.

Busse, F.H., and J.A. Whitehead, Oscillatory and collective instabilities in large Prandtl number convection, J. Fluid Mech. 66, 67-79, 1974.

Buzano, E., and Golubitsky, Bifurcation on the hexagonal lattice and the planar Bénard problem, Trans. Roy. Soc. London A 308, 617, 1983.

Golubitsky, M., J.W. Swift, and E. Knobloch, Symmetries and pattern selection in Rayleigh-Bénard convection, Physica D., in press, 1983.

Küppers, G., and D. Lortz, Transition from laminar convection to thermal turbulence in a rotating fluid layer, J. Fluid Mech. 35, 609-620, 1969.

Segel, L.A., and J.T. Stuart, On the question of the preferred mode in cellular thermal convection, J. Fluid Mech. 13, 289-306, 1962.

Silveston, P.L., Wärmedurchgang in waagerechten Flüssigkeitsschichten Forsch. Ing. Wes. 24, 29-32, 59-69, 1958.

Thompson, D., On Growth and Form, Cambridge University Press, 1942.

CHEMOTACTIC COLLAPSE IN TWO DIMENSIONS

Stephen Childress
Courant Institute of Mathematical Sciences
New York University
New York, N.Y. 10012

Summary: An asymptotic expansion describing imminent collapse in two dimensions of a symmetric aggregate of chemotactic cells is given. The result is obtained in a Keller-Segel model with constant coefficients. The analysis is consistent with the existence of a threshold cell population for chemotactic collapse.

1. Preliminary Remarks.

In this note we consider some nonlinear features of chemotaxis in two spatial dimensions. The model we use is a special case of those introduced by Keller and Segel (1970). The present calculations continue earlier work (Childress and Percus 1981, hereafter denoted by CP). It was conjectured in CP that the elementary chemotaxis model we study (eqns. (1.1), (1,2) below) leads in two dimensions to a threshold sensitivity to collapse: if an aggregate contains more than a critical number of cells, the aggregate can collapse to infinite density (in finite time). For aggregates less than this critical size, such a collapse would then never occur. We shall describe here an asymptotic result consistent with this conjecture and the other supporting arguments advanced in CP.

As explained in CP, the model equations are

$$\frac{\partial a}{\partial t} = \nabla \cdot [\mu \nabla a - \chi a \nabla b] \quad , \tag{1.1}$$

$$\frac{\partial b}{\partial t} = fa - gb + \nu \nabla^2 b \; , \tag{1.2}$$

where a and b are respectively cell and chemical density. The parameters μ, χ, f, g, ν are all positive constants. We shall deal with radially symmetric distributions within a disc of radius L in the plane so a,b will depend upon $r = (x^2+y^2)^{1/2}$ and t. The boundary conditions are homogeneous Neumann conditions on $r = L$.

It is convenient to introduce dimensionless variables defined by

$$\alpha = \frac{f\chi L^2}{\mu\nu} a\ , \quad \beta = \frac{\chi}{\mu} b\ , \quad \rho = r/L\ , \quad \tau = t\mu/L^2\ . \tag{1.3}$$

Then (1.1), (1.2) may be written, with the assumed symmetry, in the form

$$\frac{\partial\alpha}{\partial\tau} = \rho^{-1}\frac{\partial}{\partial\rho}\rho\left(\frac{\partial\alpha}{\partial\rho} - \alpha\frac{\partial\beta}{\partial\rho}\right)\ , \tag{1.4}$$

$$\delta_1\frac{\partial\beta}{\partial\tau} = \alpha - \delta_2\beta + \rho^{-1}\frac{\partial}{\partial\rho}\left(\rho\frac{\partial\beta}{\partial\rho}\right)\ , \tag{1.5}$$

where

$$\delta_1 = \mu/\nu\ , \qquad \delta_2 = gL^2/\nu\ . \tag{1.6}$$

Given $\alpha(\rho,0)$, $\beta(\rho,0) \geq 0$, and that $\partial\alpha/\partial\rho = \partial\beta/\partial\rho = 0$ on $\rho = 1$, we seek to study the evolution of α,β in the limit of small δ_1 and δ_2 . The total cell population, given by

$$\theta = \int_0^1 \rho\alpha\, d\rho = \text{constant} \tag{1.7}$$

will also be fixed to a value $4 + \epsilon$, $0 < \epsilon << 1$, where in the present variables $\theta = 4$ is the threshold value introduced in CP.

2. The Formal Limit

If δ_1 is set equal to zero in (1.5), then from (1.6) we see that, in effect, diffusion of β is rapid relative to diffusion of α and, for a given α , β will rapidly adjust to a magnitude and distribution consistent with (1.5) and the Neumann condition at $\rho = 1$. As $\delta_2 \to 0$, we must therefore include a large constant component and set

$$\beta = \frac{2\theta}{\delta_2} + \beta^*\ , \qquad \beta^* \sim O(1)\ . \tag{2.1}$$

In the limit $\delta_1,\delta_2 \to 0$, (2.1),(1.5) then imply, after one integration,

$$\int_0^\rho r\alpha(r,\tau)dr - \theta\rho^2 + \rho\frac{\partial\beta^*}{\partial\rho} = 0\ . \tag{2.2}$$

In view of (1.7) we see that the Neumann condition on β is now satisfied.

Eliminating β from (1.4) and integrating once we then obtain

$$\frac{\partial A}{\partial \tau} = 2(A-\theta\sigma)\frac{\partial A}{\partial \sigma} + 4\sigma\frac{\partial^2 A}{\partial \sigma^2} \tag{2.3}$$

where

$$A = \int_0^\rho r\,\alpha(r,\tau)dr\ , \qquad \sigma = \rho^2\ .$$

The conditions on (2.2) are

$$A(0,\tau) = 0\ , \qquad A(1,\tau) = \theta\ . \tag{2.4}$$

Equation (2.3) is an improvement on equation (4.11a) of CP, the latter being valid only within a small aggregate. Here (2.3) holds throughout the disc in the limiting case considered.

3. The Expansion

We introduce a small expansion parameter for a solution of (2.2), (2.4) by setting $\theta = 4 + \in$. According to CP, we expect collapse to occur if the initial density of α is sufficiently concentrated at $\rho = 0$. This suggests the ordering $\sigma = \in s$, s,τ of order unity. Then (2.3) may be written

$$\frac{\partial}{\partial s}\left[A^2 - 4A + 4s\frac{\partial A}{\partial s}\right] = \in\left[\frac{\partial A}{\partial \tau} + 2\theta s\frac{\partial A}{\partial s}\right]\ . \tag{3.1}$$

Suppose, therefore, that

$$A = A_0(s,\tau;\in) + \in A_1(s,\tau;\in) + O(\in^2) \tag{3.2}$$

where we allow for A_0 and A_1 to contain intermediate-order terms involving powers of $\ell n\in$. Substitution of (3.2) into (3.1) shows that A_0 satisfies (3.1) with zero right-hand side. The only solution vanishing at $s=0$ is

$$A_0 = \frac{4k(\tau;\in)s}{1+k(\tau;\in)s}\ , \tag{3.3}$$

where $k(\tau;\in)$ is an arbitrary function of τ (cf. (4.17) of CP, with k replaced by 8k). The equation for A_1 is

$$\frac{\partial}{\partial s}\left[(2A_0-4)A_1 + 4s\frac{\partial A_1}{\partial s}\right] = \frac{\partial A_0}{\partial \tau} + 2\theta s\frac{\partial A_0}{\partial s}\ . \tag{3.4}$$

Introducing $u = ks$, we can bring (3.4) into the form

$$\frac{\partial}{\partial u}\left[\left(\frac{u-1}{u+1}\right) A_1 + u\,\frac{\partial A_1}{\partial u}\right] = \left(\frac{\overset{\circ}{k}}{k^2} + \frac{2\theta}{k}\right)\frac{u}{(1+u)^2}\,Y \tag{3.5}$$

with $A_1(0,\tau) = 0$, where $\overset{\circ}{k} = dk/d\tau$. The solution is

$$A_1 = \left(\frac{\overset{\circ}{k}}{k^2} + \frac{2\theta}{k}\right)\left[\frac{u}{(1+u)^2}\right]\Bigg[(1+u)\ln(1+u) - 2u - \frac{(1+u)}{u}\ln(1+u)$$

$$+ 2\int_0^u \frac{1}{v}\ln(1+v)\,dv\Bigg] + c(\tau)\,\frac{u}{(1+u)^2}\ .$$

We now impose the condition that $A = \theta = 4 + \epsilon$ at $\sigma = 1$. Using (3.3) and (3.6) in (3.2), we obtain

$$4 + \epsilon = \frac{4k}{\epsilon + k} + \epsilon\left(\frac{\overset{\circ}{k}}{k^2} + \frac{2(4+\epsilon)}{k}\right)\left[\ln(k/\epsilon) + O[\epsilon(\ln\epsilon)^2]\right]\ , \tag{3.7}$$

or, with $k = \ln(1/\epsilon)K$,

$$\overset{\circ}{K} = K^2\,\{1 + [\ln K/\ln(1/\epsilon)] + [\ln(\ln(1/\epsilon))\,/\,\ln(1/\epsilon)]\}^{-1}$$

$$- 8K + O(\epsilon(\ln\epsilon)^2) \tag{3.8}$$

From (3.8), with $\epsilon << 1$, K will grow and become infinite in a finite time provided $K(0)$ exceeds a critical value $K^* \underset{\sim}{\sim} 8$. If $K(0) < K^*$, $K \to 0$ as $\tau \to \infty$. Returning to $A(\sigma,t)$, this implies that the threshold distribution of cells is given by

$$A = A^* = \frac{4K^*\epsilon^{-1}\ln(1/\epsilon)\sigma}{1 + K^*\epsilon^{-1}\ln(1/\epsilon)\sigma}\ . \tag{3.9}$$

4. Discussion

According to the conjecture made in CP, in the region $\theta > 4$ there is a branch of unstable equilibrium aggregates n which terminates at the first bifurcation point $\theta \sim 7.3$. As $\theta \downarrow 4$ on this branch, the amplitude of the solution as measured by $\beta(0)$ should tend to infinity (cf. Figure 3 of CP). The asymptotic structure of this unstable branch has been obtained above. Indeed, if (3.9) is used in (2.3), noting that for a steady solution

$$\int_0^1 \beta^* d\sigma = 0 \quad , \tag{4.1}$$

we obtain the asymptotic expression

$$\beta(0) \sim \frac{2\theta}{\delta_2} + 2\ln\left(K^* \frac{\ln(1/\epsilon)}{\epsilon}\right) \quad , \quad \epsilon \to 0 \quad . \tag{4.2}$$

With $\epsilon = \theta - 4$, (4.2) provides the limiting form of the unstable branch in the $(\theta, \beta(0))$ plane.

Since this result applies only for small ϵ , the possibility of collapse to a delta function of amplitude sensibly greater than 4 remains an open question. The present analysis suggests, however, that first collapse can involve an aggregate of arbitrary size > 4 . In a general two dimensional problem, collapsed aggregates can move as discrete entities, attracting further cells and thereby increasing in amplitude. An example of this process is given elsewhere in this volume by Lauffenburger.

Using equation (2.3) we can easily study the final phase of the above accretion process, given that a delta function with amplitude A_o occurs initially at $\sigma = o$, provided that A_o is taken close to θ . We may then substitute

$$A = \theta + (A_o - \theta)\, f(\sigma,\tau) \tag{4.3}$$

into (2.2) and neglect the term quadratic in $A_o - \theta$. Then

$$\frac{\partial f}{\partial \tau} \simeq 2\theta(1-\sigma)\frac{\partial f}{\partial \sigma} + 4\sigma \frac{\partial^2 f}{\partial \sigma^2} \quad , \tag{4.4a}$$

$$f(0,0) = 1 \quad , \qquad f(1,\tau) = 0 \; . \tag{4.4b}$$

One solution of (4.4) is $f = e^{-2\theta\tau}(1-\sigma)$, and others would presumably exhibit similar decay, since the convection of f in (4.4a) is now uniformly to the left, while f is monotone decreasing for all time. It remains an open question whether or not symmetric solutions of (2.3) which do not relax to the uniform state $A = \theta\sigma$ invariably form a delta function at $\sigma = 0$ of amplitude θ .

This research was supported by the National Science Foundation under Grant MCS-8301809 at New York University, and by a Visiting Fellowship Grant from the SERC at the University of Newcastle. The author is indebted to A. M. Soward for helpful discussion.

REFERENCES

Childress, S. and Percus, J.K., 1981. Math. Biosci. 56, 217.

Keller, E.F. and Segel, L.A., 1970. J. Theor. Biol. 26, 399.

SINGULAR BIFURCATION IN REACTION - DIFFUSION SYSTEMS

Thomas Erneux, Bernard J. Matkowsky and Edward L. Reiss
Department of Engineering Sciences and Applied Mathematics
Northwestern University
Evanston, Illinois 60201/USA

1. What is Singular Bifurcation?

Bifurcation Theory is a study of the branching of solutions of equations as a parameter λ, called the bifurcation parameter, is varied. The branching, or bifurcation points, are singular points of the solutions. In Bifurcation Theory solutions are analyzed near bifurcation points.

In this theory there is usually a distinguished solution which is called the basic state. Typically, it exists for all values of λ and it is usually determined by the physics, or biology that the bifurcation problem models.

The bifurcation points of the basic state are the primary bifurcation points and the solutions branching from them, other than the basic state, are the primary bifurcation states. These states are determined analytically by asymptotic methods, such as the Poincaré-Linstedt or other equivalent methods. The results of these asymptotic analyses yield a system of equations, simpler than the original bifurcation problem, to determine the amplitudes of the primary bifurcation states. They are called the primary amplitude equations. The primary bifurcation states are usually more complex spatially and/or temporally than the basic state, thus initiating a pattern formation.

Similarly, solutions of the bifurcation problem, other than the basic state, that branch from the primary bifurcation states are secondary bifurcation states and the corresponding branching points are secondary bifurcation points. The secondary bifurcation states are frequently more complex than the primary bifurcation states, thus continuing the pattern formation. More generally, there may be a sequence of bifurcations whose corresponding solutions are increasingly more complex spatially and temporally. We have referred to this elsewhere [1,2] as cascading bifurcation. The idea of cascading bifurcation has been proposed as a possible explanation of the transition from laminar to turbulent flows in certain hydrodynamic stability problems as a flow parameter λ, such as the Reynolds number, increases. Similarly, cascading bifurcations have been offered as a theoretical explanation of biological and biochemical morphogenesis [3].

In many problems the primary bifurcation points, which we denote by $\lambda_1(\underset{\sim}{p}),\lambda_2(\underset{\sim}{p}),\cdots$, depend on a vector $\underset{\sim}{p}$ of auxiliary parameters that occur in the original bifurcation (reaction-diffusion) problem. Since the work presented in [4], it has been customary to analyze secondary and cascading bifurcation by considering the additional singular situation when two or more primary bifurcation points coincide as the vector $\underset{\sim}{p} \to \underset{\sim}{p}_0$ to produce a multiple primary bifurcation point. Then asymptotic and other methods are employed to determine the cascading bifurcation of solutions, where the small parameter in the asymptotic analysis is a monotone function of $|\underset{\sim}{p} - \underset{\sim}{p}_0|$. The results of the analysis yield amplitude equations for the secondary and higher order states; see e.g. [1,2,4].

It has recently been recognized in certain reaction-diffusion problems [5,6] that the primary or secondary amplitude equations are singular, i.e. they are not uniquely solvable as $\underset{\sim}{p}$ approaches special values $\underset{\sim}{p}_1$. We refer to this situation as singular bifurcation (primary or secondary) or more generally singular cascading bifurcation. Then the asymptotic expansions that have been obtained are not uniformly valid as $\underset{\sim}{p} \to \underset{\sim}{p}_1$. New asymptotic expansions are then obtained that are uniformly valid in $|\underset{\sim}{p} - \underset{\sim}{p}_1|$. An analysis of the resulting amplitude equations reveals new mechanisms for the appearance of additional steady and periodic solutions of the original bifurcation problem. We demonstrate this phenomenon of singular bifurcation by briefly describing two examples of reaction-diffusion problems that we have studied [5,6], All details of the analysis and discussions of their physical and biological significance are omitted. The reader is referred to the original papers. The problems we consider are a Brussellator boundary value problem and a model reaction-diffusion problem whose solutions possess propagating fronts. The study of the latter model was motivated by investigations [7,8] of flame front propagation in combustible fluids.

2. Pattern Formation During Morphogenesis

The Brussellator [3] is a frequently used model of biochemical morphogenesis. In [5] we considered a boundary value problem for a one dimensional Brussellator system of length L. The dependent variables u(x,t) and v(x,t), which are the concentrations of two intermediate chemicals, satisfy the "no-flux" boundary conditions at x = 0,1. In addition the Brussellator model depends on two prescribed concentrations of reactants A and B and the ratio of the diffusion coefficients $\theta^2 \equiv D_u/D_v$ of the u and v species. The bifurcation parameter is defined by $\lambda \equiv L^2/D_v$.

The modes corresponding to the primary bifurcation points are proportional to $\cos n\pi x$. The mode corresponding to n = 1 is called a polar structure and the more complex, but symmetric, mode corresponding to n = 2 is called a duplicate structure. Since we wish to study the transition from polar to duplicate structures we have analyzed the solution of the Brussellator problem when the two primary bifurcation points corresponding to n = 1 and n = 2 coalesce. They coalesce when $B \to b_0$, where b_0 is a number defined in [5]. By expanding the solutions in a small parameter ε which is proportional to $B - b_0$, we obtain an asymptotic approximation for λ near the multiple primary bifurcation point λ_0 as a linear combination of the modes $\cos \pi x$ and $\cos 2\pi x$. The amplitudes α_1 and α_2 of these modes then satisfy the secondary amplitude equations,

$$\alpha_1(P_1 + P_2\alpha_2) = 0 \quad , \quad Q_1\alpha_2 + Q_2\alpha_1^2 = 0 \tag{2.1}$$

The coefficients P_1, Q_1, P_2 and Q_2 are defined by

$$P_1(\ell) \equiv -c(\ell - q) \quad , \quad Q_1(\ell) \equiv c(\ell + q)$$

$$c \equiv 3A^2/(4\pi^2) \quad , \quad q \equiv \beta/c \tag{2.2}$$

$$P_2(A,\theta) = (1 - A^2\theta^2)/A \quad , \quad Q_2(A,\theta) = (2 + A\theta)(1 - 2A\theta)/4A$$

where $\beta \equiv \pm 1$ depending on whether $B > b_0$ or $B < b_0$ and ℓ is defined by

$$\lambda = \lambda_0 + \ell\varepsilon + O(\varepsilon^2) \quad . \tag{2.3}$$

We observe that $\alpha_1 = \alpha_2 = 0$ is a solution of (2.1) for all values of ℓ. It corresponds to the basic state solution of the Brussellator problem. In addition, we observe that (2.1) has the solution $\alpha_1 = 0$, $\alpha_2 \neq 0$ (α_2 is arbitrary) provided that $Q_1 = 0$, i.e. that $\ell = -q$. Therefore, this corresponds to "vertical" primary bifurcation from the primary bifurcation point $\ell = -q$. The amplitude α_2 for this branch is determined by analyzing higher order terms in the asymptotic expansions, as we show in [5]. However, for $\alpha_1 \neq 0$ and $\alpha_2 \neq 0$, (2.1) have the pair of solutions

$$\alpha_1^2 = \frac{P_1}{P_2}\frac{Q_1}{Q_2} \quad , \quad \alpha_2 = -\frac{P_1}{P_2} \tag{2.4}$$

provided that the conditions

$$P_2Q_2 \neq 0 \quad \text{and} \quad P_1Q_1/(P_2Q_2) \geq 0 \tag{2.5}$$

are satisfied. Since $P_1Q_1 = -c^2(\ell^2 - q^2)$ these solutions exist only for $\ell^2 < q^2 (> q^2)$ if $P_2Q_2 > 0 (< 0)$. Furthermore, they branch from the basic solution at the primary bifurcation point $\ell = q$ because $\alpha_1(q) = \alpha_2(q) = 0$. However, at $\ell = -q$: $\alpha_1 = 0$, $\alpha_2 = \alpha_2^s \equiv -P_1(-q)/P_2 = -2\beta/P_2$ so that solutions (2.4) also branch from the primary bifurcation states at the secondary bifurcation point $\ell = -q$, $\alpha_1 = 0$, $\alpha_2 = \alpha_2^s$.

The solutions (2.4) are bounded if $P_2Q_2 \neq 0$. However, as we observe from (2.2), P_2Q_2 vanishes if

$$A = A^* = 1/(2\theta) \quad , \quad \text{or} \quad A = A^{**} = 1/\theta \quad . \tag{2.6}$$

Then if $\ell \neq \pm q$, α_1 and possibly α_2 are unbounded and the asymptotic expansions are not uniformly valid for A in a neighborhood of either A^* or A^{**}. Thus the bifurcation equations are singular for these values of A. In [5] we derive new asymptotic expansions of the solution of (2.1) near these singular values. The analysis of the resulting new bifurcation equations shows that there are other branches of steady solutions and, in addition, there is tertiary bifurcation of periodic solutions. These results and their biological significance are discussed briefly in [5].

3. Pulsating Propagating Fronts

Cascading bifurcations of solutions are observed experimentally and analytically in the propagation of laminar flame fronts in a combustible fluid, see e.g. [9]. In [6] we have considered a general reaction diffusion equation whose solutions possess propagating fronts. More specifically, there is a sequence of bifurcations from the steadily propagating planar solution(fronts),which corresponds to the basic state, to more complex propagation such as steadily propagating cellular and pulsating propagating fronts as the bifurcation parameter is increased.

In [6] it is shown that for certain ranges of parameters there are three primary bifurcation states that branch from a multiple primary bifurcation point λ_1. They represent two periodic traveling waves along the wave front and a standing wave front. The null space of the linearized theory at $\lambda = \lambda_1$ is four dimensional. In addition, the primary amplitude equations become singular at certain parameter values $\underset{\sim}{p} = \underset{\sim}{p}_1$, so that the asymptotic expansion leading to these amplitude equations is invalid near these critical parameter values. We have therefore obtained a new asymptotic expansion of the reaction diffusion problem near these singular values. The resulting amplitude equations show that there are quasi-periodic traveling waves along the front which occur by secondary bifurcation. We now briefly describe this analysis.

The steadily propagating planar front is the basic state and it is stable (unstable) for $\lambda < \lambda_1 (> \lambda_1)$, where λ_1 is the lowest primary bifurcation point and it has multiplicity four. A standard bifurcation analysis in the small parameter ε which is proportional to $\lambda - \lambda_1$ gives an asymptotic expansion of the solution of the model reaction diffusion problem as a linear combination of four waves traveling along the wave front. The amplitude of these waves, which are given in terms of two complex valued functions $\alpha_1(\tau)$ and $\alpha_2(\tau)$ of the slow time τ defined by $\tau \equiv \varepsilon^2 t$, satisfy a pair of complex valued amplitude equations. If we employ the polar notation $\alpha_1 = \rho e^{i\theta}$ and $\alpha_2 = \delta e^{i\phi}$, then the amplitude equations are:

$$\rho_\tau = \rho(P + A_1\rho^2 + B_1\delta^2) \tag{3.1a}$$

$$\delta_\tau = \delta(P + A_1\delta^2 + B_1\rho^2) \tag{3.1b}$$

$$\rho\theta_\tau = \rho(Q + A_2\rho^2 + B_2\delta^2) \tag{3.1c}$$

$$\delta\phi_\tau = \delta(Q + A_2\delta^2 + B_2\rho^2) \tag{3.1d}$$

where the coefficients in (3.1) depend on the coefficients in the reaction diffusion problem. Steady solutions of (3.1) correspond, to lowest order,to periodic solutions of the model reaction diffusion problem.

We observe that $\rho = \delta = 0$ corresponds to the basic state or plane wave front of the reaction diffusion problem. In addition, the solutions $\rho = 0$, $\delta \neq 0$, θ and ϕ constants; and $\rho \neq 0$, $\delta = 0$, θ and ϕ constants, correspond to periodic traveling wave fronts, which are primary bifurcation states. In addition, standing wave fronts are primary bifurcation states which are steady solutions, with $\rho = \delta \neq 0$ and θ and ϕ constants.

To obtain quasi-periodic solutions of the reaction diffusion problem, we first observe that Eqns. (3.1a,b) for ρ and δ are uncoupled from the phase equations (3.1c,d). Thus, we solve (3.1) by first solving the subset (3.1a,b), then insert the results in (3.1c,d) and solve for θ and ϕ.

We seek solutions of (3.1) in which ρ and δ are nonzero constants. It follows from (3.1a,b) that ρ and δ satisfy

$$\begin{bmatrix} A_1 & B_1 \\ B_1 & A_1 \end{bmatrix} \begin{bmatrix} \rho^2 \\ \delta^2 \end{bmatrix} = -P \begin{bmatrix} 1 \\ 1 \end{bmatrix} \quad . \tag{3.2}$$

If the determinant $\Delta \equiv A_1^2 - B_1^2$ of the matrix in (3.2) does not vanish then we obtain the standing wave solution previously described. However, if $\Delta = 0$, then (3.2) is singular. If $A_1 = -B_1$ it can be shown that there is no solution to (3.1). However, if $A_1 = B_1$ then the singular system (3.2) is reduced to

$$\rho^2 + \delta^2 = -P/A_1 > 0 \quad , \tag{3.3}$$

which is a one parameter family of solutions of (3.1a,b). The corresponding phases from (3.1c,d) are

$$\theta(\tau) = Z_1\tau + \theta(0) \quad ; \quad \phi(\tau) = Z_2\tau + \phi(0) \tag{3.4a}$$

where Z_1 and Z_2 are defined by

$$Z_1 \equiv Q + A_2\rho^2 + B_2\delta^2 \quad , \quad Z_2 \equiv Q + A_2\delta^2 + B_2\rho^2 \tag{3.4b}$$

and $\theta(0)$ and $\phi(0)$ are specified initial values of the phases. If $Z_1 \neq Z_2$ then the frequency modulations for the propagating modes will be different and hence lead to quasi-periodic solutions. Thus, in [6] we analyze the singular amplitude equations by another asymptotic expansion in a new small parameter that is proportional to $B_1 - A_1$. This leads to the secondary bifurcation of quasi-periodic solutions from the previously obtained periodic solutions. Conditions for the existence, stability and location of these new solutions depends on the coefficients in the reaction diffusion equations, see [6] for details.

4. Acknowledgements

This research was supported by the U.S. Air Force Office of Scientific Research under Grant No. AFOSR 80-0016A and the U.S. Department of Energy under Grant No. DE-AC02-78ERO-4650. TE is Senior Research Assistant with the F.N.R.S. (Belgium). BJM gratefully acknowledges the support of a John Simon Guggenheim Memorial Foundation Fellowship.

5. References

1. E. L. Reiss, Cascading Bifurcations, SIAM J. Appl. Math. 43, 57-65 (1983).
2. T. Erneux and E. L. Reiss, Splitting of Steady Multiple Eigenvalues May Lead to Periodic Cascading Bifurcation, SIAM J. Appl. Math. 43, 613-624 (1983).
3. G. Nicolis and I. Prigogine, Self-Organization in Nonequilibrium Systems, John Wiley, New York (1977).
4. L. Bauer, H. B. Keller and E. L. Reiss, Multiple Eigenvalues Lead to Secondary Bifurcation, SIAM Rev. 17, 101-122 (1975).
5. T. Erneux and E. L. Reiss, Singular Secondary Bifurcation, SIAM J. Appl. Math., to appear (1984).
6. T. Erneux and B. J. Matkowsky, Quasi-Periodic Waves Along a Pulsating Propagating Front in a Reaction-Diffusion System, SIAM J. Appl. Math., to appear (1984).
7. B. J. Matkowsky and D. O. Olagunju, Travelling Waves Along the Front of a Pulsating Flame, SIAM J. Appl. Math. 42, 486-501 (1982).
8. B. J. Matkowsky and D. O. Olagunju, Spinning Waves in Gaseous Combustion, SIAM J. Appl. Math. 42, 1138-1156 (1982).
9. S. B. Margolis and B. J. Matkowsky, Nonlinear Stability and Bifurcation in the Transition from Laminar to Turbulent Flame Propagation, Comb. Sci. and Tech., in press (1983).

HOW CHEMICAL STRUCTURE DETERMINES SPATIAL STRUCTURE IN FLAME PROFILES

P. C. FIFE AND B. NICOLAENKO

University of Arizona
Tucson

Los Alamos National Lab
Los Alamos

1. Introduction. Flames embody an array of highly complex physico-chemical phenomena. Their complexity is, in part, a consequence of fluid dynamical and other instabilities which render them prone to irregular behavior, or at least to complicated patterns in space-time. Another complexity is due to the large number of chemical reactions which nearly always contribute to the combustion process. Both of these difficulties are often ignored in analytical treatments, although some major advances achieved in the last decade have served to clarify a few issues in the first category. There have also been analyses illustrating the effects which multiple reactions may have on flames (e.g., see [5], [6], [7], [8], and [9], and related issues in [1]).

The purpose of the present paper is to outline a systematic scheme for incorporating the action of complex chemical systems in the study of plane steady premixed flame profiles. What our scheme amounts to is an algorithm by which knowledge about the reaction network (chemical structure) can be transformed into knowledge about the velocity of the flame (or flames) resulting therefrom, and about the spatial structure of the profile. While our objective is to be sophisticated about the chemistry, we shall strive to keep all other aspects of the problem, such as the geometry and the fluid dynamics, as simple as possible. The transport (diffusion) matrix is supposed to be constant. Our hope is that once a procedure for handling fairly general chemical networks is systemized within the scenario of plane laminar flames, this procedure, or at least the insights it provides, may be helpful in other types of combustion problems as well.

The beginnings of a general procedure were described in [2], [3], and [4], and the present paper continues in that vein. The earlier scheme was designed for model chemical networks in which every reaction was irreversible with high activation energy. Among other things, we show here how to overcome these limitations. We

---

Research supported by NSF Grant MCS 8202056, Mathematics Department, Kyoto University, and the Center for Nonlinear Studies, Los Alamos National Laboratory.

allow reversible reactions, and we merely suppose that one of the reactions (at least) has high activation energy. On the other hand, we impose a "genericity" condition on the reaction network and its various rate constants (Basic Assumption 2 below) which says, among other things, that no two reactions in the scheme should have the same or nearly the same set of Arrhenius rate constants, when measured relative to the reaction with highest activation energy. The assumption is mainly for convenience when confronting the task of developing a general algorithm. It could be relaxed to a great extent, but doing so would result in more involved definitions and procedures than those explained here.

Many analytical questions germane to the present scheme have been treated in some detail in [2] and [3] and will not be repeated here. Specifically, these include singular perturbation aspects of the flame layer analyses, the question of the relevance of the functions $H_k(T)$ in comparing reactions, and the unique solvability of the limit problem. Rather than dwelling on these points, we shall turn our attention more toward an attempt to describe the general scheme, and the ideas on which it is based, in words.

Every realistic problem in combustion theory is burdened with a large number of parameters. Their orders of magnitude may drastically affect the kind of analysis used in treating the problem. In line with our strategy of dealing with only one sophistication, we make a blanket assumption that all parameters other than the Arrhenius parameters in the various rate functions are of order unity. This assumption, vague as it is, will not be restated in the sections below. The tactic is that whenever a parameter such as a diffusion coefficient or specific heat or heat release parameter enters the analysis, we shall assume that this parameter is not of such a small or large magnitude as to interfere in any way with the activation energy asymptotics which <u>are</u> being performed.

Temperatures and activation temperatures are always nondimensionalized in such a way that the range of temperature in the flame (maximum minus minimum) is unity. Certain characteristic temperatures T^{κ} will figure in the analysis, and in line with the blanket assumption mentioned above, we assume that all differences between these quantities have magnitude $O(1)$. As before, a more painstaking study would remove most of these limitations.

<u>2. Preliminary Definitions and Assumptions.</u> As stated before, our primary purpose is to outline a general procedure by which knowledge of the chemical reaction network, together with the rate functions and the unburned state of the gas, will yield knowledge about the profile structure and velocity of the flame or flames which will bring this unburned state to chemical equilibrium (the burned state or states). The recipe we propose constitutes a generalization of that given

in [3]. The central notion, as before, is that of an "allocation" for the given initial state and given network. The term "allocation" refers to a way of allocating the resources (unburned state) to the various reactions in the process of attaining chemical equilibrium. The set of all possible allocations is determined in a straight-forward manner. A subset of these will be called "feasible"; each of the latter corresponds to an actual steady flame bringing the state of the medium to equilibrium. This subset can be found by purely algebraic means. In case there exist no feasible allocations, there typically may exist at least one allocation which satisfies a weaker criterion, which we call "succession feasible". This is a larger subset which, again, can be characterized precisely, and determined in a straight-forward way. Our claim is that the latter allocations correspond to "flame successions", rather than to single flames. By flame succession, we mean a sequence of flames, traveling at different velocities and bringing the gas to different "equilibrium" states with respect to different subnetworks of the original network. The final result of the succession is to achieve equilibrium with respect to the entire network.

We begin by defining a <u>chemical reaction network</u> $\mathbf{N}$ as a set of <u>reactions</u>

$$\mathbf{N} = \{R_j\}_1^m,$$

each R_j being an expression like

$$A_1 + A_2 \rightarrow A_3 + A_4,$$

the A_i denoting the various chemical species in our system. We let n be the total number of species, and m the total number of reactions.

The <u>state space</u> is $\mathbf{R}^{n+1}$; its elements (states) will be denoted by $U = (U_0, U_1, \ldots, U_n)$. The zeroth component U_0 represents the (nondimensional) temperature T of the gas, and the other components are the concentrations, in the gas, of the n species. Sometimes these other components will be denoted by the n-vector $Y = (Y_1, \ldots, Y_n)$, so that $U = (T,Y)$. Actually, Y_j will denote the mass fraction of the species A_j divided by its molecular weight. This choice somewhat simplifies the expressions for the reaction vectors, described below.

Associated with each reaction R_j is a <u>reaction vector</u> $K_j = (Q_j, K_{j1}, \ldots)$. Here Q_j is a certain measure of the heat released by the reaction (if the reaction is endothermic, then $Q_j < 0$). The other components K_{ji} represent the net change in number of molecules of A_i during an "event" of reaction R_j; this change is calculated by subtracting the stoichiometric coefficient of A_i on the left of the reaction arrow from the one on the right [3]. Note that if R_j is reversible, then the reverse reaction is considered to be a new reaction, say $R_{j'}$; then we simply have $K_{j'} = -K_j$. We denote by V_j the n-vector $(K_{j1}, \ldots)$, so that $K_j = (Q_j, V_j)$.

Also associated with R_j is its <u>reaction rate</u> $\omega_j(U)$, which we shall always take to be of mass action-Arrhenius form

$$\omega_j(U) = B_j M(Y) e^{-E_j/T}. \qquad (1)$$

Here M is the mass action monomial (which vanishes if and only if $Y_k = 0$ for some k for which A_k appears on the left of the reaction R_j). Also, E_j is a dimensionless activation energy or activation temperature. We make no direct assumption about the order of magnitude of the constants B_j.

The following functions H_j will serve to gauge the relative importance of the R_j at the various temperatures; most importantly, they will tell us when certain reactions may be neglected. We define

$$H_j(T) = \varepsilon\,(\ln B_j - E_j/T), \qquad (2)$$

where the normalization constant ε is chosen so that the maximum of the ranges of the functions H_j as T varies between the unburned temperature T_- and the maximum possible flame temperature, is equal to unity.

Our first basic assumption is essentially that at least one of the reactions has high activation energy; i.e., has large E_j. According (2) and the definition of ε, this means that ε must be small:

<u>Basic assumption 1:</u> $$\varepsilon \ll 1. \qquad (3)$$

The algorithm we shall propose for determining the structure of the flame profile has a lot to do with the pattern of graphs of the functions $H_j(T)$, so at this point it is appropriate to make some observations (and an assumption) about these graphs. Our procedure follows a type of asymptotic analysis based on the smallness of ε. In a sense it is the lowest order asymptotics, and we shall be logically consistent by neglecting the effect of quantities whose magnitude is $O(1)$ (as $\varepsilon \to 0$) <u>relative</u> to the magnitude of quantities which are retained. For example, if $H_k(T) < H_j(T)$ for some value of T, and the difference δ between these two functions is such that $\varepsilon = O(\delta)$, then the Arrhenius factor $B_k \exp(-E_k/T)$ of the first rate function, when divided by that of the second, does not surpass $e^{-\delta/\varepsilon} \ll 1$. Generally, this will imply that R_k will be effectively frozen at that temperature compared to R_j. (However, caution must be employed in applying this criterion; there are important examples for which the two reactions will be comparable, because of the influence of the mass action factors in the rate functions ω_j. Further details on the justification for neglecting reactions on the basis of the relative magnitude of the functions H_j can be found in [2] and [3]. Cases when $\delta > 0$ but $\varepsilon \neq O(\delta)$ are more complicated. For the sake of simplicity, we shall explicitly exclude those cases when possible, and that is the intent of our next basic assumption:

Basic Assumption 2: For any $j \neq K$, $|H_j(T) - H_k(T)| > O(\varepsilon)$ for all T except on a small interval of length $O(1)$. Furthermore, the set of all intersection points of the graphs of the H_j is isolated, and the set of values of T for these intersection points are spaced $> O(\varepsilon)$ apart.

The implication of this is that we may typically neglect one of the reactions R_j or R_k in favor of the other, except near certain critical values of T, which will be intersection points. (But there are important exceptions to this simple rule. Our procedure automatically accounts for these exceptions!) The assumption means that no two of the curves H_j lie in an ε-neighborhood of each other for any appreciable distance. Apparently in some sense, this will generically be the case if the H_j are randomly chosen subject to natural restrictions. Note also the following implication. Suppose j_0 is the index such that H_{j_0} has the maximum range of 1, and k is an index such that $E_k << E_{j_0}$. Then the graph of H_k will be relatively flat, so we approximate it by a horizontal line. The group of such reactions R_k are those with "low" or zero activation energy. Assumption 2 implies that all these horizontal graphs are spaced a distance $> O(\varepsilon)$ apart. (Actually, this is one place where Assumption 2 can be easily weakened, as we do not need all of the horizontal graphs to be so spaced; we'll not pursue the details.) This does not mean that from this group only the one with largest H is nonnegligible! In fact, typically some of these reactions, maybe with large H functions, will be inoperative at some places in the profile because they are in equilibrium. And when they are brought out of equilibrium, for example, by the sudden production of radicals at certain triggering temperatures, they may be only evanescent reactions, as the radicals are quickly recombined. Bearing in mind that the functions H are monotone nondecreasing in T, and that their slopes at a given value of T are proportional to their activation energies, it is not hard to conclude on the basis of Assumption 2 that at an intersection point, the slopes of the two intersecting H curves differ by an amount $> O(\varepsilon)$.

The equilibrium set function $\mathbf{E}(T)$ is defined to be the set of vectors Y in the nonnegative orthant of species space such that for $U = (T,Y)$,

$$\sum_{j=1}^{m} \omega_j(U)K_j = 0. \qquad (4)$$

Note that in our set-up, reversible reactions are really pairs of reactions, each with its own H function. For definiteness, let us denote a certain pair of forward and reverse reactions by R_1 and R_2. If H_1 and H_2 intersect at some temperature (say T^*), then according to Assumption 2, one or the other of them may be neglected except for temperatures near T^*. We could therefore replace these two functions by a new combined function $H_{12}(T) \equiv \max[H_1(T),H_2(T)]$ representing a reaction which goes forward (say) for $T < T^*$, and backwards for $T > T^*$. This will be consistent with our asymptotic analysis and it will, in fact, be convenient to make the replacement. (There are important cases, however, when by necessity the reaction proceeds at -- or near -- T^*, so that the forward and backward reactions rates are comparable, and equilibrium is achieved without either going fully to completion.) So in constructing the equilibrium function $\mathbf{E}$ for networks containing this reversible reaction, we require that the forward reaction be in equilibrium for $T < T^*$, and the backward one for $T > T^*$. Of course if the two functions do not intersect, then either the forward or the backward reaction will always be neglected. If there exist reaction cycles, such as $A_1 \to A_2 \to A_3 \to A_1$, then we make a similar replacement for the three relevant H functions.

The initial, unburned state will be denoted by $U_- = (T_-,Y_-)$.

The most central concept in our algorithm is that of an <u>allocation</u> α. an allocation is a nonnegative m-vector

$$\alpha = (\alpha_1,\alpha_2, \dots, \alpha_m)$$

such that if we define

$$U_+(\alpha) \equiv (T_+(\alpha),Y_+(\alpha)) = U_- + \sum_1^m \alpha_j K_j, \qquad (5)$$

then

$$Y_+(\alpha) \in E(T_+(\alpha)). \qquad (6)$$

The set of all allocations will be denoted by $\mathbf{A}$. Note that $\mathbf{A}$ depends on U_- as well as on $\mathbf{N}$. Knowing $\mathbf{E}$, it will be straight-forward to determine $\mathbf{A}$. The components α_j of an allocation α serve to specify the extent to which the various reactions R_j take part in the process by which the initial state U_- is converted to the final burned state, which we denote by U_+. This latter state is given by (5), which we may for the moment take as its definition. Examples of sets $\mathbf{A}$ were given in [3] and [4].

<u>3. Feasible and Succession-feasible Allocations.</u> In the following we shall define a subset A_f of $\mathbf{A}$; if this set is not empty, its elements will be called "feasible allocations". They are the ones, we claim, which correspond to realizable

simple flames. Another subset $\mathbf{A}_f^*$, containing $\mathbf{A}_f$, will also be defined. Its elements, called <u>succession-feasible allocations</u>, correspond to realizable flame successions, as described before.

We define $\mathbf{A}_f$ by outlining tests an allocation must pass in order to be "feasible".

Given α, we define $T_+(\alpha)$ as in (5). Since the zeroth component of K_j is the heat release coefficient Q_j, we have

$$T_+(\alpha) = T_- + \sum_1^m \alpha_j Q_j. \qquad (7)$$

We consider here only the case when $T_+(\alpha) > T_-$. We next define

$$H_+(\alpha) \equiv \min_j [H_j(T_+(\alpha)) : j \text{ such that } \alpha_j \neq 0] \qquad (8)$$

and the set-valued function

$$\mathbf{N}(T) = \{R_j : H_j(T) > H_+\}, \quad \text{for } T < T_+. \qquad (9)$$

Note that all inequalities in (9) are strict. For convenience later, we extend this definition to <u>larger values,</u> $T \geq T_+$, by replacing the inequality ">" by "≥". We also define

$$\mathbf{E}^*(T) = \mathbf{E}(T;\mathbf{N}(T)) \equiv \text{the equilibrium set function associated with } \mathbf{N}(T) \qquad (10)$$

$$= \{Y : \Sigma\omega_j(T,Y)K_j = 0\}, \text{ the summation being over } \{j : R_j \in N(T)\}.$$

We next define a sequence of temperatures $\{T^\kappa\}$, $\kappa = 1, \ldots, m'$, in increasing order, consisting of the values of T in the interval $(T_-, T_+(\alpha))$ at which $\mathbf{E}^*(T)$ is discontinuous, together with the endpoints T_- and T_+ of the interval. Thus $T_- = T^1$ and $T_+ = T^{m'}$. It turns out that by virtue of our handling of reversible reactions, the function $\mathbf{E}^*$ will be constant for values of T between the T^κ.

This is one point at which the present recipe differs from our original one, which excluded reversible reactions. There we defined the T^κ as points where the horizontal line $H = H_+$ intersects the curves $H_j(T)$. Since $\mathbf{E}^*(T)$ is discontinuous at those intersection points, they are included among our present T^κ; but we also include points like T^*, described above, where the forward and reverse rates of a reversible reaction are equal (a simple model example is shown in Figure 1); i.e., the "bending" point of the function $H_{12}(T)$. Here, neither H_1 or H_2 intersect $H = H_+$, but $\mathbf{E}^*$ is discontinuous because R_1 dominates over

R_2 for $T < T^*$, and vice versa for $T > T^*$.

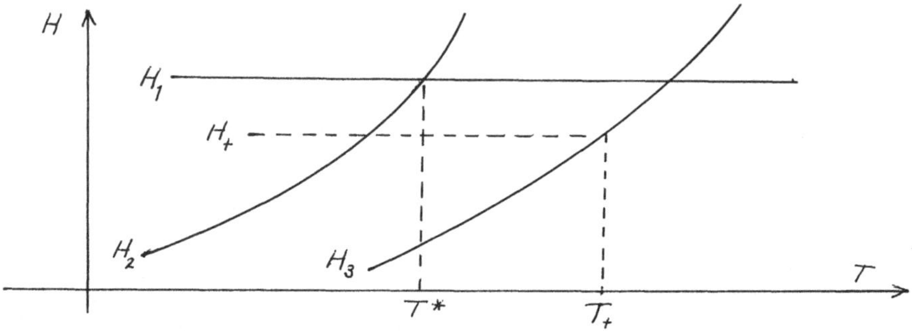

Figure 1. $H_j(T)$ for the network: $(R_1)\ A_1 \to A_2$; $(R_2)\ A_2 \to A_1$; $(R_3) A_3 \to P$.

Given an integer $\kappa = 1, 2, \ldots, m'$, the symbol α^κ will denote a suballocation for $\mathbf{N}(T^\kappa + \delta)$, i.e., a nonnegative m-vector such that

$$Y^\kappa \equiv Y_- + \Sigma\alpha_j V_j \in \mathbf{E}(T) \text{ for } T \text{ slightly larger than } T^\kappa. \qquad (11)$$

Given $\alpha^{\kappa-1}$ (for $\kappa > 1$), the suballocation α^κ is called "maximal" if among all such suballocations with $\alpha^\kappa \geq \alpha^{\kappa-1}$, it is chosen so that those components α_j^κ with the largest value of $H_j(T^\kappa)$ are maximized [3]. That is, the component with the largest value of H is maximized first; then the one with the next largest, etc. If it should happen that the H-functions for the forward and backward parts of a reversible reaction intersect at T^κ, then the two corresponding components of α^κ are exempted from this criterion (the 2 parts partially conteract each other, and so no a priori judgment can be made as to their net effect). Maximal suballocations will not necessarily be unambiguously determined this way; for example, they will not if two functions H_j intersect at T^κ.

Also given an integer $\kappa \leq m'$, we define a temperature

$$T_*^\kappa = T_- + \Sigma\alpha_j^\kappa Q_j. \qquad (12)$$

Definition. The allocation α is feasible if:

i) The subnetwork $\mathbf{N}_0 = \{R_j : H_j(T_-) \geq H_+(\alpha)\}$ is in equilibrium at the initial state U_-;

ii) there exists a sequence of suballocations $\{\alpha^\kappa\}_1^{m'}$ which satisfy $\alpha^{\kappa-1} \leq \alpha^\kappa$ for all $\kappa \leq m'$ (equality not being allowed for $\kappa = m'$), which are maximal for $\kappa > 1$, and which satisfy $\alpha^1 = 0$ and $\alpha^{m'} = \alpha$;

iii) $$T_*^\kappa < T^\kappa \text{ for all } 0 < \kappa < m'. \qquad (13)$$

(Note that equality necessarily holds here for $\kappa = 0$ or m'.) The set of all feasible allocations is denoted by $\mathbf{A}_f$. It is by no means guaranteed that $\mathbf{A}_f$ is nonempty. However, there typically will exist succession-feasible allocations, which will be defined next.

Definition. The allocation α is succession-feasible if there exists an increasing sequence of subnetworks $\{\mathbf{N}^\mu\}$, $\mu = 1, \dots, \mu^*$ with $\mathbf{N}^{\mu^*} = \mathbf{N}$, together with initial states U_-^μ and final states U_+^μ connected by suballocations $\alpha^\mu \geq 0$:

$$U_+^\mu = U_-^\mu + \Sigma\alpha_\kappa^\mu K_\kappa$$

(the summation being over κ), satisfying

i) each α^μ is feasible for $(\mathbf{N}^\mu, \underline{U}^\mu)$, with associated constants H_+^μ,

ii) $U_+^\mu = U_-^{\mu+1}$ whenever both sides are defined;

iii) $U_-^1 = U_-$,

iv) $\alpha = \Sigma_\mu \alpha^\mu$;

v) the H_+^μ are strictly decreasing in μ.

The set of all such allocations is denoted by $\mathbf{A}_f^*$; it contains $\mathbf{A}_f$ and is contained in $\mathbf{A}$.

4. Single Flames. Our claim is that every feasible allocation determines a steady flame. Our object now is to show how the basic properties of that flame can be determined: specifically, the mass flux and the structure of the temperature-concentration profile. This is done by use of a limit problem which can be derived in a formal logical manner, as has been described in previous papers [2], [3], [4]. Here, we shall only sketch the procedure and provide few details.

We first describe the basic structure of the profile. Let $\alpha \in \mathbf{A}_f$ be given. We recall the Definition (8) of H_+, as a minimum of certain numbers $H_j(T_+(\alpha))$. As explained in [3], this minimum will be attained, generically in a certain sense, at either one or two values of j. We distinguish between two possible cases:

(a) At least one of these two functions H_j realizing the minimum has nonzero slope (always, when we say something is nonzero with reference to the functions H_j, it should be understood that we mean bounded away from zero by something of larger order than ε).

(b) There is only one such function, and it has zero slope (which really means a slope which is $\leq O(\varepsilon)$).

In case (a), call the relevant index $j = j_0$. Then the reaction R_{j_0} is frozen at temperatures lower than T_+, and so does not go forward at all in the flame until that final temperature is reached. At that temperature, however, it suddenly goes to completion, according to the component α_{j_0} allocated to it. If there is a second such reaction R_{j_1} whose H function has nonzero slope and intersects the first at T_+, then it also goes to completion. The net effect will be the sudden release of thermal energy (the effect is that of an exothermic reaction; it is not too hard to see that the final temperature must be the maximal temperature if the allocation is feasible, and that the maximal temperature cannot be attained with an endothermic reaction). In the limit problem [3], [4], this sudden release is responsible for a bend (or corner) in the flame's profile.

At this point, it is crucial to rescale space (rather, the traveling wave coordinate) and mass flux so that the magnitude of this bend (i.e., the jump in the derivative of the profile) is of order $O(1)$. How to perform this scaling was described in [2], [3]; we suppose it has been done.

The next thing to do is perform a layer analysis at the corner. This involves stretching both the dependent and independent variables in its neighborhood by a factor ε^{-1}, and keeping only the lowest order terms. There results a system of autonomous differential equations, the "flame layer equations" [2], which can be solved and matched to the solution behind the flame layer to yield the value of the discontinuity in profile slope at that point. If there is only one minimizing j_0, the solution can be written down explicitly in terms of an integral. On the other hand if there are two minimizing j's, then the solution cannot be displayed explicitly; we must be content with proving its existence. This was done in [2]. In any case, the magnitude of the "bend" is determined.

At each of the other values T^κ, $1 < \kappa < m'$, certain sets of reactions again go to completion when the temperature in the profile reaches that value. Specifically, they are the reactions in the subnetwork $\mathbf{N}(T^\kappa + \delta) \setminus \mathbf{N}(T^\kappa)$. It may happen that in the process, some reactions, not in the subnetwork, which were previously in equilibrium, are brought out of it; if this is the case, then those extra reactions which have H values greater than H_+ at T^κ will again very quickly return to equilibrium; we call them <u>evanescent</u> reactions. These evanescent reactions are automatically found, and their effect accounted for, during the procedure of determining the suballocations α^κ [3].

The reactions in the subnetwork mentioned above suddenly bring the network $\mathbf{N}(T^\kappa + \delta)$ to equilibrium, with a sudden concomitant heat release or heat absorption, resulting in another corner in the flame profile. In the case of a heat release, the corner is convex; but it is possible also for it to be concave in case the net effect is endothermic, as long as the temperature profile in the neighborhood of the corner is monotonic.

Thus in case (a), the flame profile will consist of a series of corners (perhaps just one) with (in the asymptotic limit) no reactions taking place in between. In the rescaled coordinates, all these corners represent slope discontinuities of the same order of magnitude $O(1)$, by virtue of the fact that the relevant reactions all either proceed at temperatures which will guarantee their H values are the same (equal to H_+); or have higher values of H, and are matched with competing reverse reactions so that the net effective reaction has a smaller order of magnitude rate, namely one which would have an H value also equal to H_+; or have higher values of H and represent the chemical decay of evanescent species whose concentration is so low that the actual reaction rate again has an order of magnitude typical of H_+.

It turns out that the jump in slope of the profile at the temperature T^κ is an unknown scalar times a known vector (denoted by K_j, in [3], [4]). This information, together with knowledge of the diffusion matrix, suffices, as described in the aforementioned papers, for the determination of the mass flux and the entire profile, including the locations of the corners.

For one simple example in case (a), we choose a network with one reversible reaction:

$$A_1 \to A_2 \qquad (R_1)$$

$$A_2 \to A_1 \qquad (R_2)$$

$$A_3 \to P \qquad (R_3).$$

Suppose that R_1 and R_3 are exothermic, R_2 and R_3 have high activation energy, R_1 has zero activation energy, and the H-diagram is the one shown in Figure 1. We take $Y_- = (0,1,1)$, so there are unit amounts of A_2 and A_3 to burn. Assume $Q_3 > Q_1$, $T_- < T^*$, and the allocation $\alpha = (0,1,1)$, suppose T_+ is shown; then $m' = 3$: $T^1 = T_-$, $T^2 = T^*$, and $T^3 = T_+$. In the flame profile, there will be two locations where rapid combustion takes place: T^* (where the backward reaction suddenly becomes important) and T_+. The pecularity here is that even though $H_2 > H_+$ at T^*, nevertheless its rate is mitigated by the presence of the forward rate, which at that temperature is comparable. The net effect of the competing reactions is a backward reaction with (smaller) rate appropriate to H_+ rather than to $H_1(T^*)$. The endothermic nature of R_2 causes a concave corner in the temperature profile at T^*; if it were exothermic, the corner would be convex.

We shall give one more example, namely the network

$$A_1 + A_2 \to 2A_2 \qquad (R_1)$$

$$2A_2 \to P \qquad (R_2)$$

both reactions being exothermic, R_1 with zero activation energy, and R_2 with high activation energy. The H functions appear as in Figure 1, with H_3 omitted from the diagram. Assume $T_- < T^*$. With $Y_- = (1,0)$, it may be checked that $A = \{(\alpha_1,\alpha_2): 0 \leq \alpha_1 = 2\alpha_2 \leq 1\}$. The initial state is in equilibrium, so that

$\alpha = 0$ is feasible. There is another feasible or succession-feasible allocation as well, depending on whether $T_- + Q_1 + (Q_2/2) \leq T^*$ or $> T^*$. In the latter case, choose $\alpha_1 \in (0,1)$ so that

$$T_+ = T_- + \alpha_1 Q_1 + \alpha_1 Q_2/2 = T^*;$$

then $(\alpha_1, \alpha_1/2)$ is seen to be feasible. In the former case, $\alpha = (1,1/2)$ is succession-feasible with $\mathbf{N}^1 = \{R_1\}$, $\mathbf{N}^2 = \mathbf{N}$.

In case (b), the function H_{j_0} is constant in T. Since the H_j are measures of the reaction rates, we conclude that reaction R_{j_0} is active throughout the flame profile, except where it is in equilibrium. Part i) of the definition of feasibility guarantees that it is in equilibrium for temperatures near the initial temperatore T_-, i.e., far to the left in the profile. It may be brought out of equilibrium at other places in the profile, but again according to the definition of allocation, it must reach its equilibrium at the final state U_+, and is the last reaction to do so. Note that all other reactions with low or zero activation energy which actually take part in the combustion process have H values strictly greater than H_+. Their role will be to cause evanescent species rapidly to disappear when they are generated.

It follows from our assumptions that R_{j_0} must be exothermic. Having low activation energy, it reaches equilibrium, not abruptly in a corner, but rather gradually. Doing so, it causes a smooth rise in temperature to $T_+(\alpha)$ in the rear part of the flame.

The determination of the mass flux is not so simple in case (b) as in case (a), and the same is true of the locations of the corners corresponding to T^κ for $\kappa < m'$. Nevertheless, the qualitative structure of the profile and the order of magnitude of the velocity can be immediately ascertained from the above considerations. The limit problem can be formulated as well in case (b), and these qualitative features can be read off from the formulation, but the method for solving it given in [3] no longer applies. This type of limit problem will be examined in some detail in a future publication.

5. Flame successions. The simplest flame successions have already been studied in [2]. We called them F1,12 and F2,21 on pages 245-6 of that article. Both involve a pair of high activation energy sequential reactions

$$A_1 \to A_2 \to P. \qquad (14)$$

Call the first reaction R_1 and the other one R_2. There is only one equilibrium state $E(T) = 0$, and only one allocation. Hence there is only one possible T_+ and only one H_+. The flame succession F1,12 occurs if $H_1(T_+) > H_2(T_+)$ and

$T_*^2 = T_- + \alpha_1 Q_1 < T^2$. In this case, the sequence $\{N^\mu\}$ consists of $2 = \mu^*$ networks $N^1 = \{R_1\}$, $N^2 = N$. Also, $H_+^1 = H_1(T_*^2) > H_1(T^3) = H_+^2 = H_+$, so indeed the H_+^μ are strictly decreasing in μ, as required in part (v) of the definition of succession-feasibility.

In the general case, given a succession-feasible allocation, together with its subnetworks N^μ, one first simply determines the structure and velocity of each of the component flames separately on its own merits. This is the procedure described in Section 4. Thus, the flame which takes N^μ into its equilibrium state U_+^μ will have its own velocity. After this is done, one compares the various velocities. It turns out that the velocities of the component flames are ordered according to the H_+^μ, so that the first one (corresponding to N^1) has the greatest velocity and pushes ahead of the others. And in fact they all draw further apart as time goes on.

About flame successions, it must be said that in realistic terms only the first of the succession may be important. This would be the case if the velocities in the later components are too slow to be of interest to the observer (in other words, the later reactions are frozen in the observers' time scale).

6. Conclusion. We have outlined a procedure for the approximate determination of possible steady laminar premixed flame profiles, and their corresponding mass fluxes,given the chemistry. The latter includes both the reaction network and the rate functions. Our restrictions are that at least one reaction have high activation energy, plus a genericity assumption regarding the relative magnitudes of the constants in the Arrhenius rate functions. The analysis constitutes asymptotics on the simplest level, with regard to the high activation energy. Our strategy is to construct, by algebraic methods, all feasible allocations - there may be none, one, or many. For each feasible allocation, there exists a flame whose qualitative features can be easily determined on the basis of that allocation and its concomitant temperature sequence $\{T^\kappa\}$ and sequence of suballocations. This includes the order of magnitude of the mass flux. The profile's fine details and a more precise value for the mass flux are obtained by solving a limit problem. The solution is straight-forward in the case (a) described in Section 4, but is more difficult in case (b).

Note that neither the existence nor uniqueness of a flame bringing the entire network to equilibrium is guaranteed. In the case of multiple solutions (i.e., multiple feasible allocations), the flame actually occurring in an experiment would presumably depend on how the gas is ignited. In the case of no feasible allocations, no doubt there still exists a succession-feasible allocation under acceptable restrictions (that is the case in all the examples we have examined), which means that the complete burning process must occur in several stages by single flames traveling at different velocities.

Examples of our analysis, applied to various types of networks, are given in [2], [3], [4]. Those examples include cases of nonexistence and nonuniqueness.

To complete the analysis, it would be desirable to investigate more fully the connection between the limit problem and the system of differential equations taken as the original model [2], [3], [4]. This should be done by careful singular perturbation analysis, including layer analysis and matching techniques. Such an analysis has been carried out in most standard cases (see these same references), but there is more to be done, especially for reversible reactions and for flames which terminate with a low activation energy reaction.

References

1. M. A. Birkan and D. R. Kassoy, Homogeneous thermal explosion with dissociation and recombination, Combustion Sci. Technology 33 (1983), 135-166.

2. P. C. Fife and B. Nicolaenko, The singular perturbation approach to flame theory with chain and competing reactions, pp. 232-250 in Ordinary and Partial Differential Equations, W. N. Everitt and B. D Sleeman, eds., Lecture Notes in Mathematics No. 964, Springer-Verlag: Berlin (1982).

3. P. C. Fife and B. Nicolaenko, Asymptotic flame theory with complex chemistry, pp. 235-256 in Nonlinear Partial Differential Equations, J. Smoller, Ed., Contemporary Mathematics 17, Amer. Math. Soc.: Providence (1983).

4. P. C. Fife and B. Nicolaenko, Flame fronts with complex chemical networks, Proc., Conf. on Fronts, Interfaces, and Patterns, Physica D, in press.

5. A. Kapila and G. S. S. Ludford, Two-step sequential reactions for large activation energies, Combustion and Flame (1977), 167-176.

6. S. B. Margolis and B. J. Matkowsky, Flame propagation with multiple fuels, SIAM J. Appl. Math. 42 (1982), 982-1003.

7. S. B. Margolis and B. J. Matkowsky, Flame propagation with a sequential reaction mechanism, SIAM J. Appl. Math. 42 (1982)

8. A. Zebib, F. A. Williams, and D. R. Kassoy, Effects of kinetic mechanism on diffusion-controlled structure of hydrogen-halogen reaction zones, Combustion Sci. Technology 10 (1975), 37-44.

9. Ia. B. Zel'dovič, G. I. Barenblat, V. B. Librovič, and G. M. Mahviladze, Mathematical Theory of Combustion and Detonation, ed. Nauka, Moscow, 1980.

AN AGE-DEPENDENT POPULATION MODEL WITH APPLICATIONS TO MICROBIAL GROWTH PROCESSES

Mats Gyllenberg
Helsinki University of Technology, Institute of Mechanics
SF-02150 Espoo 15, FINLAND

1. INTRODUCTION. The classical linear model of age-dependent population dynamics essentially due to Von Foerster [30] and Sharpe and Lotka [28] consists of the partial differential equation

$$(1.1)\qquad \frac{\partial x(t,\tau)}{\partial t} + \frac{\partial x(t,\tau)}{\partial \tau} + f(\tau)x(t,\tau) = 0$$

derived from the requirement of population balance, together with the boundary condition

$$(1.2)\qquad b(t) = x(t,0) = \int_0^\infty m(\tau)x(t,\tau)d\tau$$

describing the birth process. Here t denotes time and τ denotes age. The unknown $x(t,\tau)$ is the population age density at time t (the integral $\int_{\tau_1}^{\tau_2} x(t,\tau)d\tau$ is the number of individuals whose age is between τ_1 and τ_2) and $b(t)$ is the birth rate at time t. f and m are prescribed functions known as the mortality and maternity function, respectively.

It is well-known that if nativity is greater than mortality (in a certain sense, which can be made precise), then the model (1.1), (1.2) predicts that the total population

$$(1.3)\qquad P(t) = \int_0^\infty x(t,\tau)d\tau$$

grows in a Malthusian way, i.e. exponentially, and hence is inadequate for describing populations which competes for resources (cf. Gurtin and MacCamy [12]). Therefore many nonlinear extensions of the classical model have been presented and investigated in recent years (e.g. Gurtin and MacCamy [11], Rorres [25, 26, 27], Haimovici [19], Gyllenberg [15, 16, 17]. The most commonly used method of generalization of (1.1), (1.2), has been to introduce nonlinearity by allowing the death and birth modulus to depend on a control variable y, which in turn depends on the age distribution of the population. This means that $f(\tau)$ and $m(\tau)$ are replaced by $f(\tau,y(t))$ and $g(\tau,y(t))$, respectively, and that the system is supplemented by an equation defining the control variable $y(t)$. It is the choice of this control variable that distinguishes the models of the authors cited above.

In [11] the total populations P given by (1.3) plays the role of the control variable. Rorres [25, 26] uses a very similar control variable, namely

$$(1.4)\qquad s(t) = \int_0^\infty c(\tau)x(t,\tau)\,d\tau ,$$

where different age groups are given different weights.

Gyllenberg [14, 15] considered age-dependent growth of continuously propagated bacterial cultures and used the food supply or substrate concentration in the reactor as growth-limiting variable. The substrate concentration S obeys an integrodifferential equation of the following type:

$$(1.5)\qquad S'(t) = \int_0^\infty g(\tau,S(t))x(t,\tau)\,d\tau + h(S(t)) ,$$

where the given functions g and h describe the consumption of substrate and input and washout of substrate, respectively.

The model of Haimovici [19] involves resources (denoted by r) and pollution (denoted by p) and can be put into our framework if one lets the control variable y be a vector valued function defined by

$$(1.6)\qquad y(t) = \begin{pmatrix} r(t) \\ p(t) \end{pmatrix} .$$

The changing of y is then represented by the equation

$$(1.7)\qquad y'(t) = g(y(t)) \int_0^\infty x(t,\tau)\,d\tau + h(y(t)) ,$$

where g and h are given R^2-valued functions.

Some authors (cf. [5]) have generalised the Lotka-Von Foester model to describe the age specific evolution of a population consisting of m different, mutually interacting classes. These classes may belong to different species or they may be well defined groups of individuals of the same species. This is easily achieved by letting x be an R^m-valued function, the components of which are the age densities of the different classes and by replacing f and g by matrix valued functions F and G.

The purpose of this paper is to give a common criterion for exponential asymptotic stability of the model for a large class of control variables. Specifially, we consider the following system:

$$(1.8)\qquad \mathrm{D}x(t,\tau) + F(\tau,y(t))x(t,\tau) = 0 , \qquad t,\tau > 0 ,$$

$$(1.9)\qquad L_1y'(t) + L_2y(t) = \int_0^\infty G(\tau,y(t))x(t,\tau)\,d\tau + h(y(t)) , \qquad t > 0$$

supplemented by the boundary condition

$$(1.10)\qquad x(t,0) = \Lambda y(t)$$

and the initial conditions

$$(1.11)\qquad x(0,\tau) = x_0(\tau) ,$$

$$(1.12)\qquad L_1y(0) = y_0 .$$

Here D is the derivative in the direction (1,1) x: $R^+ \times R^+ \to R^m$, y: $R \to R^n$ are the unknown; F: $R^+ \times R^n \to \mathcal{L}(R^m)$, G: $R^+ \times R^n \to \mathcal{L}(R^m;\ R^n)$, h: $R^n \to R^n$ L_1, $L_2 \in \mathcal{L}(R^n)$, $\Lambda \in \mathcal{L}(R^n;\ R^m)$, $x_o \in L^1(R^+;\ R^m)$ and $y_o \in R^n$ are given. Linear operators from one finite dimensional Euclidean space to another are identified with the corresponding matrices.

The birth process is now represented by the boundary condition (1.10). As is seen from (1.9) and (1.10), the model allows the birth rate to act as a growth controlling factor. Cases where y controls the birth rate are also included. In the paper of Rorres [27] y is chosen to be simply the birth rate.

We have explicity assumed that the function Λ occurring in the boundary condition is linear. Recently Gurtin and Levine [14], Di Blasio et al. [6] and Thompson et al. [29] have investigated population models of the form (1.8) - (1.12) with Λ nonlinear. However, the nonlinear function h in (1.9) makes it possible to include birth processes of this kind, too.

The matrix $F(\tau,y)$ describes how individuals of age τ are deleted from and introduced into the population (e.g. by death and immigration) and how individuals transmit from one class to another when the control variable has the value y.

We assume that the matrix L_1 has the following decomposition into blocks

$$(1.13)\quad L_1 = \begin{pmatrix} I_{j\times j} & 0 \\ 0 & 0 \end{pmatrix}$$

for some integer j, $0 \le j \le n$, where $I_{j\times j}$ denotes the $j\times j$ identity matrix. Throughout the paper j denotes the dimension of the range of L_1. If $\Phi \in \mathcal{L}(R^k,\ R^n)$. ($k$ arbitrary) then the submatrices $\Phi^1 \in \mathcal{L}(R^k,\ R^j)$, $\Phi^2 \in \mathcal{L}(R^k,\ R^{n-j})$ are defined such that

$$(1.14)\quad \Phi = \begin{pmatrix} \Phi^1 \\ \Phi^2 \end{pmatrix}.$$

In the case $k = n$, Φ is a square matrix and $\Phi^{\mu\nu}$ ($\mu,\nu = 1,2$) are defined such that

$$(1.15)\quad \Phi = \begin{pmatrix} \Phi^{11} & \Phi^{12} \\ \Phi^{21} & \Phi^{22} \end{pmatrix},$$

where $\Phi^{11} \in \mathcal{L}(R^j,\ R^j)$, $\Phi^{22} \in \mathcal{L}(R^{n-j},\ R^{n-j})$. Of course, if $j = 0$ or n, then some of the submatrices are not defined.

Whenever a matrix occurs decomposed as in (1.14) or (1.15), it will be understood that the submatrices have the dimensions indicated above.

The initial value y_o occurring in (1.12) is a given vector in R^n with $y_o^2 = 0$. If $j = 0$, then no one equation in (1.9) contains a derivative and there is no need for an initial condition. Accordingly, (1.12) reduces to the identity $0 = 0$.

It is immediately seen that all models mentioned above are special cases of (1.8)-(1.12).

The system (1.8) - (1.12) can also be used to describe predator-prey interactions with predation dependent on age of prey. In fact, if one lets one component of the control variable correspond to the size of the predator population, then the models considered by Gurtin and Levine [13], Cushing and Saleem [5] and Thompson et al. [29] fit exactly the formulation of this paper. Our model contains also some epidemic models, for instance the model (2.1) - (2.5) in the paper of Webb [31].

If y: $R^+ \to R^n$ is any continuous function, let y_h denote its translate, $y_h(t) = y(t+h)$, and let Φ_y be the fundamental matrix associated with $F(\cdot,y(\cdot))$, that is, Φ_y is the unique solution of the matrix equation

$$(1.16)\quad \frac{d}{ds}\,\Phi_y(s,s_0) + F(s,y(s))\Phi_y(s,s_0) = 0\ ,$$

$$(1.17)\quad \Phi_y(s_0,s_0) = I\ ,$$

where I is the identity matrix (we assume that F is continuous from $R^+ \times R^n$ to $\mathcal{L}(R^n)$). If one integrates eq. (1.8) along characteristics and takes (1.10) and (1.11) into account, one obtains

$$(1.18)\quad x(t,\tau) = \begin{cases} \Phi_{y_{t-\tau}}(\tau,0)\Lambda y(t-\tau)\ , & t \geq \tau \\ \Phi_{y_{t-\tau}}(\tau,\tau-t)x_0(\tau-t)\ , & t \leq \tau \end{cases}$$

(The occurrance of $y_{t-\tau}$ in the case $t < \tau$ may see curious, however, it appears only in the principal matrix $\Phi_{y_{t-\tau}}(\tau,\tau-t)$ with initial value $\tau-t$ so we need no information of y for $t < 0$).

Substitution of (1.18) into (1.9) yields

$$(1.19)\quad L_1y'(t) + L_2y(t) = \int_0^t G(\tau,y(t))\Phi_{y_{t-\tau}}(\tau,0)\Lambda y(t-\tau)d\tau$$
$$+\int_t^\infty G(\tau,y(t))\Phi_{y_{t-\tau}}(\tau,\tau-t)x_0(\tau-t)d\tau$$
$$+h(y(t))\ .$$

By a solution of the problem on an interval $I = [0,T]$ or R^+ we understand a continuous mapping (x,y): $t \to (x(t,\cdot),\ y(t))$ defined on I with values in $L^1(R^+;\ R^m) \times R^n$ such that y is a continuous (weak) solution of (1.19) and $x(t,\tau)$ is given by (1.18).

We shall not be concerned with the nontrivial problem of existence and uniqueness of solutions. Specific results on existence, uniqueness and positivity of solutions can be found in [3], [6], [11], [15], [16], [19], [31] for some important special cases.

We shall analyze the asymptotic stability of the model (1.9) - (1.12) by the method of linearization. The main idea is to use the Paley-Wiener Theorem to show that the resolvent associated with the linearized system belongs to L^1. An estimate of the kernel of the linearized system then shows that the resolvent actually decays exponentially. Recently Cushing [4] and Prüss [23] have used linearization methods in analyzing related problems from different points of view.

2. UNCONTROLLED GROWTH AND BIRTH RATE. In this case the matrices F and G are independent of y, eq. (1.9) describes solely the birth process, $n \leq m$ and the system (1.8) - (1.11) takes the form

$$(2.1) \qquad \mathrm{D}x(t,\tau) + F(\tau)x(t,\tau) = 0 \;,$$

$$(2.2) \qquad y(t) = \int_0^\infty G(\tau)x(t,\tau)d\tau \;,$$

$$(2.3) \qquad x(t,0) = \Lambda y(t) \;.$$

Let Φ be the principal matrix corresponding to F, then

$$(2.4) \qquad x(t,\tau) = \begin{cases} \Phi(\tau,0)\Lambda y(t-\tau) \;, & t \geq \tau \\ \Phi(\tau,\tau-t)x_0(\tau-t) \;, & t \leq \tau \end{cases}$$

which substituted into (2.2) gives

$$(2.5) \qquad y(t) = \int_0^t K(\tau)\Lambda y(t-\tau)d\tau + \int_t^\infty K(\tau)\phi(\tau-t)d\tau \;,$$

where $K(\tau) = G(\tau)\Phi(\tau,0)$, $\phi(\tau) = \Phi(0,\tau)x_0(\tau)$. The problem is thus reduced to the uncoupled problems of solving a system of linear ordinary differential equations (i.e. finding the principalmatrix Φ) and solving the system (2.5) of linear renewal equations (the biological interpretation of the given functions requires that both the kernel $K(\cdot)\Lambda$ and the forcing function are nonnegative). Once these problems are solved, the age distribution is obtained from (2.4). An example in §4 shows the techniques used to investigate the asymptotic behaviour of the solution.

3. EQUILIBRIUM SOLUTIONS AND STABILITY OF EQUILIBRIA. By an *equilibrium solution* of the system (1.8) - (1.10) we understand any time independent solution thereof. More exactly, an equilibrium is a global solution (x,y) such that there is a function $x^* \in L^1(\mathrm{R}^+;\, \mathrm{R}^m)$ and an $y^* \in \mathrm{R}^n$ satisfying

$$(3.1) \qquad x(t,\tau) = x^*(\tau) \;,$$

$$(3.2) \qquad y(t) = y^*$$

for all t and τ in R^+. In what follows, we shall denote an equilibrium simply by (x^*, y^*).

Observe that although all components of the age density are nonnegative in all applications to population dynamics, we do not impose any positivity conditions on x^* because this would be irrelevant to our purposes.

The equilibrium solution (x^*, y^*) is called exponentially asymptotically stable if there is an $\alpha > 0$ and for any given $\varepsilon > 0$ there exists a $\delta > 0$ such that

$x_o \in L^1(\mathrm{R}^+; \mathrm{R}^m)$, $\| x_o - x^* \|_1 < \delta$, $y_o \in \mathrm{R}^m$, $y_o^2 = 0$, $|y_o - L_1 y^*| < \delta$ imply that the problem (1.8) - (1.12) has a unique solution (x,y) existing and satisfying $\| x(t,\cdot) - x^* \|_1 < \varepsilon e^{-\alpha t}$, $|y(t) - y^*| \le \varepsilon e^{-\alpha t}$ for all $t \in \mathrm{R}^+$.

It is obvious that at equilibrium

$$(3.3) \quad \frac{d}{d\tau} x^*(\tau) + F(\tau,y^*)x^*(\tau) = 0 ,$$

$$(3.4) \quad L_2 y^* = \int_0^\infty G(\tau,y^*)x^*(\tau)d\tau + h(y^*) ,$$

$$(3.5) \quad x^*(0) = \Lambda y^* .$$

The solution of the initial value problem (3.3), (3.5) is given by

$$(3.6) \quad x^*(\tau) = \Phi_{y^*}(\tau,0)\Lambda y^* ,$$

where Φ_{y^*} is the principal matrix corresponding to $\tau \to F(\tau,y^*)$. Substitution of (3.6) into (3.4) yields

$$(3.7) \quad L_2 y^* = \int_0^\infty G(\tau,y^*)\Phi_{y^*}(\tau,0)d\tau\Lambda y^* + h(y^*) .$$

We therefore conclude that

there exists an equilibrium solution if and only eq. (3.7) *has a solution* y^* *and that the equilibrium age density is given by* (3.6).

The equation (3.7) seems quite complicated but in many important special cases it can be reformulated using the notion of the net reproduction rate in such a way that the condition for existence of equilibria is intuitively acceptable, see Gurtin and McCamy [11], Rorres [25,27], Gyllenberg [16].

We will now investigate stability of equilibrium solutions of the population problem. We will show that the corresponding linearized system of (1.8) - (1.10) can be interpreted as a perturbated linear Volterra equation (either an integro-differential equation or a pure integral equation), where the perturbation term has a quite complicated functional dependence on the unknown. We can then use well-known results and methods from the resolvent theory of linear Volterra equations.

We first introduce some notation. If ϕ is any vector or matrix valued function on $\mathrm{R}^+ \times \mathrm{R}^n$ for which the partial derivative $\phi_y(\tau,y^*)$ exists we denote the small term occurring in the definition of the derivative by $\varepsilon_\phi(\tau,\cdot)$ $\phi(\tau,y^*)$ by $\phi^*(\tau)$ and $\phi_y(\tau,y^*)$ by $\phi_y^*(\tau)$. Thus we have

$$(3.8) \quad \phi(\tau,y^*+\eta) = \phi^*(\tau) + \phi_y^*(\tau)\eta + \varepsilon_\phi(\tau,\eta) .$$

By definition,

$$(3.9) \quad |\eta|^{-1}\varepsilon_\phi(\tau,\eta) \to 0 \quad \text{as} \quad \eta \to 0$$

for each $\tau \in \mathrm{R}^+$.

Let (x^*,y^*) be an equilibrium solution of (1.8) - (1.10). Denote the deviation from the equilibrium by (ξ,η), that is

(3.10) $\xi(t,\tau) = x(t,\tau) - x^*(\tau)$,

(3.11) $\eta(t) = y(t) - y^*$

for all t and τ in R^+. If one assumes that the partial derivatives $F_y^*(\tau)$, $G_y^*(\tau)$ and $h'(y^*)$ exist for all $\tau \in R^+$, then an easy computation which uses (3.3) - (3.5) shows that the system (1.8) - (1.12) is equivalent to

(3.12) $D\xi(t,\tau) + F^*(\tau)\xi(t,\tau) + k(\tau)\eta(t) = \chi(t,\tau)$,

(3.13) $L_1\eta'(t) + A\eta(t) = \int_0^\infty G^*(\tau)\xi(t,\tau)d\tau + \psi(t)$,

(3.14) $\xi(t,0) = \Lambda\eta(t)$,

(3.15) $\xi(0,\tau) = \xi_0(\tau)$,

(3.16) $L_1\eta(0) = \eta_0$,

where $k(\tau) \in \mathcal{L}(R^n; R^m)$ and $A \in \mathcal{L}(R^n; R^n)$ are defined

(3.17) $k(\tau)\eta = [F_y^*(\tau)\eta]x^*(\tau)$,

(3.18) $A\eta = L_2\eta - \int_0^\infty [G_y^*(\tau)\eta]x^*(\tau)d\tau - h'(y^*)\eta$

and

(3.19) $\chi(t,\tau) = -\varepsilon_F(\tau,\eta(t))x^*(\tau) - [F(\tau,y^*+\eta(t)) - F^*(\tau)]\xi(t,\tau)$,

(3.20) $$\psi(t) = \int_0^\infty \varepsilon_G(\tau,\eta(t))x^*(\tau)d\tau + \int_0^\infty [G(\tau,y^*+\eta(t)) - G^*(\tau)]\xi(t,\tau)d\tau + \varepsilon_h(\eta(t)) ,$$

(3.21) $\xi_0(\tau) = x_0(\tau) - x^*(\tau)$,

(3.22) $\eta_0 = y_0 - L_1y^*$.

Let $\Phi^* = \Phi_{y^*}$, that is, Φ^* is the principal matrix corresponding to F^*. For fixed functions χ, η, ξ_0, integration of (3.12) along characteristics yields

(3.23) $$\xi(\chi,\eta,\xi_0)(t,\tau) = \begin{cases} \Phi^*(\tau,0)\Lambda\eta(t-\tau)+\int_0^\tau \Phi^*(\tau,s)[\chi(t-\tau+s,s)-k(s)\eta(t-\tau+s)]ds , & \tau \le t \\ \Phi^*(\tau,\tau-t)\xi_0(\tau-t)+\int_0^t \Phi^*(\tau,s+\tau-t)[\chi(s,\tau-t+s)-k(\tau-t+s)\eta(s)]ds , & \tau \ge t . \end{cases}$$

Substituting (3.23) into (3.13), one obtains the following "functional-integro-differential equation":

(3.24) $L_1\eta'(t) + A\eta(t) = (K*\eta)(t) + \Psi(\chi,\eta,\xi_0)(t)$.

Here $*$ denotes convolution (i.e. $(K*\eta)(t) = \int_0^t K(t-s)\eta(s)ds$), K is an n by n matrix-valued and Ψ an R^n-valued function. They are defined by

(3.25) $K(t) = G^*(t)\Phi^*(t,0)\Lambda - \int_0^\infty G^*(\tau+t)\Phi^*(\tau+t,\tau)k(\tau)d\tau$,

(3.26) $\Psi(\chi,\eta,\xi_0)(t) = \int_0^t\int_0^\infty G^*(\tau+t-s)\Phi^*(\tau+t-s,\tau)\chi(s,\tau)d\tau ds$

$$+\int_0^\infty G^*(\tau+t)\Phi^*(\tau+t,\tau)\xi_0(\tau)d\tau + \psi(\chi,\eta,\xi_0)(t) ,$$

where $\psi(\chi,\eta,\xi_0)$ is, for every χ, η and ξ_0, the function defined by (3.20) with ξ replaced by $\xi(\chi,\eta,\xi_0)$ defined by (3.23). We have thus reduced the problem to the study of the equation (3.24) supplemented by

(3.27) $\chi(t,\tau) = -\varepsilon_F(\tau,\eta(t))x^*(\tau) - [F(\tau,y^*+\eta(t)) - F^*(\tau)]\xi(\chi,\eta,\xi_0)(t,\tau)$.

It should be noted that the functional Ψ defined by (3.26) is nonanticipative in the sense that the value of $\Psi(\chi,\eta,\xi_0)(t)$ depends only on the restrictions of χ and η to the interval $[0,t]$. Hence (3.24) can be rewritten in an equivalent variation of constants form.

If $L_1 = I$, let R be the differential resolvent associated with the pair (A,K), that is, R is the solution of the matrix eqution

(3.28) $R'(t) + AR(t) = (K*R)(t)$, $R(0) = I$.

Then (3.24) is equivalent to

(3.29) $\eta(t) = R(t)\eta_0 + \int_0^t R(t-s)\Psi(\chi,\eta,\xi_0)(s)ds$

(cf. Grossmann and Miller [9]).

In applications to population dynamics $L_1 \neq I$, since at least one of the components in eq. (1.9) describes the birth process, which has an undifferentiated form (cf. (1.2)). Suppose therefore that L_1 has the decomposition into blocks with $j < n$. Let e denote the function $t \to e^{-t}$, $t \in R^+$. We have $e(0) = 1$, $e' = e$; e, $e' \in L(R^+)$. It is easily seen that for a continuously differentiable function a on R^+, $e*a'$ exists and

(3.30) $e*a' = a - a(0)e - e*a$.

Hence, if both sides of equation (3.24) are convoluted by $\begin{pmatrix} eI & 0 \\ 0 & \delta I\end{pmatrix}$ (δ is the Dirac measure) and some terms are transferred from the left to the right side, one obtains the equation

(3.31) $\tilde{A}\eta = \tilde{K}*\eta + \tilde{\Psi}(\chi,\eta,\xi_0,\eta_0)$,

where

(3.32) $\tilde{A} = \begin{pmatrix} I & 0 \\ A^{21} & A^{22} \end{pmatrix}$,

(3.33) $\tilde{K} = \begin{pmatrix} eI - eA^{11} + eI*K^{11} & -eA^{12} + eI*K^{12} \\ K^{21} & K^{22} \end{pmatrix}$

and

(3.34) $\tilde{\Psi} = \begin{pmatrix} \eta_o^1 + eI*\Psi^1 \\ \Psi^2 \end{pmatrix}$

If $\tilde{A}$ is invertible, which happens if and only if the $(n-j)\times(n-j)$ submatrix A^{22} is invertible, then (3.31) is equivalent to

(3.35) $\eta = \tilde{A}^{-1}K*\eta + \tilde{A}^{-1}\tilde{\Psi}$.

Let r be the integral resolvent associated with the kernel $\tilde{A}^{-1}\tilde{K}$, i.e. r is the unique solution of the matrix equation

(3.36) $r = \tilde{A}^{-1}\tilde{K} + \tilde{A}^{-1}\tilde{K}*r$

on R^+. In this case (3.35) can be rewritten as

(3.37) $\eta = \tilde{A}^{-1}\tilde{\Psi}(\chi,\eta,\xi_o,\eta_o) + r*\tilde{A}^{-1}\tilde{\Psi}(\chi,\eta,\xi_o,\eta_o)$.

We do not have to treat the cases $L_1 = I$ and $L_1 \neq I$ separately; in both cases eq. (3.24) is equivalent to (3.35). But the case $L_1 = I$ is simpler because there $\tilde{A} = I$ and we need no extra hypothesis on the invertibility of a submatrix of A to transform the equation into the form (3.37).

In the proofs of the stability results the exponential decay of the resolvent r is of crucial importance. By a well-known theorem of Paley and Wiener [22], assuming $K \in L^1(R^+; \mathcal{L}(R^n, R^n))$ (which clearly implies $\tilde{A}^{-1}\tilde{K} \in L^1(R^+; \mathcal{L}(R^n, R^n)))$, $r \in L^1(R^+; \mathcal{L}(R^n, R^n))$ if and only if

(3.38) $\det(I-\tilde{A}^{-1}\hat{\tilde{K}}(z)) \neq 0$, $\quad \mathrm{Re}\, z \geq 0$,

where $\wedge$ denotes Laplace transform. The condition (3.38) is easily seen to be equivalent to

(3.39) $\det(zL_1+A-\hat{K}(z)) \neq 0$, $\quad \mathrm{Re}\, z \geq 0$.

It is also a simple matter to prove that if in addition to (3.39) $K(t)$ decays expotentially, that is, if there are constants $\alpha > 0$ and $M < \infty$ such that $\| K(t) \| < M e^{-\alpha t}$, then the same is true of the resolvent.

In the case $L_1 = I$ one could alternatively base the stability proof on the form (3.29) and use the integrability criterion of R (due to Grossmann and Miller [10]), which also is equivalent to (3.39).

For our first theorem we need the following hypotheses on the smoothness of the given functions:

(S_1) There is a neighborhood V of y^* such that the mapping $y \to F(\cdot,y)$ belongs to $C(V; L^\infty(R^+; \mathcal{L}(R^m, R^m)))$.

(S_2) The partial derivatives $F_y^*(\tau)$, $G_y^*(\tau)$ exist for all $\tau \in R^+$. If $j < n$, there is a neighborhood V of y^* such that the partial derivative $G_y^2(\tau,z)$ exist for all $\tau \in R^+$ and all $z \in V$ and $z \to G_y^2(\cdot,z)$ is a continuous mapping from V into $L^\infty(R^+; \mathcal{L}(R^n, \mathcal{L}(R^m, R^{n-j})))$.

(S_3) $\lim_{\eta \to 0} \sup_{\tau \in R^+} \{|\eta|^{-1} \| \varepsilon(\tau,\eta) \| \} = 0$ for $\varepsilon = \varepsilon_F, \varepsilon_G$ (cf. (3.8), (3.9)).

(S_4) h is differentiable at y^*. If $j < n$ there is a neighborhood V of y^* such that h is continuously differentiable in V.

We also need the following assumptions concerning the bigness of the data:

(B_1) There are constants $\alpha > 0$ and $M < \infty$ such that $\| \Phi^*(t,s) \| \le Me^{-\alpha(t-s)}$, $0 \le s \le t < \infty$.

(B_2) $G^* \in L^\infty(R^+; \mathcal{L}(R^m, R^n))$.

(B_3) $F_y^* \in L^\infty(R^+; \mathcal{L}(R^n, \mathcal{L}(R^m, R^m)))$,
$G_y^* \in L^\infty(R^+; \mathcal{L}(R^n, \mathcal{L}(R^m, R^n)))$.

The smoothness hypotheses (S_1) - (S_5) are of quite technical character. They seem acceptable as do the bigness hypotheses (B_2), (B_3). The hypothesis (B_1) may appear as too restrictive, however, it is very natural as the following biological interpretation shows.

It is a well-known fact in the theory of linear ordinary differential equations that (B_1) is equivalent to the seemingly weaker condition that the zero solution of the equation

$$(3.40) \quad u'(s) + F(s,y^*)u(s) = 0$$

is uniformly asymptotically stable (cf. Hale [20. p. 84]). Suppose the value of the control variable is kept fixed at y^*. Then by (1.8), $u(s) = x(t+s,s)$ satisfies (3.40) and the uniform asymptotic stability of this equation means, roughly speaking, that any sufficiently small group of individuals, which are of age τ at time t, will remain small for all time and will eventually extinct. (The concept of sufficient smallness is independent of τ and t). For systems where population loss is due only to deaths this condition is automatically satisfied.

It should be noted that the hypotheses imply that $\| K(t) \| \le Me^{-\alpha t}$.

<u>THEOREM 3.1.</u> *Let* (x^*,y^*) *be an equilibrium solution of the system* (1.1) - (1.3) *and suppose* (S_1) - (S_4), (B_1) - (B_3) *are satisfied. Then, if condition* (3.39) *holds and*

the matrix A^{22} *is invertible (if* $j < n$*), there are constants* $\delta > 0$, $\alpha > 0$ *and* $M < \infty$ *such that* $\| x_0 - x^*\|_1 + |y_0 - L_1 y^*| < \delta$ *implies that*

$$(3.41)\quad \| x(t,\cdot) - x^*\|_1 + |y(t) - y^*| \le M\{\| x_0 - x^*\|_1 + |y_0 - L_1 y^*|\}\, e^{-\alpha t}$$

as long as the solution (x,y) *satisfying* $x(0,\tau) = x_0(\tau)$, $L_1 y(0) = y_0$ *exists. Moreover,*

$$(3.42)\quad |x(t,\tau) - x^*(\tau)| \le M\{\| x_0 - x^*\|_1 + | y_0 + L_1 y^*|\}\, e^{-\alpha t}$$

for $t \ge \tau$ *as long as the solution exists.*

Thus if one knows a priori that to each initial value (x_0, y_0) corresponds a unique solution existing for all time, then the conclusion of Theorem 2.1. is that the equilibrium is exponentially asymptotically stable in the $L^1(\mathrm{R}^+;\ \mathrm{R}^m)\times\mathrm{R}^n$-topology and in addition that $x(t,\tau)$ tends to $x^*(\tau)$ uniformly in τ and exponentially in t as $t \to \infty$.

If one imposes slightly more smoothness on F, G and h in some neighborhood of y^* one can prove local existence of solutions with initial values sufficiently near (x^*,y^*) using a fixed point argument. Standard continuation techniques then yield a global solution and hence by the preceding theorem exponential asymptotic stability of the equilibrium. We need:

(S_5) There are constants $\delta > 0$ and $L < \infty$ such that

$$\| \varepsilon_{G^1}(\tau,\eta_1) - \varepsilon_{G^1}(\tau,\eta_2)\| \le L|\eta_1 - \eta_2| ,$$

$$|\varepsilon_{h^1}(\eta_1) - \varepsilon_{h^2}(\eta_2)| \le L|\eta_1 - \eta_2|$$

when $|\eta_1|$, $|\eta_2| < \delta$, $\tau \in \mathrm{R}^+$.

(S_6) To every $\varepsilon > 0$ corresponds a $\delta > 0$ such that

$$\| \varepsilon_{F^2}(\tau,\eta_1) - \varepsilon_{F^2}(\tau,\eta_2)\| \le \varepsilon|\eta_1 - \eta_2| ,$$

$$\| \varepsilon_{G^2}(\tau,\eta_1) - \varepsilon_{G^2}(\tau,\eta_2)\| \le \varepsilon|\eta_1 - \eta_2| ,$$

$$|\varepsilon_{h^2}(\eta_1) - \varepsilon_{h^2}(\eta_2)| \le \varepsilon|\eta_1 - \eta_2|$$

when $|\eta_1|$, $|\eta_2| < \delta$ $\tau \in \mathrm{R}^+$.

<u>THEOREM 3.2.</u> *Let* (x^*,y^*) *be an equilibrium solution and suppose the conditions* (S_1) - (S_6), (B_1) - (B_3) *are satisfied. Then* (x^*,y^*) *is exponentially asymptotically stable if* (3.39) *holds and if (in the case* $j < n$*)* A^{22} *is invertible. Moreover, the estimate* (3.42) *holds for* $0 \le \tau \le t < \infty$ *if* (x,y) *is a solution corresponding to initial data sufficiently close to* (x^*,y^*).

The proofs of results slightly less exact than those in Theorems 3.1 and 3.2 for the case of only one species ($m = 1$) can be found in [17]. The proofs of theorems 3.1 and 3.2 will appear in a forthcoming paper.

In [17] it is shown that the stability results of the papers [11], [16], [26], [27] are essentially special cases of Theorems 3.1 and 3.2.

4. EXAMPLES. In this section we illustrate the application of the theory presented in the preceding sections by two examples.

A. Batch cultivation of *Saccharomyces cerevisiae*

Saccharomyces cerevisiae reproduces by budding. Each fission leaves a scar on the wall of the mother cell. The population is thus divided into well defined scar classes according to the number of bud scars (cf. [2], [21]). Let $x_0(t,\tau)$ be the age density of virgin cells (i.e. cells without bud scars) and let $x_1(t,\tau)$ be the density of cells which have divided at least once (i.e. cells with one or more bud scars). Here τ does not mean chronological age but cell cycle age, which represents the phase in the cell cycle, in which the cell is. Thus the mother cell obtains the age $\tau = 0$ after each fission (cf. [18]).

If the cells are cultivated in an excess of substrate and if there is no crowding, then there is no growth controlling factor, mortality can be neglected and the evolution of the system can be modelled using (2.1) - (2.3) with $m = 2$, $n = 1$,

$$(4.1) \qquad F(\tau) = \begin{pmatrix} f_0(\tau) & 0 \\ 0 & f_1(\tau) \end{pmatrix}$$

$$(4.2) \qquad G(\tau) = (f_0(\tau) \quad f_1(\tau))$$

and

$$(4.3) \qquad \Lambda = \begin{pmatrix} 1 \\ 1 \end{pmatrix} .$$

(2.5) is then a scalar equation and takes the form

$$(4.4) \qquad y(t) = \int_0^t p(\tau)y(t-\tau)d\tau + \phi(t) ,$$

where

$$(4.5) \qquad p_i(\tau) = f_i(\tau)\exp(-\int_0^\tau f_i(s)ds) , \qquad p = p_0 + p_1$$

and

$$(4.6) \qquad \phi_i(\tau) = \int_t^\infty f_i(\tau)\exp(-\int_{\tau-t}^\tau f_i(s)ds)x_{io}(\tau-t)d\tau , \qquad \phi = \phi_0 + \phi_1 .$$

The sizes of the scar classes are obtained by integration of eq. (2.4) over all ages, which in this case yields

$$(4.7) \qquad X_i(t) = \int_0^\infty x_i(t,\tau)d\tau = \int_0^t \pi_i(\tau)y(t-\tau)d\tau + \psi_i(t)$$

where

$$(4.8) \quad \pi_i(\tau) = \exp(-\int_0^\tau f_i(s)ds)$$

and

$$(4.9) \quad \psi_i(t) = \int_t^\infty \exp(-\int_{\tau-t}^\tau f_i(s)ds)x_{io}(\tau-t)d\tau \ .$$

We shall assume that the functions p_i, π_i, ϕ_i and ψ_i have compact support, an assumption which corresponds to the biological requirement of the existence of a maximum length of the cell cycle. Then the characteristic equation

$$(4.10) \quad \hat{p}(z) = 1$$

has infinitely many roots, of which one, μ, is positive and the others appear as pairs of complex conjugates with real part less than μ. It then follows easily (for details, see [18]) that the different generations ultimately grow exponentially with the same specific growth rate (or Malthusian parameter) which equals μ, that the relative frequencies $X_o(t)/(X_o(t) + X_1(t))$ and $X_1(t)/(X_o(t) + X_1(t))$ of the different generation approach constant values and that they perform damped oscillations until demographic equilibrium is attained. This behaviour of populations of *S. cerevisiae* has been empirically known for a long time (see Beran [2], Hartwall and Unger [21]), but previous growth models (e.g. that of Gani and Saunders [7]) of *S. cerevisiae* have failed to show such oscillations. The model of Adams et al. [1] predicted damped oscillations, but the model was investigated only for special choices of initial distributions. As shown in [18], the oscillations of our model is a consequence of the asymmetrical division pattern and the age specific behaviour of the population.

B. Continuous culture of bacterial populations

As a second example we consider the following model of age-dependent growth of continuously propagated bacterial cultures, the cells of which reproduce by binary fission.

$$(4.11) \quad Dx(t,\tau) + [f(\tau) + m(\tau) + D]x(t,\tau) = 0 \ ,$$

$$(4.12) \quad S'(t) = -\gamma\int_0^\infty \mu(\tau,S(t))x(t,\tau)d\tau + D[S_{in} - S(t)] \ ,$$

$$(4.13) \quad b(t) = x(t,0) = \int_0^\infty f(\tau)x(t,\tau)d\tau + \int_0^\infty \mu(\tau,S(t))x(t,\tau)d\tau \ .$$

Here the scalar function $x(t,\tau)$ is the age density of *biomass* (*not* individuals), f and m are the age specific fission and mortality modulus, respectively, D is the dilution rate, S is the concentration of growth-limiting substrate in the growth chamber, S_{in} the input substrate concentration, γ is the inverse of the yield constant, and μ is the age specific growth rate of the population. All these functions are positive. The system (4.11) - (4.13) is clearly a special case of (1.8) - (1.10) with $m = 1$,

$n = 2,\ y = \begin{pmatrix} S \\ b \end{pmatrix},\ F(\tau) = f(\tau) + m(\tau) + D,\ L_1 = \begin{pmatrix} 1 & 0 \\ 0 & 0 \end{pmatrix},\ L_2 = \begin{pmatrix} D & 0 \\ 0 & 1 \end{pmatrix},$

$G(\tau,S) = \begin{pmatrix} -\gamma\mu(\tau,S) & 0 \\ 0 & f(\tau)+\mu(\tau,S) \end{pmatrix}$ and $h = \begin{pmatrix} DS_{in} \\ 0 \end{pmatrix}$

The absence of μ from eq. (4.11) and its occurrence in the boundary condition (4.13) means that it is assumed that the cells grow instantaneously to their final size at the time of fission. Although this can never be entirely correct, it seems to be a fairly good approximation when the growth phase of the cells is short compared with the length of the whole cell cycle (cf. [24]).

Equation (4.12) includes the assumption that the production of one mass unit of new biomass always consumes the same amount of substrate. We think this is quite plausible, too.

Using the concept of the net reproduction rate defined by

$$(4.14)\quad R(S,D) = \int_0^\infty [f(\tau) + \mu(\tau,S)]\exp\left(-\int_0^\tau (f(u)+m(u)+D)\,du\right)d\tau$$

it is easily seen that in order for the equilibrium condition (3.7) to be satisfied, it is necessary that

$$(4.15)\quad R(S^*,D) = 1 .$$

Let $p(\tau) = \mu(\tau,S^*)\exp(-\int_0^\tau [f(u)+m(u)]du)$, $q(\tau) = f(\tau)\exp(-\int_0^\tau [f(u)+m(u)]du)$. After some manipulations the characteristic equation $\det(zL_1+A-\hat{K}(z)) = 0$ can be written in the form

$$(4.16)\quad (z+D)\left[1 - \frac{\hat{p}(z+D)}{1-\hat{q}(z+D)}\right] = -\gamma b^* \frac{\partial R}{\partial S}(S^*,D) .$$

For $\operatorname{Re} z \geq 0$

$$(4.17)\quad |\hat{p}(z+D)| + |\hat{q}(z+D)| \leq \hat{p}(\operatorname{Re} z+D) + \hat{q}(\operatorname{Re} z+D) = R(S^*,\operatorname{Re} z+D) \leq 1 ,$$

where the last inequality follows from (4.15) and the fact that $R(S^*,\cdot)$ is the Laplace transform of a positive function. (4.17) implies that

$$(4.18)\quad |\hat{p}(z+D)|/|1 - \hat{q}(z+D)| \leq 1$$

and hence that the left-hand side of (4.16) cannot be a strictly negative number. Thus, if $\frac{\partial R}{\partial S}(S^*,D) > 0$, the characteristic equation (4.16) has no root in the right half-plane. On the other hand, it is easily seen that if $\frac{\partial R}{\partial S}(S^*,D) \leq 0$ the characteristic equation has a root in the right half-plane (in fact a positive root). Thus the stability criterion (3.39) is equivalent to the condition

$$(4.19)\quad \frac{\partial R}{\partial S}(S^*,D) > 0$$

and we have arrived at the following result concerning the model (4.11) - (4.13):

A nontrivial equilibrium solution of the model (4.11) - (4.13) *is exponentially asymptotically stable if the derivative of the net reproduction rate is positive at the equilibrium substrate concentration.*

For a more detailed study of this model, see [16], where, however, the derivation of the above statement is incorrect.

REFERENCES

1. J. Adams, E.D. Rothman and K. Beran, The age structure of populations of *Saccharomyces cerevisiae*, Math. Biosci. 53 (1981), 249-263.

2. K. Beran, Budding of yeast cells, their scars and ageing, Advan. Microb. Physiol. 2 (1968), 143-171.

3. M. Chipot, On the equations of age-dependent population dynamics, Arch. Rational Mech. Anal. 82 (1983), 13-25.

4. J.M. Cushing, Model stability and instability in age structured populations, J. Theor. Biol. 86 (1980), 709-730.

5. J.M. Cushing and M. Saleem, A predator prey model with age structure, J. Math. Biology, 14 (1982), 231-250.

6. G. Di Blasio, M. Iannelli and E. Sinestrari, Approach to equilibrium in age structured populations with an increasing recruitment process, J. Math. Biology, 13 (1982), 371-382.

8. K. Gopalsamy, Age-specific coexistence in two species competition, Math. Biosci. 61 (1982), 101-122.

9. S.I. Grossman and R.K. Miller, Perturbation theory for Volterra integrodifferential systems, J. Differential Equations 8 (1970), 457-474.

10. S.I. Grossman and R.K. Miller, Nonlinear Volterra integrodifferential systems with L^1-kernels, J. Differential Equations 13 (1973), 551-566.

11. M.E. Gurtin and R.C. MacCamy, Non-linear age-dependent population dynamics, Arch. Rational. Mech. Anal. 54 (1974), 281-300.

12. M.E. Gurtin and R.C. MacCamy, Population dynamics with age dependence, in Non-linear Analysis and Mechanics: Heriot-Watt Symposium, Vol. III (R.J. Knops, Ed.), Pitman, San Fransisco, 1979, pp. 1-35.

13. M.E. Gurtin and D.S. Levine, On predator-prey interactions with predation dependent on age of prey, Math. Biosci, 47 (1979), 207-219.

14. M.E. Gurtin and D.S. Levine, On populations that cannibalize their young, SIAM J. Appl. Math. 42 (1982), 94-108.

15. M. Gyllenberg, Age-dependent population dynamics in continuously propagated bacterial cultures, Report No 8, Helsinki University of Technology, Institution of Mechanics, (1981).

16. M. Gyllenberg, Nonlinear age-dependent population dynamics in continuously propagated bacterial cultures, Math. Biosci. 62 (1982), 45-74.

17. M. Gyllenberg, Stability of a nonlinear age-dependent population model containing a control variable, SIAM J. Appl. Math. 43 (1983), to appear.

18. M. Gyllenberg, The age structure of populations of cells reproducing by asymmetric division, Mathematics in Biology and Medicine, An International Conference, Bari, July 18-22, 1983, to appear.

19. A. Haimovici, On the growth of a population dependent on ages and involving resources and pollution, Math. Biosci. 43 (1979), 213-237.

20. J. Hale, Ordinary differential equations, Wiley, New York, 1969.

21. L.H. Hartwell and M.W. Unger, Unequal division in *Saccharomyces cerevisiae* and its implications for the control of cell division, J. Cell. Biol. 75 (1977), 422-435.

22. R.E.A.C. Paley and N. Wiener, Fourier transforms in the complex domain, Amer. Math. Soc. Colloquium Publications. (1934).

23. J. Prüss, Equilibrium solutions of age-specific population dynamics of several species, J. Math. Biology, 11 (1981), 65-84.

24. J. Ranta, On the mathematical modelling of microbial age dynamic and some control aspects of microbial growth processes, Acta Polytech. Scand., Ma35, Helsinki (1982).

25. C. Rorres, Stability of an age specific population with density dependent fertility, Theor. Popul. Biol. 10 (1976), 26-46.

26. C. Rorres, Local stability of a population with density dependent fertility, Theor. Popul. Biol. 16 (1979), 283-300.

27. C. Rorres, A nonlinear model of population growth in which fertility is dependent on birth rate, SIAM J. Appl. Math. 37 (1979), 423-432.

28. F.R. Sharpe and A.J. Lotka, A problem in age distribution, Phil. Mag. 2. (1911), 435-438.

29. R.W. Thompson, D. Di Blasio and C. Mendes, Predator-prey interactions: Egg-eating predators, Math. Biosci, 60 (1982), 109-120.

30. H. von Foerster, Some remarks on changing populations, in The kinetics of cellular proliferation, Grune and Stratton, New York (1959), 382-407.

31. G.F. Webb, An age-dependent epidemic model with spatial diffusion, Arch. Rational Mech. Anal. 75 (1980), 91-102.

CELL TRACTION AND THE GENERATION OF ANATOMICAL STRUCTURE

Albert K. Harris
Department of Biology
Wilson Hall (046A)
University of North Carolina
Chapel Hill, North Carolina 27514
USA

SUMMARY

The component cells of the body exert traction forces by which they can propel themselves and rearrange extracellular materials, in particular the fibrous protein collagen. Because the compression and alignment created by these cellular forces can, in turn, affect cell behavior, positive feedback cycles of several kinds arise, and these cycles are capable of spontaneously generating regular geometric patterns of cells and matrix. In this way, the mechanical activities of cells can themselves accomplish the morphogenetic functions usually attributed to the diffusion and reactions of chemical "morphogens".

INTRODUCTION

When humans build something, the process of laying out the planned geometry (drawing of blueprints, surveying and so on) normally precedes and is kept separate from the actual physical or mechanical process of construction. In general the personnel are even separate, usually with some diminution in social status between the architect and the bricklayer.

Likewise, in our attempts to comprehend how anatomical structures are brought into existence in the embryo, it has come to be assumed that there is a similar sequence of separate steps: First the geometry is laid out in the form of chemical diffusion gradients (to provide "positional information"). Then, subsequently, the cells respond to the local concentrations of these chemicals, either by differentiation or by movement (or by some combination of both) so as to create the desired anatomical pattern in physical form. So firmly have these assumptions taken hold, that many recent publications on pattern formation actually confine their considerations to mechanisms for generating systems of diffusion gradients, rather than for generating actual anatomical patterns or structures. Likewise, the successful coordination of cell behavior over any considerable

A single fibroblast cell from an embryonic chicken crawling on a thin sheet of silicone rubber. The strong traction forces exerted by the cell have distorted the rubber sheet into a complex pattern of compression and tension wrinkles. The same traction forces exerted on fibers of the protein collagen in the developing body may be responsible for creating ligaments, tendons, muscles and organ capsules, as well as controlling where parts of the skeleton will form. This cell is about 80 micrometers wide.

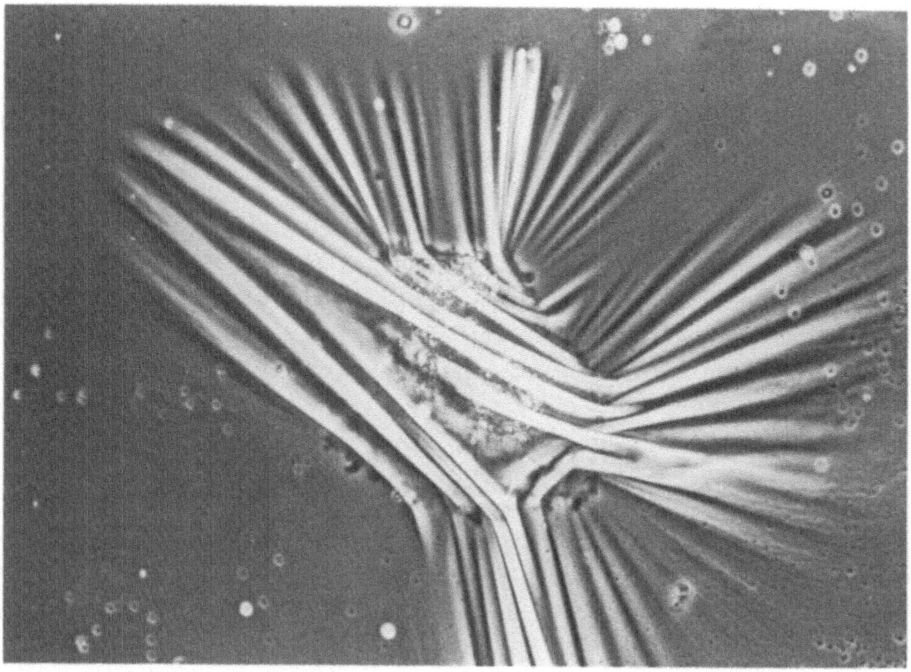

distance has itself come to be taken as evidence or even as proof that these postulated gradients really exist - as if diffusion were the only possible mechanism for action at a distance.

Contrary to these assumptions, recent studies of the physical forces exerted by embryonic cells have shown that these forces are themselves capable of generating and coordinating long range effects - without the need for control by any intervening system of diffusion gradients (Harris, Stopak and Wild, 1981). In other words, mechanical interactions themselves can accomplish the "surveying" functions, for which diffusion gradients have been postulated. This morphogenetic capability has been shown perhaps most dramatically in the case of ligaments, tendons and muscles (Stopak and Harris, 1982), where it is not difficult to bring about the creation of some reasonable facsimiles of such structures in organ culture simply by controlling the application of fibroblast traction to gels of reprecipitated collagen.

Biological evolution proceeds by adjustment of whatever processes are already present. Therefore, considering that mechanical forces are necessarily present in the embryo to rearrange cells into structures anyway, then if these forces can themselves perform the coordination role for which morphogen gradients have traditionally been postulated, perhaps neither we nor the embryos have much need for all these hypothetical diffusion gradients. Of course, there is no doubt that anatomical patterns are ultimately determined by genes, and thus by chemistry, so that the two contrasting viewpoints can perhaps be expressed by the following alternative causal sequences:

CHEMISTRY -> GEOMETRY -> MECHANICS

as opposed to

CHEMISTRY -> MECHANICS -> GEOMETRY

The top sequence represents the currently dominant "positional information" viewpoint, as well as the motivation for most work on reaction-diffusion systems capable of generating the diffusion gradients. The bottom sequence represents the viewpoint of my collaborators and myself and will be further developed here.

Our approach is not really all that novel, however, and might fairly be described as an attempt to generalize mechanisms of embryological pattern generation to include mechanical (vector and tensor) properties, instead of just the simple (scalar) property of chemical concentration, to which students of morphogenesis have tended to restrict themselves (perhaps somewhat artificially) in the past.

CELL LOCOMOTION AND TRACTION

Most of what is known about the locomotion of the body's component cells has been learned from observing their behavior in tissue culture (see Trinkaus, 1984). By culturing the cells to be studied outside the body, in artificial media and on artificial surfaces (substrata), one achieves not only optimum visibility but also the ability to manipulate experimentally the cells' chemical and physical environments to a degree which is limited only by one's imagination - rather than by the resistance of the body's homeostatic mechanisms.

The locomotion of the component cells of multicellular animals is commonly referred to as being "amoeboid", and the propulsive forces responsible for locomotion are believed to be generated (both in amoebae and in tissue cells) by the interactions of cytoplasmic proteins very similar to actin and myosin - the proteins responsible for muscle contraction. On the other hand, there are many important differences between the locomotion of tissue cells and organisms like *Amoeba proteus*. Actually, the various organisms lumped together as amoebae use many different propulsive mechanisms.

In contrast to the type of massive forward cytoplasmic flow which one sees in *Amoeba proteus*, most kinds of tissue cells lack cytoplasmic flow (a different sort of flow does occur in some white blood cells). Instead, the most evident correlate of locomotion in tissue cells is a complex set of outward and rearward movements of the plasma membrane (Harris, 1983). Tissue cell locomotion also differs from that of *Amoeba* in being very much slower (less than one um. per minute for most cell types) so that time lapse photography is needed even to observe it. When simply observed through the microscope, tissue cells appear to be standing still.

An important property of tissue cell locomotion is its dependence upon adhesion to a substratum. Most cell types remain approximately spherical in the absence of adhesion to a substratum and can extend themselves only by attaching and pulling themselves along the surface of some external object. Indeed their locomotion could reasonably be described as pulling themselves along by means of their adhesions - in contrast to *Amoeba proteus*, which gives the impression of pushing itself along.

In order to study the pulling force ("traction") by which tissue cells propel themselves, methods have been developed for preparing extremely thin layers of silicone rubber upon which cells can be cultured and which are sufficiently deformable (elastic) so that visible wrinkling can be produced, even by the traction forces exerted by individual cells (Harris, Wild and Stopak, 1980). As cells spread on these silicone rubber substrata, the traction they exert produces two sorts of wrinkles in the substratum: (1) Compression wrinkles, which form directly beneath the cells where the forces are being exerted and which are oriented transversely to the axis of maximum tension, and (2) Tension wrinkles which are aligned along the axis of maximum tension and radiate out beyond the cell margins. When cells are killed or detach from the substratum, the elastic tension relaxes almost instantaneously with the disappearance of both classes of substratum wrinkles.

Although the number and size (amplitude and length) of the wrinkles produced by a given cell are presumed to be proportional to the magnitude of the traction forces exerted, it has proven difficult to make this method truly quantitative. Part of the difficulty is that of making the rubber layers with precisely the same thickness and elastic moduli (or to know whether one has done so). The mathematical methods of tension field theory ought to be applicable to the calculation of forces from observed wrinkle patterns but suggest that the degree of wrinkling caused by a given force should be highly sensitive to the thickness of the rubber layer (to its second or third power). This expectation is not really in accord with my observations of the effects of varying rubber thicknesses, however.

Based simply on qualitative observations of wrinkling, including such things as the population density of a given type of cell needed to produce a given degree of substratum distortion, there seem to be very large differences (of as much as 2 or 3 orders of magnitude) in the strengths of the traction forces exerted by different cell types (Harris, Stopak and Wild, 1981). The strongest are the fibroblasts, glial cells and blood platelets. The weakest are the polymorphonuclear leucocytes and the growth cones of nerves (neither of which is strong enough to produce a detectable wrinkling of our rubber layers) as well as the macrophages, which are only barely able to distort these substrata. Epithelial cells are intermediate in strength between these two extremes, and cells which have been transformed to a cancerous state seem consistently to have a weakened contractility relative to their untransformed equivalents.

Using a cell line which can be converted back and forth between a normal and transformed morphology by exposure to a drug, we were able to follow the strengthening and weakening of their traction (Leader, Stopak and Harris, 1983). It is somewhat counterintuitive that weak traction should be associated with greater motility and invasiveness, but this is the clear pattern in regard to both leucocytes and cancerous cells. On the other hand, trauma apparently stimulates a gradual increase in the tractional strength of fibroblasts, a phenomenon which seems to be related to the constrictive closure of wounds (see Gabbiani, Majno and Ryan, 1973).

EFFECTS OF CELL TRACTION ON COLLAGEN

Collagen is the most important of the extracellular structural proteins of the body. Tendons, for example, consist mostly of collagen. The molecules of this protein have a strong tendency to align with one another and form lateral cross-links. Their behavior is analogous to nematic liquid crystalline substances, except that the individual collagen molecules are 3 um. long, relatively stiff rods with regularly staggered arrays of lateral bonding sites (Linsenmeyer, 1981). Aligned, cross-linked masses of collagen are physically very strong and provide most of the structural reinforcement of the body. The problem for the embryologist is to understand how the molecules come to be arranged and aligned into the proper geometric patterns and how the same protein can form such different structures as tendons and the sheet-like capsules which closely surround most internal organs.

By using the right combination of salt concentration and acidity, collagen molecules can be redissolved from certain tendons and can then be reprecipitated to form gels (Elsdale and Bard, 1972). Gelatin, in fact, is a form of collagen extracted by more destructive means. Gels of reprecipitated collagen are initially homogeneous and isotropic, but when fibroblasts (or other cells which exert strong traction) are cultured on or in these gels, the effect of the imposed forces is to realign and rearrange the collagen fibers. When cells are cultured on thin layers of collagen, this distortion takes the form of wrinkles very much like those produced by cell traction on silicone rubber substrata. When thicker layers of collagen are used, and particularly when the cells lie down in the gel itself, surrounded by fibers, the effect of traction is to realign these fibers along the axis of cellular extension. The resulting patterns are somewhat reminiscent of the arrangement of iron filings around a bar magnet (except that here direct mechanical forces are responsible, rather than action at a distance).

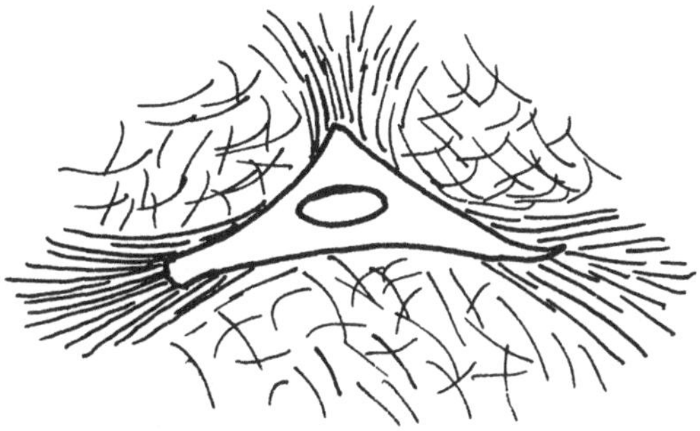

When many hundreds or thousands of cells are concentrated together in one of these gels, the mechanical effects of their traction are cumulative and can produce relatively large scale effects which even become visible to the naked eye. Surrounding an individual tissue fragment or other mass of cells, the cumulative effect of the cells' traction is to draw collagen fibers centripetally onto the surface of the cellular mass to form a tight, dense surface wrapping which grows progressively thicker as more collagen is pulled in. In the surrounding area, the effect of the traction forces is to reorient the collagen fibers radially around the mass of cells - that is, along the axis of the inward directed traction forces. When two or more masses of cells are placed in the same collagen gel, even stronger alignment occurs along straight lines connecting masses. Linear alignment also results when small "fixed points" are placed in the gel to prevent its displacement (Stopak and Harris, 1982). Fibroblast traction can easily align collagen over distances of several centimeters - across the width of an entire petri dish.

There seem to be several reasons for the magnitude of these alignment effects. One is simply the inherent tendency for any long stiff molecules to align parallel with one another, such as one sees in the case of nematic liquid crystals. Added to this is the ability of collagen molecules to form lateral bonds with one another, which tends to lock the molecules into parallel arrays, once these are established. In addition, the cells have a strong tendency to align themselves along fibers as well as to exert most of their own traction along their own long axes. Thus, a positive feedback cycle is created, in which tension in a given direction has the effect of aligning collagen fibers in that direction, which in turn stimulates cells to align in the same direction, which means that their traction will be concentrated in this direction also, which further increases the alignment of fibers, and so on. This is one of several such positive feedback cycles which can be identified in the interactions of cells and extracellular matrix materials such as collagen. These cycles have powerful capacities for magnifying initial heterogeneities or anisotropies to produce large-scale patterns and may even be able to initiate the formation of such patterns spontaneously. I shall call these cycles "anti-diffusive effects" and enumerate some of them in a later section. Their morphogenetic capacity is comparable to that of chemical autocatalysis as it is employed in reaction-diffusion systems.

The similarity of the compressed sheets of collagen to normal organ capsules, and of the aligned tracts of collagen to ligaments and tendons, led to the proposal of the "tractional structuring hypothesis" (Stopak and Harris, 1982). According to this hypothesis, capsules, ligaments, tendons and other anatomical structures, including muscles, are given their structure by the effects of cell traction exerted on extracellular matrix materials, especially collagen. In other words, we proposed that traction in the embryo produces the same sorts of rearrangements and alignments as those we observed in tissue culture and that these are the normal mechanisms of connective tissue morphogenesis in the body. For example, one specific proposal was that skeletal muscles are formed by the pull of cell traction exerted at attachment sites on the skeletal surface, stretching both collagen and muscle cells into alignment between these sites. In this way, muscles could be created by their attachments, rather than forming separately and then somehow finding these attachments, as had been believed.

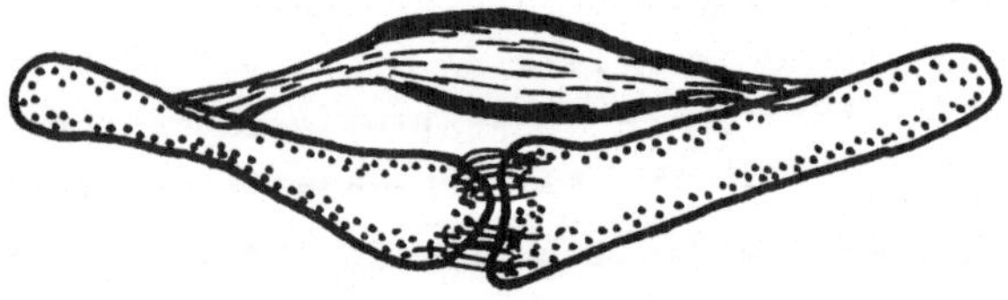

As a further extension of this basic hypothesis, it has also been proposed that the locations and shapes of the skeletal elements themselves could be controlled by the effects of cell traction on the mechanical arrangements of matrix material and cells. Because the differentiation of cells to form skeleton (cartilage) can be stimulated by increased concentrations of cells and matrix (see Newman, 1977), the compressive effects of traction should be able to make cartilage form in one place rather than another. Bone (actually cartilages, in the embryo) would form wherever cell traction caused the most compression. Complex patterns, such as those which form in the skins of some animals, can also be generated by mechanical interactions between cell traction and rheological properties of the matrix. Theoretical as well as experimental treatments of these questions will soon be published (Oster, Murray and Harris, 1983; Harris, Stopak and Warner, 1984).

MECHANICAL FACTORS WHICH CAN CONTROL TISSUE CELL LOCOMOTION

In a long series of tissue culture studies, a number of different guidance or controlling mechanisms have been discovered which are capable of influencing the directions and sometimes the speeds of tissue cell locomotion. The most important of these guidance mechanisms are the following:

1) Contact Guidance: This is the orientation of cell locomotion and elongation parallel to either fibers or grooves in the substratum upon which the cells are crawling. Several possible mechanisms have been proposed (reviewed in Harris, 1983).

2) Contact Inhibition: This is the slowing of locomotion or change in its direction as a result of intercellular contact (reviewed in Harris, 1983). This phenomenon does not involve a total paralysis of locomotion, but merely a tendency not to continue in the direction in which contact has been made with another cell. It tends to cause dispersal and to inhibit overlapping, and is of special interest because the susceptibility to contact inhibition is characteristically reduced both in leucocytes and in cancerous cells - as if this reduction were perhaps the explanation for their invasiveness.

3) Chemotaxis: Positive chemotaxis (attraction) has been proven for leucocytes and was recently demonstrated in the case of nerve fibers (steering the tips of growing nerves) but the many reports of chemotaxis in other cell types are based upon experimental criteria which actually cannot distinguish between the non-directional

stimulation of locomotion as opposed to true chemotaxis, which is a directional or steering response (reviewed in Harris, 1983).

4) Haptotaxis: This is the tendency of cells to move preferentially to areas of greater substratum adhesiveness (Carter, 1965; Harris, 1973). This tendency seems to result from a sort of perpetual tug-of-war between different parts of the cells' peripheries.

"DIFFUSIVE" VERSUS "ANTIDIFFUSIVE" EFFECTS

Some of these activities tend to equalize the spatial distributions of cells, while others can contribute to the spontaneous formation of aggregations. Simple random locomotion will cause cells to occupy space evenly, and contact inhibition will strengthen this effect, as would mutual negative chemotaxis (if it occurred), both of which would effectively "push" cells away from areas of high density toward less crowded areas. We can call such effects "diffusive" in that their result is like that of diffusion, in contrast to "antidiffusive" effects which tend in the opposite direction - toward clumping and the autocatalytic magnification of density differences. For example, mutual positive chemotaxis would be a potent anti-diffusive effect (if it occurred). But probably much more important are haptotaxis and contact guidance, due to adhesion and alignment by extracellular matrix materials which have themselves been compressed and aligned by cells, so as to create positive feedback cycles between matrix alignment and cellular alignment, matrix compaction and cell aggregation (Oster et al, 1983).

Cell properties or combinations of properties, the effects of which should be diffusive:

1) Mutual negative chemotaxis
2) Random or non-directional cell locomotion
3) Elasticity of extracellular matrix (especially if the Young's modulus increases with increasing strain)
4) Inhibition of cell contractility by intercellular contact or crowding

Properties or combinations of properties whose effects should be anti-diffusive:

1) Mutual positive chemotaxis
2) Convection, cells being carried bodily along with the matrix toward areas of concentrated traction

3) Haptotaxis, in the particular case of cells migrating up a gradient of matrix density

4) Stimulation of cell contractility by intercellular contact or crowding

5) Stimulation of chondrogenesis by compression of matrix or cells (since the secretion of additional cartilage matrix further increases the local density of matrix)

6) Plasticity of the extracellular matrix (and elasticity, if Young's modulus decreases with increasing strain)

7) Surface curvature, in the particular case of cylindrical masses of cells with contractile surfaces, as will be discussed below in relation to the embryonic phenomenon of somite segmentation.

Each of these anti-diffusive effects, alone or in combination, can serve the autocatalytic function for which autocatalytic chemical reactions are hypothesized in reaction-diffusion systems. The diffusive effects could serve either a damping role or for controlling spacing.

A PARALLELISM BETWEEN EMBRYONIC CAUSE AND STRUCTURAL FUNCTION

Notice that the mechanism proposed to explain the formation of ligaments, tendons and muscles (all structures serving to resist, transmit and exert tensile stress) is the exertion of this same class of force, namely tensile stress. The skeleton, on the other hand, serves to resist and convey stresses which are predominantly compressive. And what class of force have we been led to propose as the principal determinant of skeletal morphogenesis? None other than compressive stress.

There seems to be a pattern here. For one thing, there may be the basis of a useful heuristic: if a structure serves to resist shear, torque or whatever class of imposed force, then perhaps we should look for embryonic forces of the same class as likely explanations for that structure's morphogenesis. Probably more important, however, is the automatic propensity of such mechanisms to generate structures whose components are optimally arranged for bearing the forces which will be imposed upon them during life.

The near perfection of many anatomical structures, relative to their mechanical functions, is a part of the beauty of nature which has been much admired and analyzed (Wainwright et al., 1976). In external shape, as well as internal arrangement and alignment of components, the load-bearing structures of the body have frequently been found (when analyzed from the point of view of a mechanical engineer) to represent optimal solutions to the problem of resisting the maximum load, using the minimum of material. Simply to credit evolution with this mechanical optimization is to beg the question - unless one were to imagine that natural selection has chosen the optimal configurations out of the totality of all possible configurations, ignoring entirely the developmental mechanisms used to construct them. A more attractive possibility is that evolution has yielded morphogenetic mechanisms which, for some reason, inherently tend to generate mechanically optimal structures. May I suggest that this is the reason for the intriguing parallelisms (noted above) between the forces which create a given type of structure and the forces which that structure will serve to bear. Let us consider why optimization should follow from these mechanical parallelisms.

The optimization of mechanical design consists, more than anything else, in putting the load-bearing elements exactly where the imposed forces are and aligning these elements properly relative to the directions of the forces. An optimal design will amount virtually to a map of the spatial distribution and orientations of whatever forces the structure will be called upon to resist. Because the distribution of load is itself a function of the placement of load-bearing elements, it can be very difficult to calculate where these elements should be put to withstand the greatest load with the least material. Engineers often have to model the distribution of loads with computers or by using analog methods such as photo-elasticity. Where the forces are greatest, there they put the steel or concrete, etc. What I would propose is that the optimization of anatomical structures may be accomplished in much the same way, by a comparable analog calculation - that is, by creating (within the embryo itself!) patterns of forces which model those subsequently to be borne or exerted by the structures of the body. If the shapes and relative positions of these structures develop as local responses to the forces (tension creating tendons, compression causing chondrogenesis, etc.) then the optimization will be automatic.

SOMITES AND TRACTIONAL STRUCTURING

In the case of the formation of muscles by the tractional structuring mechanism just described, we are not dealing with pattern generation in the same sense as Turing's mechanism, for example, can generate patterns de novo from an initial state of homogeneity. Rather, the alignment of collagen and muscle cells depends upon pre-existing patterns of traction exertion at the skeletal surfaces, and so the mechanism might be thought of more as pattern execution, as opposed to pattern generation. Nevertheless, the basic mechanism of tractional structuring, when combined with a sufficient degree of positive feedback factors of the type which I have referred to above as "anti-diffusive", actually is capable of generating regular patterns, even from initial homogeneity. Indeed, there are certain parallels to the operations of chemical mechanisms like Turing's, it is merely that the morphogens are physical properties. Here I shall consider the specific case of somite formation from this point of view, but there are many other examples. In particular, there is the condensation of skin cells to form feather precursors (feather germs) in birds. That topic is treated in a pair of forthcoming papers (Oster, Murray and Harris, 1983; Harris, Stopak and Warner, 1984).

The somites are an important, though transient, set of embryonic structures which control the segmentation of the body in chordates. Somites form following gastrulation but then disperse - with their constituent cells giving rise to axial skeleton, to skeletal muscles, and to dermis. Even though the somites themselves have disappeared, their spatial periodicity continues to dictate the periodicity of the vertebrae and ribs, together with much of the axial musculature, and even the spacing of sensory ganglia and nerve outgrowth along the neural tube. The process by which the somites are formed consists of the subdivision of two, initially continuous, columns of cells (one on either side of the midline) each into a row of separate, usually squarish, blocks of cells. Each of these blocks of cells is a somite. The segmentation process is much as if some invisible knife were cutting a pair of long sausages into regular pieces; except that, of course, the "cutting" machinery resides in some unknown property of the cells being separated into these regular blocks, rather than in any sort of external "knife". This spontaneous process of self-cutting is what we wish to explain in mechanical terms.

Because these "cuts", and thus the separated somites, ordinarily form in a regular temporal, as well as spatial, sequence (from anterior to posterior), most of the eariler theories were based on the rearward movement or propagation of various sorts of somite centers. It long seemed as though the essential question to be resolved was the nature of the propagated signal. A recent variant on this theme was the proposal that segmentation was due to the combination of a rearward-moving "wave front" and a "clock" or temporal oscillator intrinsic to the pre-somitic tissue. Speculation on these issues continues, ignoring the experimental demonstration by Lipton and Jacobson (1974) that somite segmentation is perfectly capable of occurring simultaneously instead of sequentially. It is only necessary to cut the physical connection between the pre-somite columns and the adjacent mesoderm for the whole row to segment simultaneously. A theory of somite formation ought therefore to be able to account for either sequential or simultaneous segmentation, and not be inherently sequential - like the "clock and wave-front hypothesis" and most of its precursors. It would also be nice if a theory were to provide some explanation for the physical process of segmentation, and not confine itself to the generation of an unspecified chemical prepattern, from which some unspecified physical mechanism is then expected to take its orders.

A hypothetical mechanism which answers both of these criteria was suggested long ago by Waddington and Deuchar (1953), but has since been neglected. Their proposal was that somite formation should be considered as mechanically analogous to the separation of a column of water into droplets under the action of "surface tension". The instability of a liquid cylinder has a long and distinguished career of mathematical analysis (see Rayleigh, 1892), and although the breakup of actual streams from faucets depends upon intermolecular forces and physical excitations with no direct analogs in the case of embryonic cells, it is essential to realize that different physical systems can obey equivalent rules or equations even though the actual physical forces responsible are quite different. The diverse instances of simple harmonic motion are a good example of this principle, and the historic fallacy that heat must be a substance (because it obeys the same equation as diffusion) is a good example of how natural it is to equate causes when results are similar. In view of the great successes achieved by Steinberg's analysis of histotypic cell sorting in terms of its mechanical analogy to surface tension phenomena (Steinberg, 1970), it is surprising that more has not also been done along these lines with somite formation.

In order for a cylinder to break up spontaneously into separate droplets, it is sufficient that the cylinder's surface be isotropically contractile. Whether or not this contractility is literally due to surface tension (the maximization of intermolecular bonding) is not relevant. Thus it would be sufficient if cell surfaces (cortexes) were to contract, and for this contraction to become concentrated over the "exposed" parts of the cells' surfaces, where they are not adhering or in contact with one another. Such a concentration of contractile elements has recently been observed in other cases (Sobel, 1983) and can be expected to produce the effect of a "surface tension" (Harris, 1976). When a liquid cylinder's length exceeds its circumference, the contraction of its surface pinches it into segments. This is because the circumferential element of tension, acting along a sharper curve, produces an inward component which is stronger than the countervailing outward component of the longitudinal contraction. Thus the "antidiffusive element" in this case would be the imbalance in the circumferential and longitudinal components of surface contractility, but others of these elements might well also contribute. Indeed, their contribution might help to explain some of the puzzling phylogenetic variability one finds in the case of somite formation.

Waddington's mechanism for somite formation fits well with the proposed mechanisms (discussed above) for determining the locations of cartilages (by compression) and muscles (by tension). This is because the somite-derived cartilages (vertebral arcualia) form at the boundaries between the somites, where the constricting effects of the "surface tension" would be maximal. The muscle precursor cells become aligned, as the "myotome", between these.

OUTLOOK FOR THE APPLICATION OF MATHEMATICS TO EMBRYOLOGY

As we have seen, the problem of explaining how complex structural patterns are formed in the embryo, rather than being fundamentally a difficult problem in chemical kinetics, as has become widely believed, may be instead an (even more difficult!) problem in continuum mechanics. As more has been learned about the cellular forces responsible for the mechanical construction of organs, it has become clear that these forces need not be subservient to chemical diffusion gradients, but can themselves generate complex geometric patterns.

The approach to embryonic morphogenesis in terms of mechanical stability and instability is far from new, of course. D'Arcy Thompson

(1942) based a whole treatise on this approach, and more recently Steinberg (1970) developed his theory of cell sorting based on the idea of the maximization of adhesive interactions between cells. However, such mechanical explanations have been sought almost only in cases where there is a direct analogy to some familiar phenomenon of the organic world - analogies to surface tension phenomena having been by far the most popular. But what are called the "laws of surface tension" are merely the consequences to be expected from isotropic contractions of constant strength in the plane of a flexible surface. That is, they are an idealization of the behavior of simple liquids. Why we should necessarily expect the contractility of more complex materials to obey these simple rules is unclear; and when the contraction in question is that of cells, the resulting "laws" might be very different and even change with time. Those considering possible chemical systems for pattern generation have felt free to explore the consequences of a wide range of possible equations governing their reactions, so it makes little sense for those considering mechanical control to narrow their attention to the equations obeyed by soap bubbles. The various experimental observations reported above seem to suggest that cellular forces can have a much broader range of consequences.

Mathematicians have productively expended much thought on what is known as "Plateau's problem". This is the problem of calculating, for a given boundary, the surface shape having the minimum total area. Because a soap film contracts isotropically and homogeneously, it automatically adopts these minimum-area shapes. And so, if we find a cell or mass of cells which likewise adopts one of these shapes, it is reasonable to see an explanation in terms of surface contractions which are also isotropic and homogeneous. (The assumption of a minimization of free energyis unwarranted, however.) But what if biological shapes depart from those which a soap film would adopt, does this imply that surface contraction is no longer responsible - or merely that the contractions are not isotropic and homogeneous? Suppose, for example, that the contraction of a surface varies as some specified function of the local curvature in any given direction. The reulting "laws" of surface tension, and thus the shapes spontaneously adopted by such a surface, will differ accordingly.

Perhaps it is time for mathematicians to consider a problem which is, in a sense, a converse of Plateau's: Given an observed surface shape (even one as simple as the cylinder, there are many cylindrical structures in biology), what rules of variation would the contractility of the surface need to obey in order that this particular shape will arise spontaneously as a result of these forces? It has been said that Halley once asked Newton, "By what law would gravity need to vary with distance in order for the planetary orbits to be ellipses having the sun at one focus?" Newton's answer was the inverse square law (and the Principia!), and he and later mathematicians also calculated the orbit shapes which would result from other laws of variation with distance. For the inverse cube, it is an interesting spiral, for example. The shapes we observe in anatomy pose a series of analogous problems to the mathematicians of today. Except in the few cases where there happens to be an evident similarity to the shapes generated by surface tension, the unaided biologist will not even know what laws of contractility he should seek as possible explanations for the stability (and loss of stability) of the shapes he observes.

Embryologists do not yet think of morphogenesis in terms of transitions from one stable state to another. Until they do, they will not be much helped by discussions of whether such transitions can be categorized into some certain number of topologically distinct classes. What is needed are criteria for identifying such states, and sophistication about their capacities for creating and maintaining shapes determined by the physical properties of their component cells and molecules. For example, not a few still labor under the misconception that asymptotically stable configurations are necessarily states of minimum free energy.

ACKNOWLEDGEMENTS

I thank Patricia Warner, David Stopak and Elizabeth Harris for their help, and the following for their stimulating discussion and counter-arguments: James Damon, Graham Dunn, Ladnor Geissinger, Julian Lewis, James Murray, Fred Nijhout, George Oster, Nancy Preyer, James Stasheff, J.P. Trinkaus, Steve Wainwright, and Lewis Wolpert.

The research described here is supported by a grant from the National Institutes of Health, Institute of General Medicine.

REFERENCES

Carter, S.B. (1965). Principles of cell motility: the direction of cell movement and cancer invasion. Nature 208, 1183-1187.

Elsdale, T., and Bard, J. (1973). Collagen substrata for studies on cell behavior. J. Cell Biol. 54, 626-637.

Gabbiani, G., Majno, G., and Ryan, G.B. (1973). The fibroblast as a contractile cell: The myofibroblast. In: The Biology of Fibroblasts, ed. E. Kulonen and A. Pikkarainen, Academic Press, New York, pp. 137-154.

Harris, A.K. (1973). Behavior of cells on substrata of variable adhesivness. Exp. Cell Res. 77, 285-297.

Harris, A.K. (1976). Is cell sorting caused by differences in the work of intercellular adhesion? A critique of the Steinberg hypothesis. J. Theor. Biol. 61, 267-285.

Harris, A.K. (1983). Cell migration and its directional guidance. In: Cell Interactions and Development: Molecular Mechanisms, ed. K. Yamada, John Wiley and Sons, New York, pp. 123-151.

Harris, A.K., Stopak, D., and Wild, P. (1981). Fibroblast traction as a mechanism of collagen morphogenesis. Nature 290, 249-251.

Harris, A.K., Stopak, D., and Warner, P. (1984). Generation of spatially periodic patterns by a mechanical instability: a mechanical alternative to the Turing model. J. Embryol. Exp. Morphol., in press.

Harris, A.K., Wild, P., and Stopak, D. (1980). Silicone rubber substrata: a new wrinkle in the study of cell locomotion. Science 208, 177-179.

Leader, W.M., Stopak, D., and Harris, A.K. (1983). Increased contractile strength and tightened adhesions to the substratum result from reverse transformation of CHO cells by dibutyryl cyclic adenosine monophosphate. J. Cell Sci. 64, 1-11.

Linsenmeyer, T.F. (1981). Collagen. In: Cell Biology of the Extracellular Matrix, ed. E.D. Hay, Plenum, New York, pp. 5-37.

Lipton, B.H., and Jacobson, A.G. (1974). Experimental analysis of the mechanisms of somite morphogenesis. Devel. Biol. 38, 91-103.

Newman, S.A. (1977). Lineage and pattern in the developing wing bud. In: Vertebrate Limb and Somite Morphogenesis, ed. D.A. Ede, J.R. Hinchliffe, and M. Balls, Cambridge University Press, Cambridge, pp. 181-197.

Oster, G., Murray, J., and Harris, A.K. (1984). Mechanical aspects of mesenchmyal morphogenesis. J. Embryol. Exp. Morphol., in press.

Rayleigh, (J.W. Strutt) (1892). On the instability of a cylinder of viscous liquid under capillary force. Philosophical Magazine xxxiv, 155-173 (Volume 3 of Rayleigh's collected works, Cambridge University Press, Cambridge, 1902.)

Sobel, J.S. (1983). Cell-cell contact modulation of myosin organization in the early mouse embryo. Devel. Biol. 100, 207-213.

Steinberg, M.S. (1970). Does differential adhesion govern the self assembly of tissue structures? Equilibrium configurations and the emergence of a hierarchy among populations of tissue cells. J. Exp. Zool. 173, 395-434.

Stopak, D., and Harris, A.K. (1982). Connective tissue morphogenesis by fibroblast traction I. Tissue culture observations. Devel. Biol. 90, 383-398.

Thompson, D'A. (1942). On Growth and Form. Cambridge University Press, Cambridge.

Trinkaus, J.P. (1984). Cells Into Organs. Prentice Hall, New York.

Waddington, C.H., and Deuchar, E.M. (1953). Studies on the mechanism of meristic segmentation I. The dimensions of somites. J. Embryol. Exp. Morphol. 1, 349-356.

Wainwright, S.A., Biggs, W.D., Currey, J.D., and Gosline, J.M. (1976). Mechanical Design in Organisms. John Wiley and Sons, New York.

EIGHT DIFFERENT DEFINITIONS OF A CIRCLE

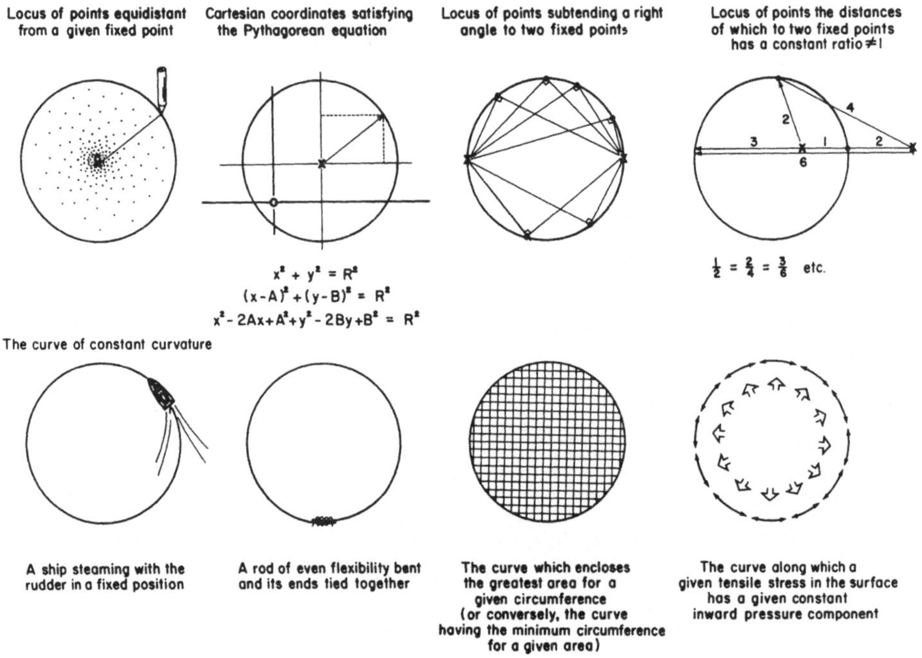
Locus of points equidistant from a given fixed point
Cartesian coordinates satisfying the Pythagorean equation
Locus of points subtending a right angle to two fixed points
Locus of points the distances of which to two fixed points has a constant ratio ≠ 1
3
1
2
2
4
6
$x^2 + y^2 = R^2$
$(x-A)^2 + (y-B)^2 = R^2$
$x^2 - 2Ax + A^2 + y^2 - 2By + B^2 = R^2$
$\frac{1}{2} = \frac{2}{4} = \frac{3}{6}$ etc.
The curve of constant curvature
A ship steaming with the rudder in a fixed position
A rod of even flexibility bent and its ends tied together
The curve which encloses the greatest area for a given circumference (or conversely, the curve having the minimum circumference for a given area)
The curve along which a given tensile stress in the surface has a given constant inward pressure component

WHAT IS THE APPROPRIATE GEOMETRY FOR COMPREHENDING MORPHOGENESIS?

It is often assumed that cells must determine their position relative to some external coordinate system, both cartesian and radial systems having been considered. It is important to recognize, however, that shape and pattern can be defined in other ways, either with reference to external positions or alternatively by local or intrinsic properties. To illustrate this point, a series of different ways of defining (or creating) circles are shown; those in the top row are extrinsic, those in the bottom row intrinsic. It may be that these intrinsic definitions are closer to those responsible for the shaping of anatomical structure, and therefore that differential geometry is more applicable than coordinate geometry to our attempts to understand morphogenesis.

Would anyone maintain that bubbles are spherical because the surface film somehow determines its distance from the point at the center and keeps this distance constant?

A hysteresis model for bacterial growth patterns

F.C. Hoppensteadt
Department of Mathematics
University of Utah
Salt Lake City

W. Jäger and C. Pöppe
Sonderforschungsbereich 123
Universität Heidelberg
Im Neuenheimer Feld 293
D-6900 Heidelberg

Growth patterns in bacterial cultures have been observed and studied by several scientists. In 1978, Hoppensteadt, Jäger, Roth and Schmid observed ring structures in cultures of histidine auxotrophic salmonella typhimurium. Similar to the classical experiment for Liesegang phenomena in chemical precipitation, concentrical growth rings are formed in response to a diffusing front of histidine spreading from the center of a Petri dish to its boundary. In many examples for growth patterns in cell populations the cells are mobile; the systems can be modelled by reaction diffusion equations showing diffusive instabilities (see e.g. the papers [1], [9], [11] in this volume). However, in this case it is impossible to observe mobility of the population: the bacteria are fixed on an agar gel containing all chemicals necessary for growth except the missing amino acid. Therefore, the spatial interaction is caused only by the diffusion of the nutrients and the buffer neutralized by acids produced as by-products of the cell growth. It has been impossible to find a mathematical model explaining the ring structures without assuming a growth inhibiting change of the pH-value. This assumption was justified experimentally. There are strains of salmonella needing other amino acids for growth, e.g. proline or histidine and proline. The experiments in case of two missing substances and two diffusion centers show lens shaped growth structures which can be explained by the same mechanisms.

Periodic growth phenomena in spatially organized microbial systems have been intensively studied by Wimpenny et al. ([12], [13], [14], [15]). Their important investigations are described in the paper by Wimpenny, Jaffe and Coombs [15] contained in this volume. Their experimental and numerical results using a similar mathematical model agree totally with those obtained for the growth patterns in salmonella typhimurium. In particular, their experimental analysis of the changes of the pH-value proves independently that the assumption of Hoppensteadt and Jäger in [6] is correct.

The following hypotheses are made in the mathematical model used in this paper.

H(1) There is no diffusion of the bacteria, the chemicals (amino acid, buffer) necessary for growth are diffusing.

H(2) The uptake of the chemicals is fast compared to their diffusion.

H(3) The adjustment of the bacteria to their environmental conditions (e.g. change of pH-value) is even faster.

H(4) There exist thresholds for growth expressed in terms of the concentrations of the chemicals. The metabolism of the cells is regulated such that there is no growth (and no consumption of chemicals) until a certain threshold is reached ("switch on"). The growth when started continues even as the system is falling "below" this threshold until a second threshold is reached where the growth stops ("switch off").

These hypotheses lead to equations of the following type (see [4,6]):

$$\frac{\partial}{\partial t} B = \alpha VB \quad ,$$

(1) $$\frac{\partial}{\partial t} H = D_H \Delta H - \beta VB \quad ,$$

$$\frac{\partial}{\partial t} G = D_G \Delta G - \gamma VB \quad ,$$

plus initial conditions and no-flux boundary conditions. Here B is the concentration of the bacteria, H the concentration of the amino acid (histidine), G the concentration of the buffer. V is a function describing the metabolic activity of the bacteria. α, β, γ are constants large compared to the diffusivities D_H and D_G of H and G. In [6] it was assumed that V has the form of a Michaelis-Menten kinetic:

$$V = \frac{H}{K+H} \cdot \frac{G}{K'+G} \quad .$$

It seems more appropriate to consider V as a function determined by a system of ordinary differential equations modelling the metabolic processes of the cells. The simplest situation is given by a differential equation for V of the following type:

(2) $$\varepsilon \frac{\partial}{\partial t} V = g(G,H,V) \quad .$$

Hypotheses H(3),(4) imply that in (2) ε is a "small" number and the equation

(3) $\qquad 0 = g(G,H,V)$

has in general multiple solutions. This means that the system is showing hysteresis. The following figure illustrates the situation in the state space given by (G,H,V).

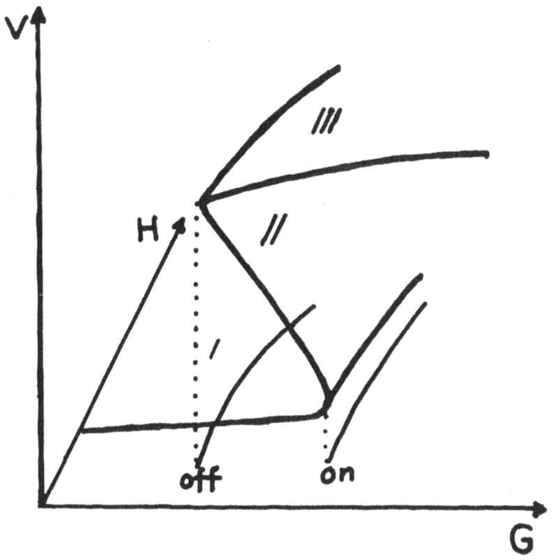

Fig. 1. The solution set of (3) is in general not a graph on the (G,H) plane. It consists here of three leaves I, II, III.

Whereas for ordinary differential equations the singular perturbation limit $\varepsilon \to 0$ is analytically well studied, there is not much known in case of diffusion systems like (1) controlled by a solution to an equation of type (2). A van der Pol oscillator with diffusion illustrates the difficulties arising in such a system for small ε.

In this paper we use a model involving a hysteresis functional instead of a solution V of an ordinary differential equation, as described in [7]. If there is no diffusion at all, this functional is obtained in the limit $\varepsilon \to 0$. For simplicity we assume that this functional only depends on G and H and its value is either 1 or 0 ("on" or "off"). It is a Volterra functional, that means for fixed (t,x) it is determined by the function $(G(\cdot,x), H(\cdot,x))$ restricted to the interval [0,t] and an initial condition. According to hypothesis H(4) it is defined using thresholds, i.e. switch curves Γ_{on} and Γ_{off} in the phase plane (G,H). These curves correspond to the boundaries of the projections of resp. leaf I and leaf III to the (G,H) plane in figure 1.

The following example is used in the numerical computation of this paper.

(4). Example

Γ_{on} (Γ_{off}) is the set of zeros of ψ_{on} (ψ_{off}) defined by

$$\psi_{on}(G,H) = \min\left(H-1,\ G-\frac{a_{on}}{H}-b_{on}\right)$$

$$\psi_{off}(G,H) = \min\left(H-1,\ G-\frac{a_{off}}{H}-b_{off}\right)$$

$$(\psi_{off} \geqq \psi_{on})$$

Here, essentially, a mass action kinetic is assumed. The cut-off at H = 1 has been introduced only for computational reasons and is not important.

The first quadrant is decomposed by the switch curves into three sets M_{off}, M_{on} and M_{on-off} (see figure 2). In case of the example:

$$M_{off} = \{(G,H),\ \psi_{off}(G,H) < 0\}$$

$$M_{on} = \{(G,H),\ \psi_{on}(G,H) > 0\}$$

$$M_{on-off} = \{(G,H),\ 0 \leqq \psi_{off}(G,H),\ \psi_{on}(G,H) \leqq 0\}$$

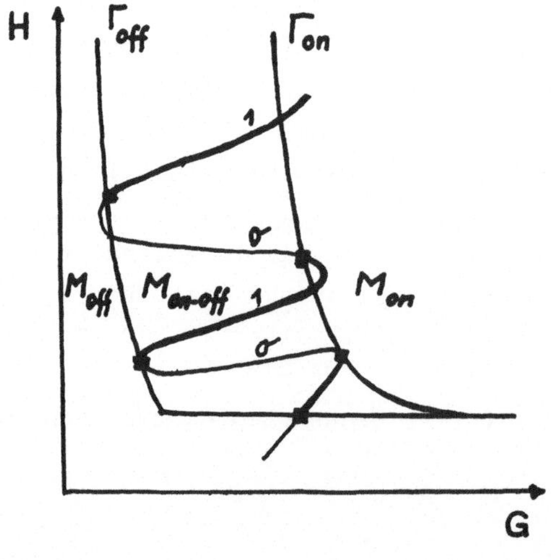

Fig. 2. Switch curves in the phase plane, definition of the hysteresis functional along a trajectory (G(·,x), H(·,x)). Switch points are marked by ▫. Numbers 0 (thin line) and 1 (bold line) refer to values of the functional.

Consider continuous functions $u = (G,H)$ of time t with values in the first quadrant. Assume that an initial switch functional s_o is given with values 1 or 0 (growth "on" or "off") satisfying

$$(5) \qquad s_o(u) = \begin{cases} 1 & \text{if } u(0) \in M^*_{on} := M_{on} \cup \Gamma_{on} \\ 0 & \text{if } u(0) \in M^*_{off} := M_{off} \cup (\Gamma_{off} \setminus \Gamma_{on}) \end{cases}$$

Define a Volterra functional s setting initially

$$(6) \qquad \begin{cases} s(u)(0) = s_o(u) \quad , \quad \text{and} \\ s(u)(t) = 1 \quad \text{if } u(t) \in M^*_{on} \text{ or if there exists a } t_o < t \text{ such that } s(u(t_o)) = 1 \text{ and } u(\tau) \notin M^*_{off} \text{ for all } t_o \leqq \tau \leqq t, \\ s(u)(t) = 0 \quad \text{otherwise.} \end{cases}$$

Because of the discontinuities on the switch curves, this definition is partially arbitrary, to what extent will not be discussed here. Setting

$$(7) \qquad V(t,x) := S(u(\cdot,x))(t)$$

in (1), one obtains a system showing all difficulties of free boundary problems. The free boundaries are the sets of points in time and space where growth is switched on or switched off. Fronts of concentration of histidine and buffer are moving through the Petri dish. Active zones are developing in the overlap of these fronts and travelling along.

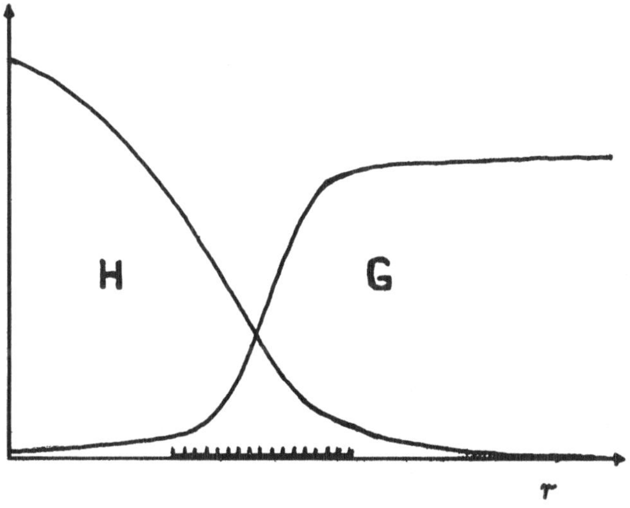

Fig. 3. Travelling fronts of histidine and buffer. Active zones (marked by small bars) arise in the overlap where the product of the concentrations exceeds the ψ_{on} threshold.

Some of the analytic problems arising have been considered in [7]. Under natural conditions for g, there are no difficulties for a system of type (1),(2) as far as existence and uniqueness of solutions are concerned. This is no longer true for the system (1),(7) because of the discontinuity of the switch functional. In this paper a discretization is used for numerical computations. The discretization of the functional s acts like introducing a time lag in the reaction term. The solutions of the corresponding equations can be determined step by step. Biologically a time lag means that consumption does not imply instant growth, which seems to be the case anyhow. Therefore, the discretized system might be considered as a good approximation.

It is clear how to define a discretization of the functional s. Consider a sequence (u_m) in the first quadrant, representing a time discretization of a function u. Define the time discretization of the functional s by

$$(8) \qquad s_m(u) := \begin{cases} 1 & \text{if } u_{m-1} \in M^*_{on} \quad \text{or} \\ & \text{if } s_{m-1}(u) = 1 \text{ and } u_{m-1} \notin M^*_{off} \\ 0 & \text{otherwise.} \end{cases}$$

Before describing the numerical results the following comments have to be made. The threshold models for the microbial growth-patterns correspond to models for Liesegang phenomena using Ostwald's hypothesis of supersaturation. The diffusing chemicals needed for cell growth correspond to the diffusing ions building up the precipitating complex. The growing cells correspond to the precipitating substance. The nucleation process in precipitation and the metabolic process in the cells should play similar roles in both systems. However, due to recent studies by Kai, Müller and Ross (see [8], [10]) the evolution of nucleation in time and space is quite different from the evolution of the precipitation patterns. Therefore, Ostwald's hypothesis has to be replaced and a more refined model has to be developed. The interesting paper of Müller and Venzl in this volume [10] contains first results for an improved theory. Whereas in a simple reaction diffusion model for Liesegang phenomena as used in the paper of Flicker and Ross [3] the pattern formation cannot be explained by diffusive instabilities, this might be possible in a refined model including the nucleation process. Concluding from the present knowledge about the biochemical processes in a growing cell population, however, there is nothing really similar to the nucleation process. Despite the similarity of the observed patterns the mechanisms and therefore their mathematical models are different. However, as in precipitation phenomena, more experimental investigations on the metabolism of the cells have to be made and their results have to be included into a mathematical model. This paper would reach an important aim if the rather crude model presented could help to ask good questions for new experiments.

The numerical result presented now is an example of a whole series. Pöppe produced a film showing the growth process for different parameters.

Let r be the radius of the dish. Assuming radial symmetry of the initial values the reduced system

$$\left.\begin{aligned} \frac{\partial}{\partial t} B &= \alpha VB \\ \frac{\partial}{\partial t} H &= D_H \frac{1}{r} \frac{\partial}{\partial r} r \frac{\partial}{\partial r} H - \beta VB \\ \frac{\partial}{\partial t} G &= D_G \frac{1}{r} \frac{\partial}{\partial r} r \frac{\partial}{\partial r} G - \gamma VB \end{aligned}\right\} \text{for } t > 0 \text{ and } 0 < r < 1, \text{ and}$$

$$\frac{\partial}{\partial r} H = 0 \, , \, \frac{\partial}{\partial r} G = 0 \qquad \text{for } r = 0,1 \text{ and } t > 0$$

is considered. Choosing definition (7) the equation can only hold in distributional sense. In general, V and therefore B will have jumps. Only outside the set of the discontinuities of $V \cdot B$ the first derivative with respect to time and the second derivatives with respect to the radius will exist. This set contains the boundary of the growing zones.

It is reasonable but not straightforward to trace the free boundary numerically. This will be done by Pöppe in another paper. Here the system is discretized in a standard way, the functional according to (8). The discretization of the operator

$$\frac{1}{r} \frac{\partial}{\partial r} r \frac{\partial}{\partial r} u$$

is as usually given by

$$\frac{1}{h^2} (u^{j+1} - 2u^j + u^{j-1}) + \frac{1}{r_j} \frac{1}{2h} (u^{j+1} - u^{j-1}) \, ,$$

where $r_j = jh$ and u^j is the value at $r = r_j$. The step size in space used here is $h = \frac{1}{100}$. The boundary conditions are treated in a standard way introducing "false boundaries". For the solution of the discrete system a hopscotch technique is used, an algorithm which combines the speed of explicit methods with the stability of implicit techniques. The accuracy, however, is not greater than that of an explicit method (see [2], [4], [5]). The algorithm used is very fast and simple and gives at least a good qualitative picture. Let h^* be the step size in time discretization. It is recommended to choose

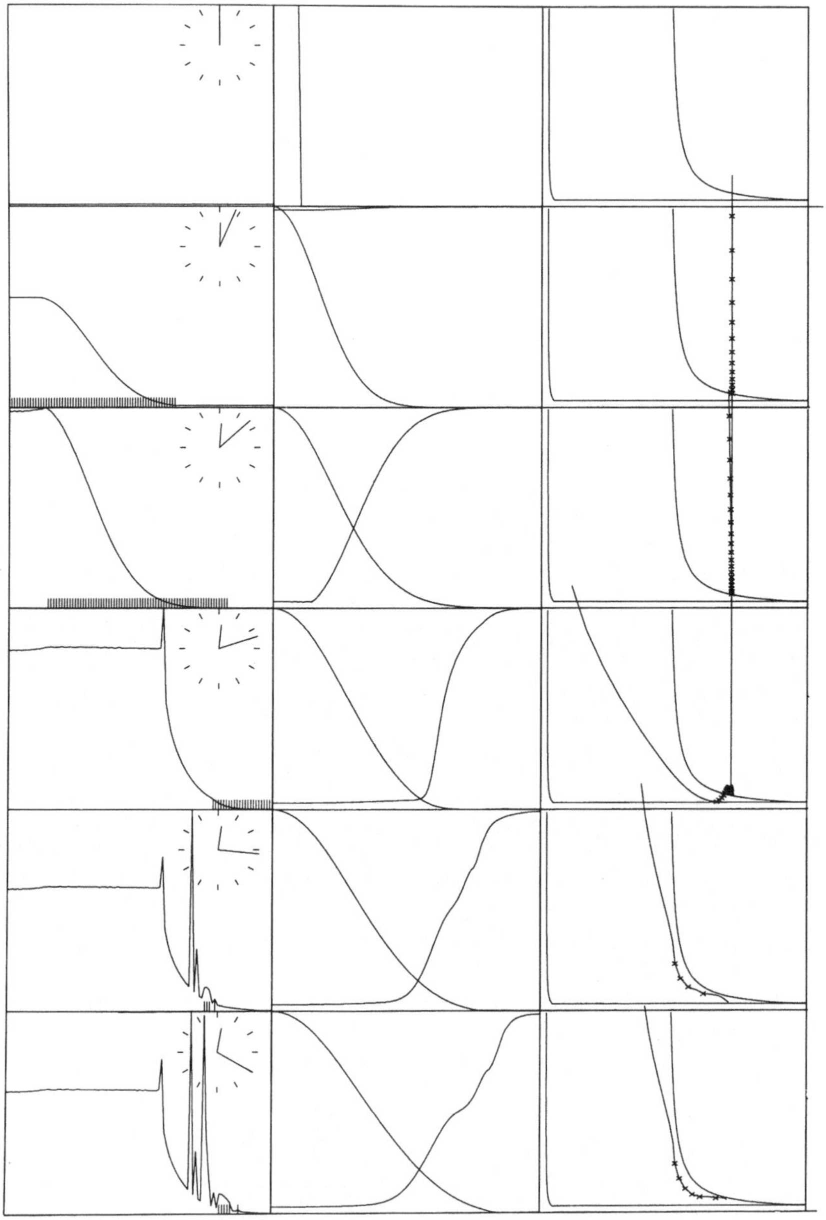

Fig. 4. Numerical simulation of bacterial growth pattern formation. (Details are explained in the text.) Scales of B and H are given below for each time.

time t	scale of B	scale of H
0	1	24 000
4	1	4 132
8	29.54	2 016
12	36.66	1 324
16	47.93	969
20	47.93	754

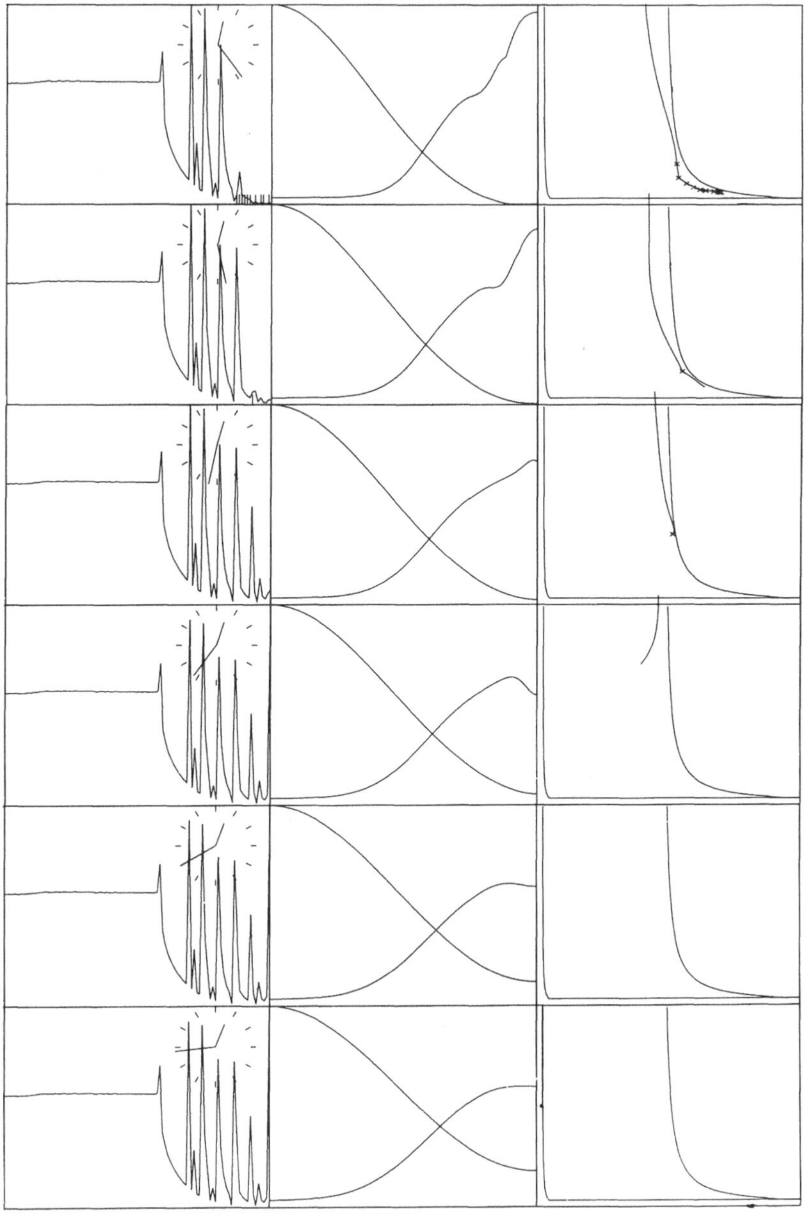

Fig. 4 (continued)

time t	scale of B	scale of H
24	47.93	613
28	47.93	514
32	47.93	441
36	51.71	384
40	51.71	340
44	51.71	304

$$h^* \leq \frac{1}{2}\,\frac{h^2}{\max(D_G, D_H)}\ .$$

The values of the parameters for the computations shown in figure 4 are as follows:

<u>differential equation</u>

D_H	D_G	α	β	γ	h	h^*
0.003	0.0005	1	5	5	0.01	0.002

<u>initial condition</u>

$$B(0,\cdot) = B_o = 0.01 \qquad\qquad G(0,\cdot) = G_o = 150$$

$$H(0,r) = \begin{cases} H_o = 24000 & \text{for } r \leq r_o = 0.1 \\ 0 & \text{elsewhere} \end{cases}$$

($H(0,\cdot)$ represents a drop of histidine at the center of the dish)

<u>switch curves</u>

a_{on}	b_{on}	a_{off}	b_{off}
100	100	5	5

For times $t = 0,4,8,\ldots$ the state of the system is displayed as a collection of three frames.

Number I shows B as a function of r. Active meshpoints are marked by small bars. A clock in the upper right corner shows the "time". One "hour" corresponds to about two days in real experiments. Number II shows G and H as functions of r. B and H are scaled by their maxima, G by G_o in order to fill the frames properly. This rescaling during time evolution has to be taken into account. The scaling of B and H is listed below the fig. 4. Number III shows the first quadrant in the phase plane, especially the switch curves and for fixed time t the trace of the curve $(G(t,\cdot), H(t,\cdot))$. Here, the scaling does not depend on the G and H values at time t, but only on the parameters of the problem. It is chosen such that the most relevant parts of the curve are shown as clearly as possible. Active points are marked by a cross.

It is this curve in the phase plane which illustrates the dynamics of the system in the best way. The initial values for G and H are represented by a vertical line in the phase plane intersecting the set M_{on}, if the drop of H in the center is "large" enough. The part of the curve belonging to active sites is driven down by reaction where it is faster than diffusion until the switch-off curve is hit. Then diffusion may drive it up again until the switch-on curve is hit again and consumption and growth are reactivated. This game may be repeated several times and in different spatial regions until the concentrations of G and H have become so small such that after switch-off a reactivation is no longer possible. Whereas G and H tend to constants, finally, the spatial oscillations remain in the distribution of B. Note that for this mechanism of pattern formation $D_G = D_H$ is allowed. Diffusive instability does not play any role.

Looking at the B-diagrams one observes at first that a central plateau of bacteria is building up. A very narrow ring is growing on the shoulder of this plateau while the active zone is travelling as a ring towards the boundary. Further rings of bacteria are developing from interior to exterior. Depending on the parameters and the selection of the step sizes the active zones may consist of very few meshpoints thus producing a very narrow growth ring. Growth need not to proceed "monotonely" from interior to exterior, in the diagrams from left to right. "Backfiring", as observed in experiments, may occur in the computations. The asymmetry of the growth peaks seems to agree with observations in experiments, too.

Whereas the algorithm appears to be quite stable against variations of the time step, variation of the spatial stepsize affects the results in a way that explicit tracing of moving bounderies including spatial stepsize control appears appropriate. Work in this direction is in progress.

References

1. Boon, J.P., Herpigny, B.: Formation and evolution of spatial structures in chemotactic bacterial systems. This volume.

2. Eilbeck, J.C.: Numerical studies of solitons. In: A.R. Bishop, T. Schneider (eds.): Solitons in condensed matter physics. Proceedings of the Symposium on Nonlinear (Soliton) Structure and Dynamics in Condensed Matter, Oxford, June 27-29, 1978, Springer Verlag 1978.

3. Flicker, M., Ross, J.: Mechanism of chemical instability for periodic precipitation phenomena. Journal of Chemistry and Physics 60, 3458-3465 (1974).

4. Gourlay, A.R.: Hopscotch: A fast second-order partial differential equation solver. J. Inst. Math. Appl. 6, 375-390 (1970).

5. Gourlay, A.R., McGuire, G.R.: General hopscotch algorithm for the numerical solution of partial differential equations. J. Inst. Math. 7, 216-227 (1971).

6. Hoppensteadt, F.C., Jäger, W.: Pattern formation by bacteria. Lecture Notes in Biomathematics 38, 68-81 (1980).

7. Jäger, W.: A diffusion reaction system modelling spatial patterns. In: Equadiff 5. Proc. 5. Czech. Conf. Diff. Eq. and Applications, Bratislava 1981. Teubner, Leipzig 1982, p. 151-158.

8. Kai, S., Müller, S.C., Ross, J.: Measurements of temporal and spatial sequences of events in periodic precipitation processes. J. Chem. Phys. 76, 1392-1406 (1982).

9. Lauffenburger, D.A.: Chemotaxis and cell aggregation. This volume.

10. Müller, S.C., Venzl, G.: Pattern formation in precipitation. This volume.

11. Schaaf, R.: Global branches of one dimensional stationary solutions to chemotaxis systems and stability. This volume.

12. Wimpenny, J.W.T., Whittaker , S.: Microbial growth in gel stabilised nutrient gradients. Society for General Microbiology Quarterly 6, 80 (1979).

13. Wimpenny, J.W.T.: Responses of microorganisms to physical and chemical gradients. Phil. Trans. R. Soc. Lond. B 297, 497-515 (1982).

14. Wimpenny, J.W.T., Lovitt, R.W., Coombs, J.P.: Laboratory model system for the investigation of spatially and temporally organised microbial ecosystems. Symposia of the Society of General Microbiology, 34, 67-117 (1983).

15. Wimpenny, J.W.T., Jaffe, S., Coombs, J.P.: Periodic growth phenomena in spatially organized microbial systems. This volume.

CHAOS IN SIMPLE THREE- AND FOUR-VARIABLE CHEMICAL SYSTEMS

J.L. Hudson
Department of Chemical Engineering
University of Virginia
Charlottesville, VA 22901/USA

and

O.E. Rössler
Institute for Physical and Theoretical Chemistry
University of Tübingen
7400 Tübingen/W. Germany

Two new 3-variable model reaction systems with chaotic behavior are presented. They each possess the property of containing a 2-variable subsystem with a Hopf bifurcation. The added third variable then allows this subsystem to alternate between its stable and its unstable regime in an autonomous fashion. The resulting overall motion is that on a torus that is strongly distorted (generating chaos). A new 4-variable reaction system based on the same general principle is also presented. It contains a 3-variable subsystem that itself can be chaotic. The added fourth variable allows this subsystem to alternate between two different regimes (a small-amplitude and a large-amplitude chaotic attractor) in an autonomous fashion. The resulting overall motion is that on a hypertorus that is strongly distorted (generating hyperchaos). Such behavior has yet to be observed experimentally.

1. Introduction

The well-stirred Belousov-Zhabotinsky reaction readily produces chaotic oscillations (cf. [1] for a preview and [2] for a review). It can be described by a system of about 25 coupled ordinary differential equations - for which many of the rate constants are still not known precisely (cf. Noyes [3]). Nonlinear systems of coupled O.D.E.s, however, do possess a wide range of possible types of qualitative behavior in dependence on dimensionality. Systems possessing four or more variables have only begun to be investigated (cf. [4]).

The types of chaotic behavior found so far in the Zhabotinsky reaction have all been explained in terms of a chaotic attractor of the usual type - that is, one that can in principle be embedded in 3 dimensions.

For example, no evidence for the presence of more than one positive Lyapunov characteristic exponent (that is, more than one direction of exponential trajectorial divergence on the attractor) has been found to date, even though no more than 4 variables are theoretically required [5,6].

In the following, first a simple (abstract) 3-variable chemical reaction system which shows chaos will be presented. Then, the underlying mechanism of chaos generation will be encountered again in another 3-variable system that happens to be related to a model of the Belousov-Zhabotinsky reaction. Finally, the second system will appear again as a partial system in a simple four-variable system. The behavior of this third system will turn out to be "more chaotic" than is possible in principle with three variables.

2. A 3-Variable Example

Consider the following abstract reaction system:

$$\begin{aligned} \dot{a} &= k_1 - k_2(a - c) - k_3ab \\ \dot{b} &= k_3ab - k_4\frac{b}{b+K} + k_5 \qquad (1) \\ \dot{c} &= k_2(a - c) \;. \end{aligned}$$

The corresponding reaction scheme is as follows:

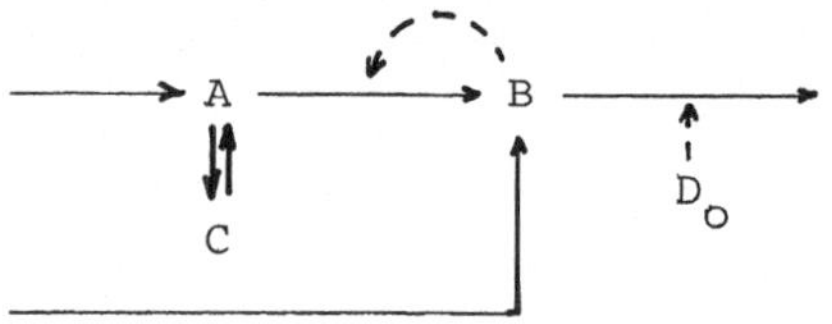

There is a constant influx to A , which reacts reversibly with C . The reaction from A to B is autocatalytic (of quadratic mass-action type). B has an additional small constant influx (k_5). Its outflux is catalyzed in a Michaelis-Menten fashion by D_o which is assumed constant (as a factor in k_4). This latter reaction is not of mass-action type itself, but can be reduced to two mass-action type elementary reactions in principle [7].

In order to understand the qualitative behavior of the system of Eq. (1), let us first turn to a 2-variable subsystem obtained from it by holding c constant, viz.,

$$\dot{a} = (k_1 + k_2 c) - k_2 a - k_3 ab$$
$$\dot{b} = k_3 ab - k_4 \frac{b}{b+K} + k_5 \quad . \tag{2}$$

This subsystem can undergo a Hopf bifurcation as the parameter c is varied upwards starting from zero. This is shown qualitatively in Figure 1.

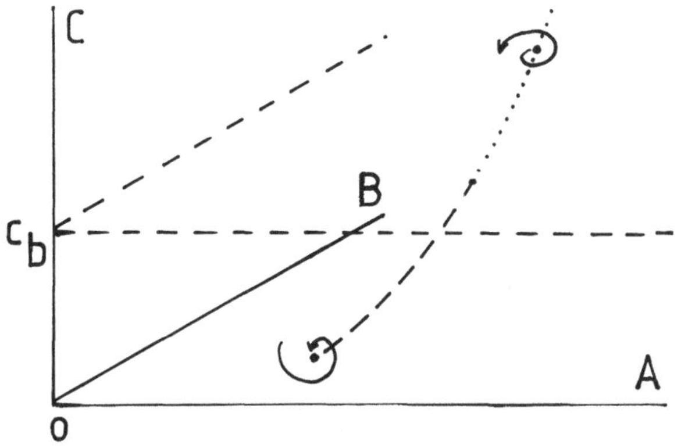

Figure 1 Hopf bifurcation in Eq.(2) in dependence on the parameter C . Schematic drawing. – – – = stable focus, ... = unstable focus, O = origin, c_b = bifurcation plane.

For values of C less than c_b , trajectories spiral into the steady state (see the dashed one-dimensional manifold in the lower part of Fig.1). For values of C larger than c_b , this steady state is unstable (dotted manifold).

Now let us turn to the case where C is not constant and the system is thus governed by Eq.(1). Consider a trajectory with all three chemical species initially small, as shown in Fig.2 (next page). In this case, the stable "steady state" of the subsystem A,B will still be approached as long as C grows slowly. After C (whose only influx is k_1 , via A) has reached a sufficient height, an at first small limit-cycle oscillation will develop between A and B . As long as the attendant consumption of A by B during part of the oscillatory cycle is less than the constant influx (k_1) to A , C will still grow on the average. As a consequence, a level of C will be reached soon where the "limit cycle" of the A,B subsystem is not small any longer. As this occurs, B is bound to consume most of A during

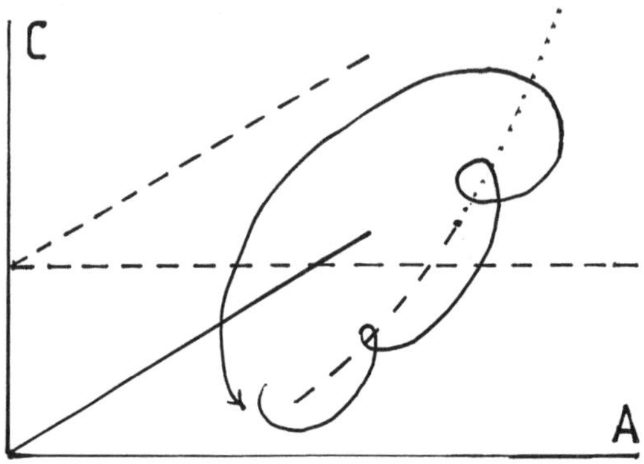

Figure 2 A variant to the situation of Fig.1, obtained by coupling C to the A,B subsystem in accordance with Eq.(1). Compare the text.

part of the oscillatory cycle, such that - due to the coupling between A and C - the concentration of C too has to diminish. Thereafter the whole process repeats.

The key to understanding what happens lies in the strong asymmetry of the A,B suboscillation as soon as its amplitude is no longer small. If this suboscillation stayed rotation-symmetric even at large amplitudes, then the up and down motions of C could be made rotation symmetric too (by supplying the right, if complicated, coupling terms). The overall motion in this case would be torus shaped (like a smoke ring around an attracting closed thread). In the present case of strong asymmetry, however, the shape of the overall motion can only resemble a "strongly distorted torus" at best. Strongly distorted tori, however, are known to produce nonperiodic behavior in dissipative systems for a long time [8].

A computer simulation confirms this qualitative reasoning: compare the result (Fig.3) with the prediction (Fig.2). The generality of the preceding argument means that, if it works in one case, many further examples (based, for instance, on the Brusselator [9] in place of the closely related Eq.(2)) must exist.

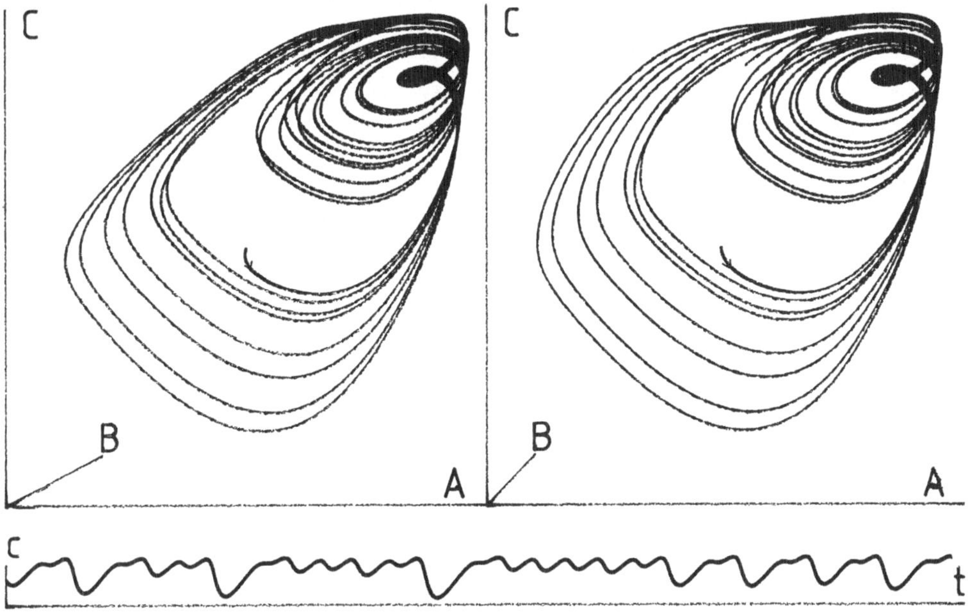

Figure 3 Chaos in Eq.(1). Numerical simulation using a standard Runge-Kutta-Merson integration routine. Stereoplot (two parallel projections) plus one time trace. Parameters assumed: $k_1 = 0.01$, $k_2 = 0.02$, $k_3 = 1$, $k_4 = 0.11$, $K = 0.08$, $k_5 = 0.0005$. Initial conditions: 0.8, 0.01, 1, respectively. Axes: 0.35 to 1.3 for A, 0 to 0.07 for B, 0.8 to 1.2 for C, 0 to 2800 for t.

3. A Second 3-Variable Example

The following abstract reaction system functions in an analogous manner:

$$\begin{aligned} \dot{a} &= k_1ac - k_2a - k_3b\,\frac{a}{a+K} + k_4c \\ \dot{b} &= k_2a - k_5b \\ \dot{c} &= k_6 - k_1ac - k_4c\,. \end{aligned} \tag{3}$$

Its reaction scheme is as follows:

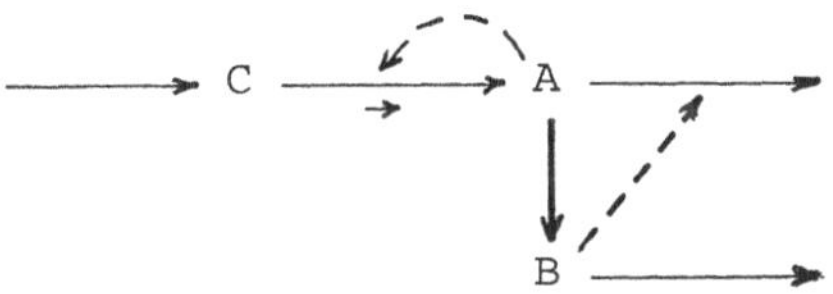

There is a small constant influx to C (k_6). C is (apart from a minor side reaction) consumed by A in an autocatalytic fashion. A reacts to give B . Another outflux of A is catalyzed by B in a Michaelis-Menten type reaction as before.

Again, we would have a 2-variable limit-cycle oscillator (A,B) if C were held constant. The equation of this chemical oscillator was first indicated by Turing [10], although he only looked at parameter values that produce a stable steady state. With the "parameter" C allowed to vary autonomously, the qualitative situation depicted in Figs. 1 and 2 applies again. The result (for one particular set of parameter values) is shown in Fig. 4. The present chaotic flow looks almost like a mirror-inverted image of that of Fig. 3.

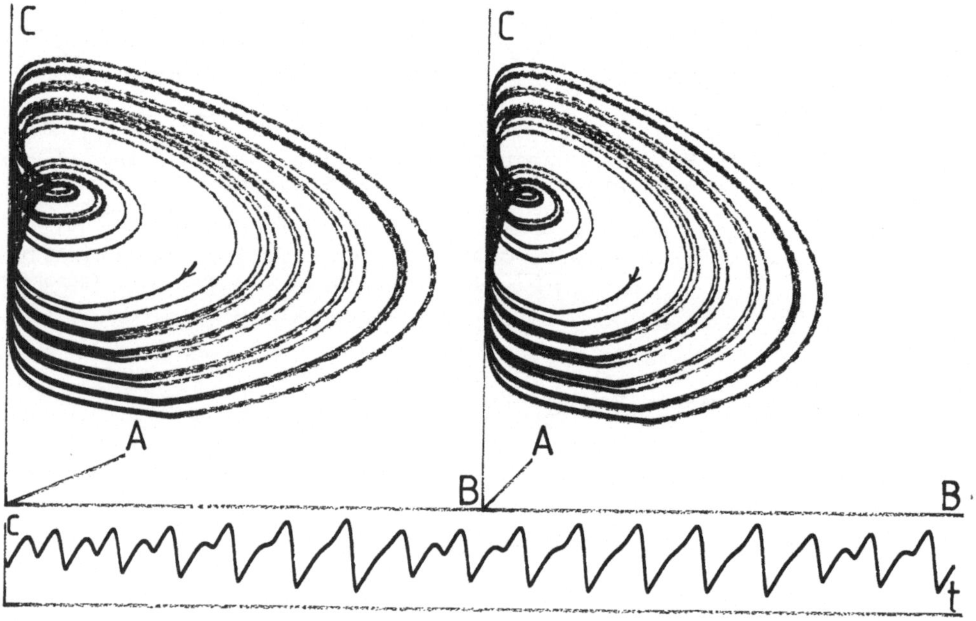

Figure 4 Chaos in Eq.(3). Compare Fig. 3. Parameters: $k_1 = 1.9$, $k_2 = 0.4$, $k_3 = 1$, $K = 0.0001$, $k_4 = 0.00006$, $k_5 = 0.5$, $k_6 = 0.005$. Initial conditions: 0.016, 0.013, 0.53, respectively. Axes: 0 to 0.03 for A , 0 to 0.05 for B , 0.4 to 0.7 for C , 0 to 1000 for t .

Even though the present system was selected for its abstract simplicity only, there exists a closely related (6-variable) mass-action system which was proposed as a possible model of the Belousov-Zhabotinsky reaction in 1972. In the simulations reported in [11], no attempt at finding anything other than a stable limit cycle was made, however.

4. A 4-Variable System

The following abstract reaction system was obtained from the preceding one (Eq.3) by the addition of a fourth variable (D):

$$\begin{aligned} \dot{a} &= k_1 ac - k_2 a - k_3 b \frac{a}{a+K} + k_4 d \\ \dot{b} &= k_2 a - k_5' b + k_6 \\ \dot{c} &= k_7 - k_1 ac - k_8 c \\ \dot{d} &= k_9 b - k_4 d \, . \end{aligned} \qquad (4)$$

Its reaction scheme is

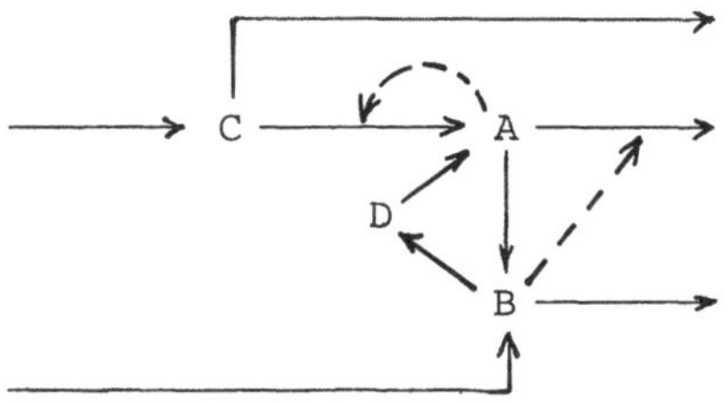

The new variable, D , is produced by B and reacts to give A in a formally 1st-order reaction. In addition, a new side-outflux to C and a constant influx to B have been assumed - two features which do not interfere with the chaotic capabilities of the A,B,C subsystem.

The present 4-variable system was devised with the same idea in mind that underlied the previous two examples. That is, the variable D , taken as a parameter, was to affect the qualitative behavior of the subsystem A,B,C . It indeed moves the A,B,C subsystem into and out of a large-amplitude chaotic behavior. At large D (so at $D > 0.135602$ if the rest of the parameters are as in Fig. 5 below), there is a small-amplitude chaotic regime. As D decreases below this value, the average amplitude of the chaotic motion starts to grow sharply. If taken as a variable, D performs a slow autonomous up and down movement in accordance with the philosophy of Fig. 2, although the underlying bifurcation (possibly a whole sequence) is more complicated this time. If everything were rotation symmetric, the new variable would merely add a periodic ("toroidal") component to the overall motion. In the absence of ideal symmetry, however, something more complicated can be expected.

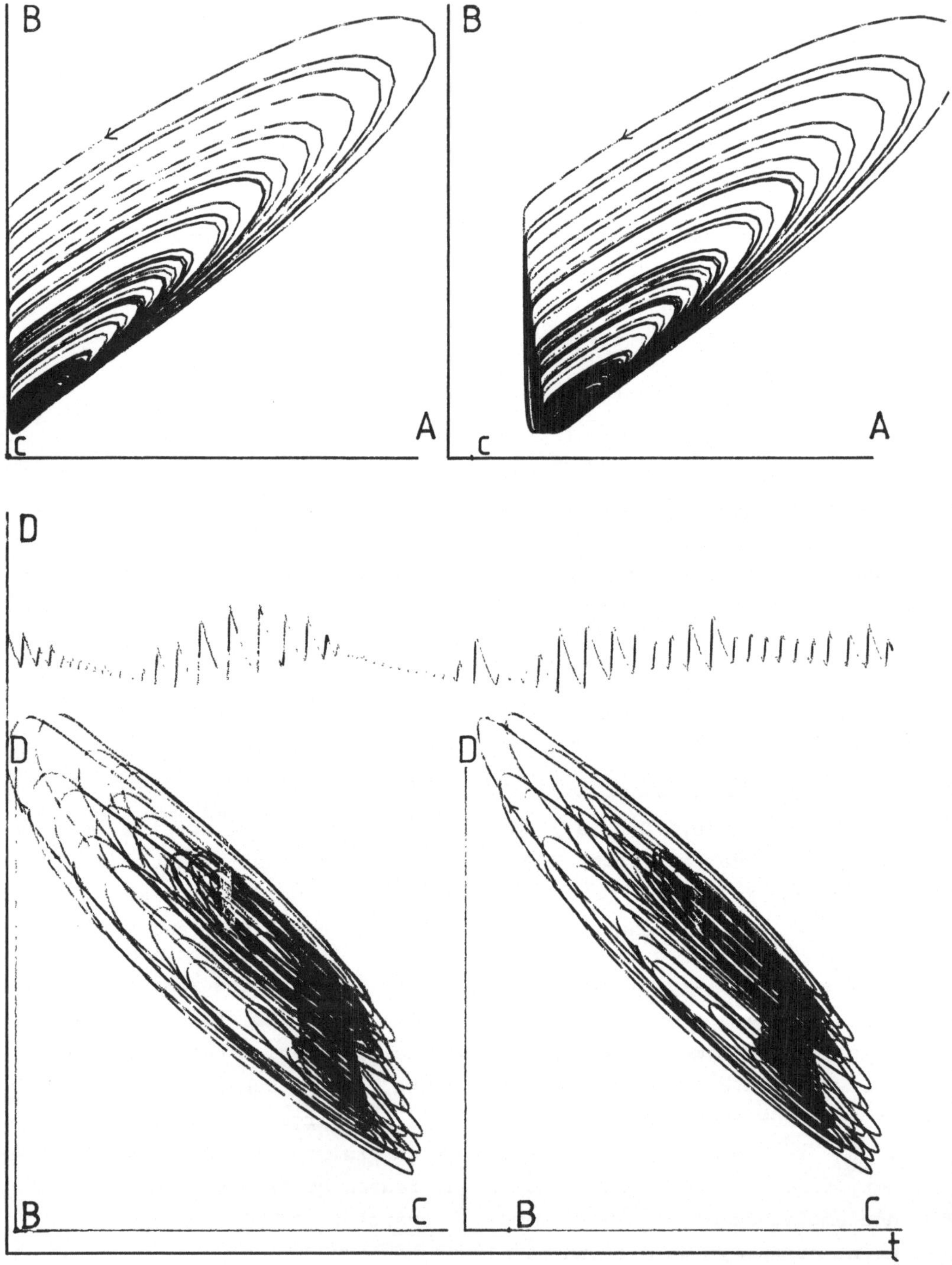

Figure 5 Higher chaos in Eq.(4). Compare Fig.3. Parameters: $k_1 = 2$, $k_2 = 0.4$, $k_3 = 1$, $K = 0.0001$, $k_4 = 0.002$, $k_5' = k_5 + k_9 = 0.5$, $k_6 = 0.0002$, $k_7 = 0.005$, $k_8 = 0.0068$, $k_9 = 0.245$. Initial conditions: 0.0043, 0.0039, 0.564, 0.138, respectively. Axes: 0 to 0.0125 for A, 0 to 0.01 for B, 0.24 to 0.32 for C (top) and 0.51 to 0.62 (bottom), 0 to 0.168 for D (middle trace) and 0.125 to 0.145 (bottom), 0 to 6100 for t.

In the simulation presented in Figure 5, two 3-D projections are shown. The A,B,C projection (top) shows a typical "chaotic" picture. The C,D,B projection (bottom) by contrast looks "torus like." When focusing on the C,D projection (lower-left picture), one sees that the "same" motions coexist in several parallel layers, so to speak. After a strong excursion of B (a large-amplitude spike of A), everything continues shifted somewhat toward the top (that is, toward a larger value of D). At such larger mean values of D , however, the amplitude of the A,B,C subsystem has to go down eventually. The motion then condenses into the more central (and to the right) bundle. At such smaller amplitudes of B, D (whose only input is B) is bound to decrease again. And so forth. The outcome is a "distorted torus," however with the special feature that one of the constituent submotions is not periodic but chaotic. The resulting object in 4-space resembles a "periodically forced chaotic motion" (cf. [4]). The time trace of D (middle portion of Fig. 5) reveals that the toroidal up and down component is not strictly periodic itself.

6. Discussion

Adding a "trivially behaving" third variable to a standard two-variable chemical oscillator proved sufficient for obtaining 3-variable chaos in two instances. We expect this design principle ("chemical glow-tube oscillator with added linear RC element" [12]) to be of rather universal validity [13]. It seems that at least one (rather complicated) abstract chaotic reaction system already contained in the literature - [14] - can be interpreted as falling into the same category.

The 4-variable example deserves a longer discussion. There are not many explicit 4-variable dynamical systems with a type of behavior specific to the fourth dimension available in the literature (cf. [4]). A recipe for a rather complicated 4-variable reaction system (with two autocatalytic reactions) producing hyperchaos was indicated in [5]. A 4-variable mechanical model of the dripping faucet exhibiting noodle-map chaos was recently devised by Shaw [15](cf. [16] for a model map).

The qualitative behavior shown by Eq.(4) has yet to be analyzed in detail. Generally speaking, four basic types of typically 4-variable behavior are known to date: 1) hypertoroidal behavior; 2) mixed chaotic and toroidal behavior; 3) higher chaos governed by a folded-towel map as a cross section [5]; 4) higher chaos governed by a noodle map [16] as a cross section. The attractor of Eq.(4) is probably of the third type (hyperchaos), but the fourth type is also a possibility dependent on parameters. Numerical detection of more than one (that is, two) positive Lyapunov characteristic exponents will establish type 3 [6]. (Type 4 will be easiest to see in a cross section [15].) At the time being, no other 4-variable system with hyperchaos is known in which the added fourth variable is stable when taken alone.

At this point it is perhaps worth mentioning that there exist many ways to generate higher chaos in 4-variable chemical systems other than the perturbation mechanism proposed above. For example, adding a bistable subsystem (single-variable or not) to a chaotic system in such a way that there is a back coupling is also sufficient in principle. But this mechanism requires additional "exotic" reactions. Simpler still is the possibility of taking a bistable chaotic system (that is, one which possesses a second bounded attractor in state space) and letting it alternate, by way of an added slow variable, between the basin of the chaotic and that of the other attractor. This possibility (singular-perturbation hyperchaos; cf. [4]) actually constitutes one mathematical prototype to the mechanism presented above. However, the above mechanism is even simpler since no hysteresis threshold is involved. Just as chaos does not depend on a hard (hysteresis type) bifurcation in a subsystem (but may for example rely on a "soft" bifurcation in a subsystem, as in Figs. 3 and 4), so can higher chaos be expected to arise through a simpler mechanism. Fig. 5 is only a first example.

The three systems presented above suggest a general conclusion. According to common experience, embedding an exotic reaction system (a switch, a limit-cycle oscillator, a chaotic oscillator, etc.) into a network of non-exotic variables is innocuous in general. Side variables (almost unavoidable in reality) are therefore usually either "lumped together" or omitted, if but a "reasonable approximation" to the actual behavior is at stake. Nevertheless, in each of the above three examples, a single added variable opened up the whole "Pandora's box" of the next higher dimension. Chemical experimentation too may thereby gain a new dimension - of difficulty.

Acknowledgments

J.L.H. acknowledges support by the Fulbright Commission, and by the National Science Foundation through grant No CPE 80.21950 . Thanks to J.R. for stimulation.

References

1. O.E. Rössler, Chaotic behavior in simple reaction systems. Z. Naturforsch. 31 a, 259-264 (1976).
2. J.L. Hudson and O.E. Rössler, Chaos and complex oscillations in stirred chemical reactors. In "Dynamics of Nonlinear Systems" (V. Hlavacek, ed.), New York: Gordon and Breach 1984.
3. R.M. Noyes, The interface between mathematical chaos and experimental chemistry. In "Stochastic Phenomena and Chaotic Behavior in Complex Systems" (P. Schuster, ed.), New York: Springer-Verlag 1984.
4. O.E. Rössler, The chaotic hierarchy. Z. Naturforsch. 38 a, 788-801 (1983).
5. O.E. Rössler, An equation for hyperchaos. Phys. Lett. 71 A, 155-157 (1979).
6. H. Froehling, J.P. Crutchfield, N.H. Packard and R.S. Shaw, On determining the dimension of chaotic flows. Physica 3 D, 605-617 (1981).
7. F.F. Seelig, Activated enzyme catalysis as a possible realization of the stable linear chemical oscillator model. J. theor. Biol. 30, 497-514 (1971).
8. M.L. Cartwright and J.E. Littlewood, On nonlinear differential equations of the second order, I. The equation $\ddot{y} - k(1-y^2)\dot{y} + y = b\lambda k \cos(\lambda t+\alpha)$, k large. J. Lond. Math. Soc. 20, 180-189 (1945).
9. I. Prigogine and G. Nicolis, On symmetry-breaking instabilities in dissipative systems. J. Chem. Phys. 46, 3542-3550 (1967).
10. A.M. Turing, The chemical basis of morphogenesis. Phil. Trans. Roy. Soc., Ser. B, 237, 37-72 (1952).
11. O.E. Rössler and D. Hoffmann, Repetitive hard bifurcation in a homogeneous reaction system. In "Analysis and Simulation of Biochemical Systems" (H.C. Hemker and B. Hess, eds.), pp. 91-102. Amsterdam: North-Holland 1972.
12. O.E. Rössler, Chaos and strange attractors in chemical kinetics. In "Synergetics - Far from Equilibrium" (A. Pacault and C. Vidal, eds.), pp. 107-113. New York: Springer-Verlag 1979.
13. O.E. Rössler and J.L. Hudson, Higher chaos in simple reaction systems. In "Chemical Applications of Topology and Graph Theory" (R.B. King, ed.), pp. 358-363. Amsterdam: Elsevier 1983.
14. T. Schulmeister, Chaos in a Lotka scheme with deposit (in German). Studia biophysica (Berlin) 72, 205-206 (1978).
15. R.S. Shaw, The dripping faucet (Tentative title). Preprint February 1983.
16. J.L. Hudson and O.E. Rössler, A piecewise-linear invertible noodle map. Physica D 1984. (In press.)

SPATIAL STRUCTURES INDUCED BY CHEMICAL REACTIONS AT INTERFACES: SURVEY OF SOME POSSIBLE MODELS AND COMPUTERIZED PATTERN ANALYSIS

M.L. Kagan[a], S. Peleg[b], E. Meisels[a] and D. Avnir[a]
Departments of Organic Chemistry[a] and Computer Science[b], The Hebrew University of Jerusalem, Jerusalem 91904, Israel.

According to the theories of far-from-equilibrium thermodynamics three types of homogeneity breaking are possible: spatial, temporal, and spatio-temporal.[1] The latter two are well known in purely chemical systems, e.g. bromate oscillators[2] and wave oxidations[3], both having their parallels in biochemistry, e.g. the glycolytic cycle[4], and slime-mold aggregation[5]. However the most abundant class of biological structures, namely spatial structures, are relatively unexplored in chemistry[6,7] (Liesegang precipitation phenomenon being an exception [8]). We present the following experimental results for a system that couples a chemical reaction to physical parameters to produce spontaneous pattern formation from initially homogeneous states. A complete model of the mechanism is still under investigation but it is clear at this preliminary stage that one is dealing with a system that is open to an analysis involving instabilities, perturbations and bifurcations. Some partial models are discussed below, some of which were also experimentally tested.

In 1977 P. Möckel reported[9] observing pattern formation during photolysis of a solution of starch, KI, and CCl_4 in water, and diphenylamine/CBr_4 in carbon tetrachloride. We began by investigating the generality of this phenomenon.[10,11] Tables 1(a) and 1(b) show some of the chemical systems that we tested for similar pattern formation. Nearly all of the reactions involve oxidation of aromatic amines by photolytically cleaved halogen species, to produce aniline type dyes (Fig. 1).

The following three conditions were found to be necessary for the formation of these types of structures: 1) A chemical reaction at an interface. 2) The product is soluble in the solvent and does not form an insoluble thin film or precipitate. 3) The solution is at rest before the reaction starts.

From the observations in Table 1, and from the works of Micheau et al.[12] and of Zinkovskaya[13], it became clear that the type of chemical reaction was of secondary relevance. However, we failed to obtain patterns by irradiating solutions of dye-product without the starting materials present.

The first possible model we discuss is that we are tracing, by means of a chemical reaction, already existing evaporative patterns, as is perhaps suggested by the similarity of Fig. 2 to typical hydrodynamic patterns of the Benard-type

Table 1(a). Structure formation during Photolysis[a] of Various Aniline Derivatives.

Substrate	Solvent and Halogen Source	Colour of Structure
Aniline	$CHCl_3$	yellow
N,N-Dimethylaniline	CCl_4	black
p-Chloroaniline	Cyclohexane	brown
Anthranilic acid	$CHCl_3$	black
Indole	CCl_4	mauve
Iminostilbene	CCl_4	green

[a] Photolysis conditions: 1% substrate, 1% halogen source (except when it is the solvent) were photolysed in Petri dishes by a rectangle array of four 15W, 254 nm U.V. lamps at a distance of 25 cm. Solution depth was 10 mm.

Table 1(b). Structure Formation during Photolysis[a] of Water Soluble Substrates.

Substrate	Halogen Sources
Diphenylamine Sulphonate	KBr; KI; Chloralhydrate; Iodoacetic acid
Tryptophan	Chloralhydrate
Anthranilic acid	Tetrabutyl ammonium iodide
4-Aminophenol hydrochloride	KI; KBr; Chloralhydrate; Tetra-ethyl ammonium iodide
Starch	KI/Chloralhydrate (9)

[a] Conditions and concentrations as in Table 1(a).

effect[14]. It seems to us that this model, which is favored by Micheau et al.[12] is quite unlikely, in view of the following experiments: 1(a) A reaction mixture of aminophenol hydrochloride and KBr in water was covered by a 3 cm layer of cyclohexane. 1(b) A reaction mixture of diphenylamine in carbon tetrachloride was covered by a 3 cm layer of water. In both cases, when irradiated from above, patterns formed at the interface of the two liquid phases.

2(a) The same mixture as in 1(a) was covered by a thin film of silicon oil.[15]
2(b) The same mixture was mixed with a surface tension reducing reagent such as a soap. Again neither prevented pattern formation.

3(a) For the reaction mixture of oxalic acid, ferric chloride and potassium ferricyanide, which is photolysed to Turnbull's blue by visible light[16], the reaction vessel was sealed with a glass cover retaining, however, a liquid/air interface. By adding a small amount of aluminium dust and by illuminating the side of the vessel by a laser beam, a Tyndall-type effect was produced which clearly showed whether there were any convective movements in the solution. Careful observation showed that before the reaction took place there were no visible hydrodynamic currents; however only once the reaction had started and patterns had formed were convection currents detected. This experiment was made more sensitive by using a binary solution of 25% methanol.[17] When the cover was removed immediate evaporation currents were generated which very effectively destroyed any previously formed chemical patterns. 3(b) A definitive test for the exclusion of evaporative

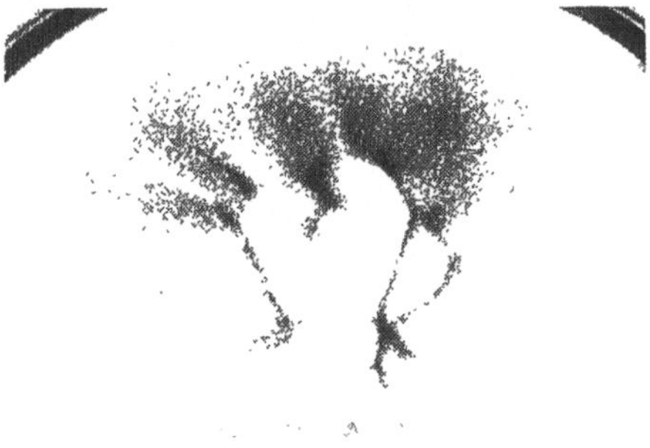

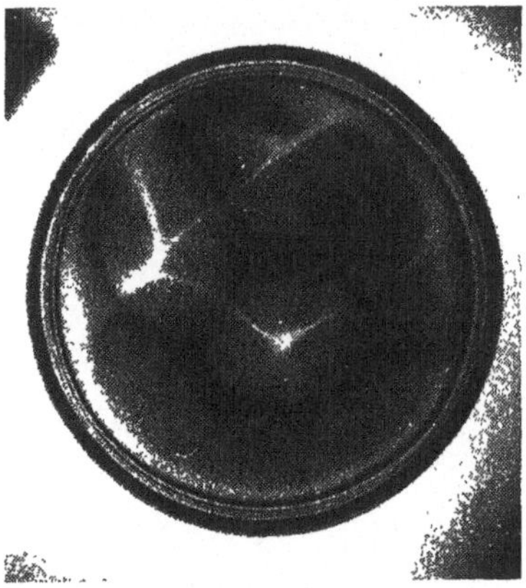

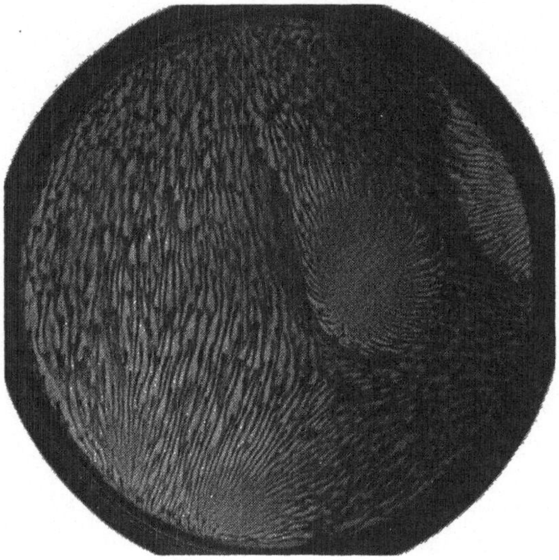

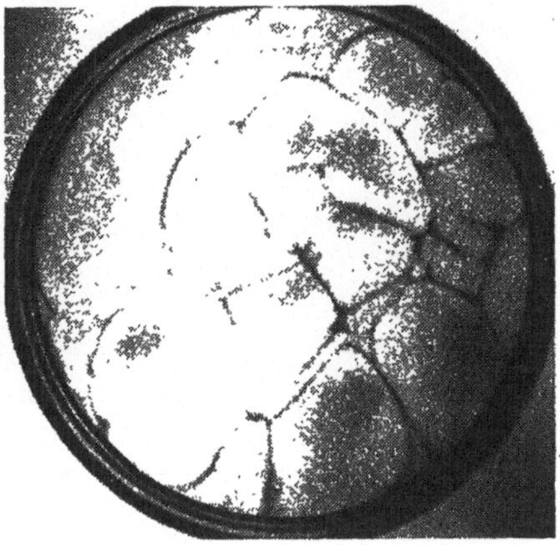

Patterns formation by: Fig. 1. Photolysis of iodoacetic acid/anthranilic acid at liquid/air interface. Fig 2. Photolysis of starch/KI/chloralhydrate at liquid air interface. Depth of solution approx. 1 cm. Fig. 3. Photolysis of oxalic acid/ $K_3Fe(CN)_6$/$FeCl_3$ at liquid/glass interface. Fig 4. Reaction of ammonia vapours with a solution of phenolphthalein.

pre-patterns was made with the same photolytic solution as in the previous test, this time, however, covering the solution without an airgap i.e. a liquid/glass interface. Patterns again formed in the liquid surface, Fig. 3.

By careful observation it was clear that the shapes in Fig. 2 were formed by the product descending along the lines of the surface pattern, and then, in shallow solutions, forming ascending cells, the white lines being the areas between neighbouring cells. Thus we have a chemical reaction inducing hydrodynamic movement and not a chemical reaction tracing hydrodynamic movement. This is illustrated in Scheme 1.

We discuss now a second possible model following the work of Ross and Nitzan[18]*. They have shown that it is possible for a reaction to be driven far-from-equilibrium by light and exhibit homogeneity breaking. In our context light would serve a dual role: as energy input for the reaction, and as the driving energy for the feedback mechanism, e.g.*

$$A \xrightarrow{h\nu} B \text{ (coloured)}$$

$$B + h\nu \longrightarrow B^*$$

$$B^* \longrightarrow B + \text{Heat}$$

In other words, a coloured product B is formed by photolysis of A, and itself absorbs light to reach an excited state B. This then decays non-radiatively. If A is of the type that diffuses to warmer regions then a fluctuation in the concentration of A or B, causing a localized 'hot-spot', will initiate, at a bifurcation point, a positive feedback loop. This would lead to growing areas of higher concentrations of B. One of the main difficulties in applying Ross' model was the discrepancy between the generality of our observation and the narrowness of the parameter window required for this model.*

While trying to determine the exact role of the light we found another general network of processes leading to the formation of dissipative structures: reactions at gas/liquid interface[19]*. Fig. 4 shows an example. The generality of the phenomenon is evident from Table 2. The above model was thus discarded.*

Our findings of pattern formation during reactions at gas/liquid interfaces led us to attempt reactions between two liquids. In order to prevent pre-reaction mixing or patterns due to extraction, the reagents in the same solvent were separated by a dialysis membrane[21]*. Table 3 shows again the generality of this interfacial phenomenon, and Fig. 5 and 6 are two examples of the resulting patterns.*[20]

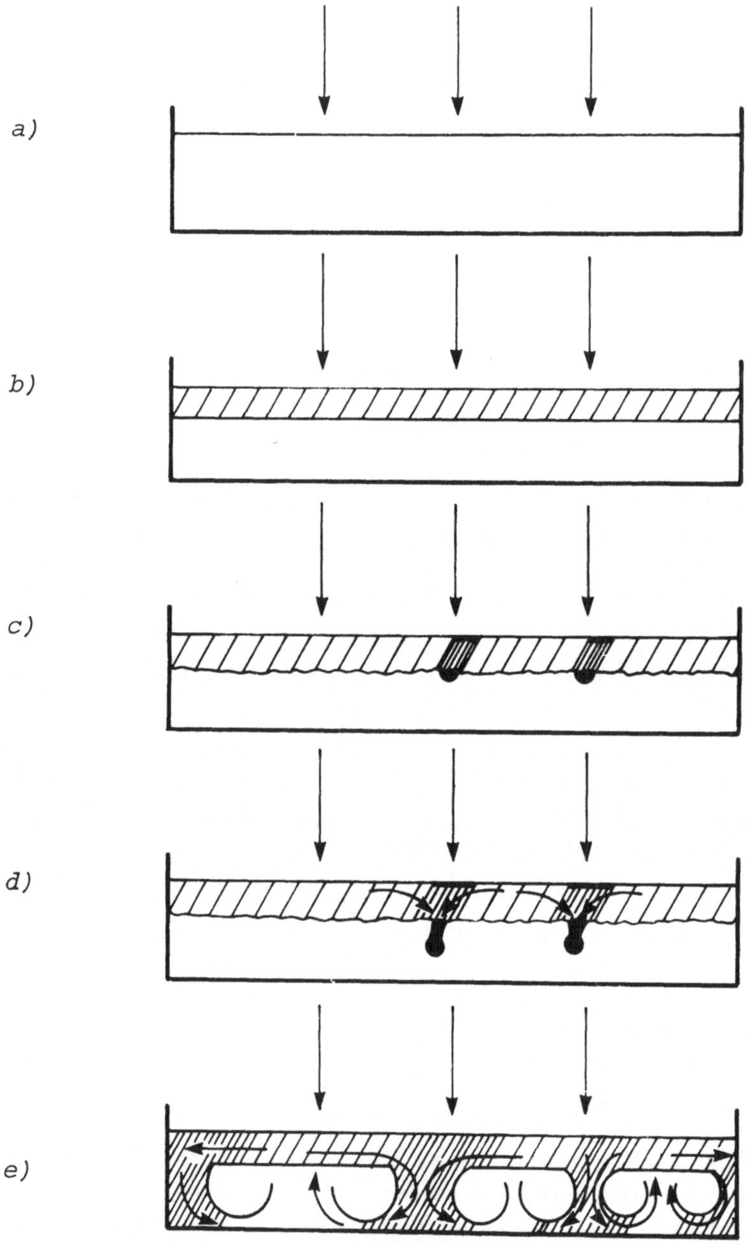

Scheme 1: a) Before reaction starts, the solutions are homogeneous and at rest; b) The reaction starts over the entire interface; c) After about 60 seconds lines of higher concentration begin to appear; d) The product descends along the pattern lines; e) If the solution is shallow enough (ca. 15 mm) the descending product form hydrodynamic-like rolls; f) (Not shown) The solution eventually becomes homogeneous again.

Table 2. Structure Formation of Gas/Liquid Interfaces[a].

Solution	Gas or Vapour (source)	Colour Change
0.5% Starch	Iodine (crystals	blue
0.5% Starch, 0.2% KI	Bromine (5% aq. sol.)	blue
0.1% Phenolphthalein	Ammonia (25% sol.)	pink
0.02% Methyl Orange	HCl (32% sol.)	red
2% Potassium hydroxide/1.2% glucose/methylene blue	Oxygen (air)	blue[b]

[a] The reactions were carried out in Petri dishes (7x1.5 cm) filled to a depth of 1 cm. The reacting gas was held in a Pyrex container of 200 ml that could be closely fitted over the dish after being filled up. [b] This structure, which is formed by a reversible redox reaction, persists for days.

Table 3. Structure Formation at Membrane Surface.

Solution[a] above membrane	Solution[a] below membrane	Reaction type	Colour of Structure
0.01% Iodine	0.5% Starch	complexation	blue
0.1% Bromine	0.5% Starch/1% KI	oxidation followed by complexation	blue
0.5% Starch/1% KI	0.1% Bromine		blue
0.01M HCl	0.02% Methyl Orange	protonation	red
0.01M KOH	0.02% Phenolphthalein	de-protonation	pink
0.5% Acetaldehyde	20% Morpholine 5% Sodium nitroprusside	oxidation	blue
1% $FeCl_3$	0.1% $K_2Fe(CN)_6$	complexation	blue
1% $FeCl_3$	0.1% Tannic acid	complexation	grey
1% $FeCl_3$	0.1% KSCN	complexation	red
0.01% Methylene blue	Water		no structure

a - All solutions in water.

It should be noted that scheme 1 applies equally to the three sets of interfaces, the most important point being that, before the onset of pattern formation, the solutions are at rest and the reaction takes place homogeneously over the whole interface. If the patterns were caused by extraction, osmosis or evaporation there would be no induction time but the immediate appearance of inhomogeneities.

A third mechanism has been suggested to us, namely, that the observed pattern could be due to double-diffusion.[22] This model has been used to explain the presence of salt inhomogeneities or 'salt-fingers' in seas and oceans.[23,24] The model involves a coupling of temperature and density in a solution containing a dissolved solute in which an upper layer of a higher temperature and a higher density (by evaporation) than the bulk, becomes unstable. At a point of fluctuation there is a small descent from this layer into the bulk. Since temperature dissipation is much faster than diffusion the density of this descending parcel increases thus accelerating its rate of descent. This feedback leads to the rapid descent of the surface layer down the path of the initial fluctuation. This model *was attractive since it explained the rapid descent of the coloured product in our experiments (∿cm/min). However, since the model is gravity dependent, it must therefore be unidirectional. For the chemical reaction in the liquid/membrane/liquid system we have shown that there is pattern formation whether the reaction occurs on the lower side of the membrane by diffusion of, for example, bromine down into* **starch**/*KI, or on the upper side of the membrane by bromine diffusing up into the same solution. Similarly, patterns were formed in the photochemical system by illuminating the petri dish from below.*

Another model involving the coupling of concentration and surface-tension resulting in Marangoni-type surface driven currents,[19b] *as in the Benard phenomenon,*[25,26] *has been tested. According to this model, as the concentration of the reaction product increases at the surface of the solution, an instability point is reached in which a small fluctuation in the concentration will cause a local change in the surface tension, which will then draw-in neighboring product, thereby initiating a positive feedback loop. The momentum resulting from this horizontal movement propels the product down into the bulk. However, we have managed to obtain typical patterns while prventing surface deformations. This was achieved, as described earlier, by forming a glass/liquid interface and irradiating through the glass, Fig. 3.*[27]

In a further series of experiments we have shown that the patterns scale with boundaries changes.[28] *In other words, as the area of the reaction vessel decreases so the number of lines per unit area increases. This is vividly shown in Fig. 7 in which a single drop of the oxalic acid/*$FeCl_3$/$K_3Fe(CN)_6$ *solution*[16] *was irradiated under a microscope. The diameter of the drop was about 5 mm. Microscopic structures were formed similar to the gas/liquid reaction of Fig. 6, the ratio of the boundary diameters being, however, 1:20. A similar result was achieved in which patterns were formed in a solution between two glass plates covering an area of diameter 50 mm but with a depth of only 0.95 mm !*

In dealing with spatial patterns we are faced with a number of unprecedented problems, for instance: the kinetics of pattern growth, changes in the complexity of patterns, and comparison of one pattern to the next. Image analysis is proving to be a very useful tool in solving these questions.[14a,29] *In the past, image analysis and pattern recognition in chemistry has been confined to analyses of analogue spectral output.*[30] *Scheme 2 outlines the various steps involved in transferring an image from the Petri dish to the computer for analysis: The pictures are recorded in situ either by a still camera (for kinetic studies it is connected to a timer) or by a Vidicon T.V. camera which is connected to a digitizer. In the former case the developed negatives are then rephotographed by the T.V. camera.*

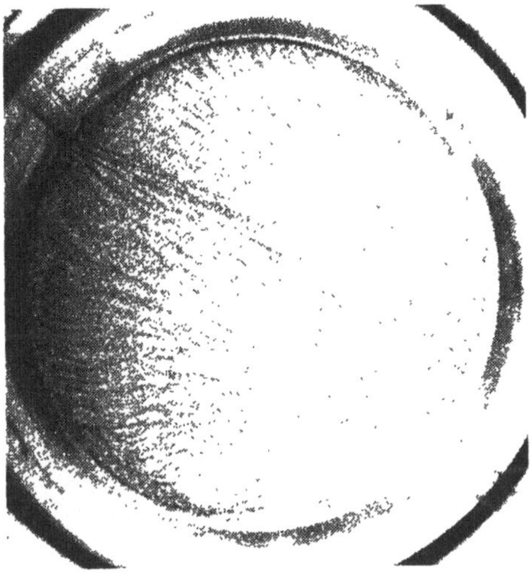

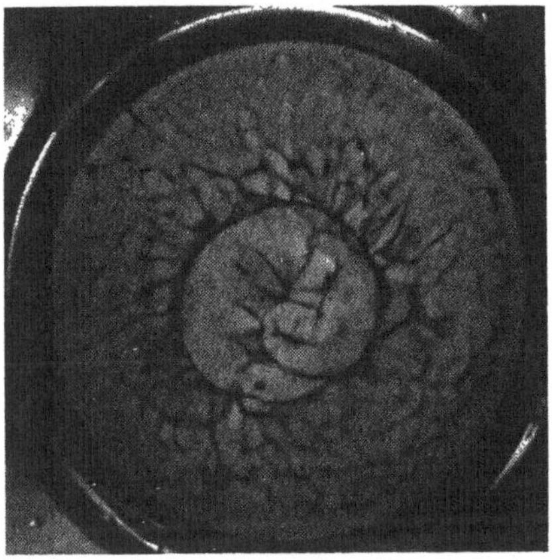

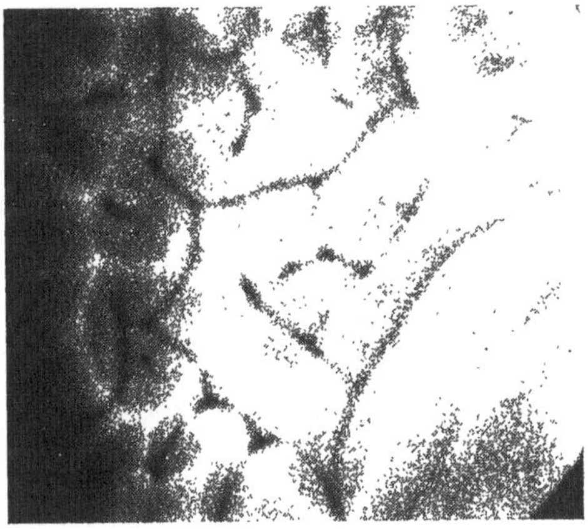

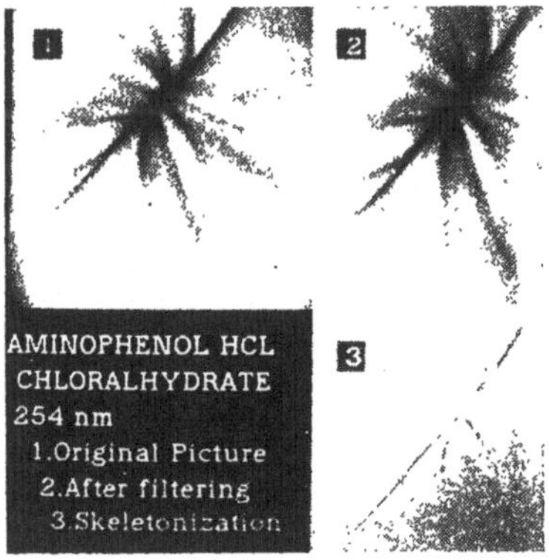

Patterns formation by: Fig. 5. Reaction of $K_3Fe(CN)_6$ with $FeCl_3$ at the surface of a dialysis membrane; Fig. 6. Reaction of bromine with starch/KI at the surface of a dialysis membrane; Fig. 7. Photolysis of oxalic acid/$K_3Fe(CN)_6$/$FeCl_3$ in drop under a microscope. Scale 1:0.025. Compare with Fig. 2. Fig. 8. Three main steps of Image Analysis.

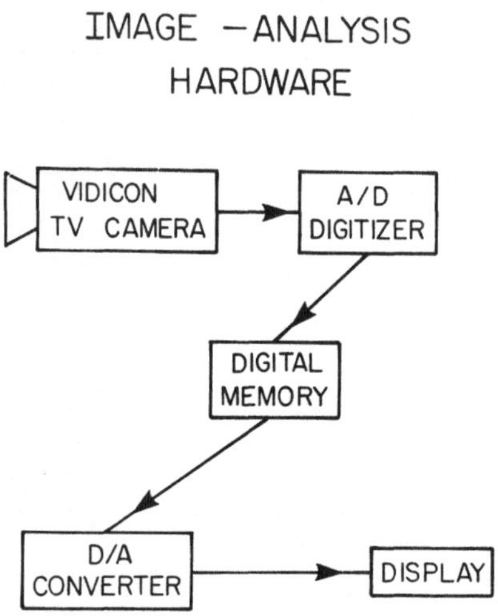

Scheme 2: Steps in transferring a structure image from a Petri dish to the computer, for analysis.

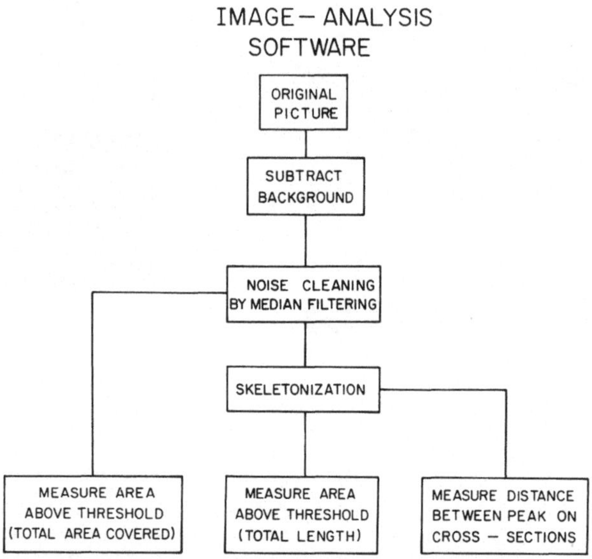

Scheme 3: The software used for processing the images.

The digitized picture is a 512 x 512 pixel array with each pixel having one of 256 possible grey levels. The software used for processing the images is shown in Scheme 3 and consists of the following steps:

Step 1. Elimination of non-random noise by subtracting the picture to be analyzed, pixel by pixel from a picture of the reaction system at time zero.

Step 2. Reduction of the size of the picture by quarter to 256 x 256. This is achieved by a general averaging algorithm that not only reduces the bit size of the image, thus making it easier to handle, but also acts as an initial filter.

Step 3. Filtering of random noise produced in the experiment itself and accummulated in the various photographic stages. This is done by a localized median averaging algorithm.

At this point a summation of the number of pixels above an optimum grey level yields the total amount of product in the picture.

Step 4. The pattern is skeletonized[31] *leaving lines of only 1-2 pixels wide but retaining their length. A summation of pixels at an optimum grey level at this stage yields the total length of the structure (the result from step 3 minus this result gives the average line width). In this way the rate of line growth can be determined, as well as the average distance between the lines as a measure of complexity needed for a scaling law. Fig. 8 shows the main stages as viewed on the image analyser television screen for patterns formed by photolysis of aminophenol HCl/chloralhydrate in water.*

Step 5. (Not shown). At this stage the option exists of using a broken line gap-filling algorithm. Such gaps sometimes appear as an artefact of the above steps.

As mentioned above, another specific problem that we are presently solving with this technique[29] *is to give a quantitative definition to patterns in order to determine how different one pattern is from another. This requires parametrizations such as line length, line spacings, fractal-signature texture analysis*[32]*, area enclosed by lines, spacing of junction points, Fourier analysis.*

Acknowledgments: *Sponsored by the Volkswagen Foundation. Assisted by the F. Haber Research Center for Molecular Dynamics. Most helpful discussions with Professor B. Hess, Professor T. Plesser, and Dr. S. Muller are acknowledged. D.A. is an M. Richter Fellow. Thanks are due to Y. Karni for programming assistance.*

References and Notes

1. *Glansdorff, P. and Prigogine, I., "Thermodynamic Theory of Structure, Stability and Fluctuations" (Wiley, N.Y., 1971); Nicolis, G. and Prigogine, I., "Self-Organization in Non-equilibrium Systems" (Wiley, N.Y., 1977).*
2. *e.g. Tyson, J.J., Lecture Notes in Biomathematics, 10, (Springer-Verlag, 1976).*
3. *Winfree, A.T., Sci. Am., 82, June (1974).*
4. *Hess, B., Goldbeter, A. and Lefever, R., Adv. Chem. Phys., 38, 363 (1979).*

5. Hess, B., Boiteaux, A., Busse, H.G. and Gerisch, G., ibid., 29, 152 (1975); Gerisch, G., Naturwiss., 58, 430 (1971).
6. Appearance of mosaic patterns during bromate oxidations: Showalter, K., J. Chem. Phys., 73, 3735 (1980); Orban, M., J. Am. Chem. Soc., 102, 4311 (1980); Zhabotinskii, A.M. and Zaikin, A.N., J. Theor. Biol., 40, 45 (1973).
7. E.g.: Hlavarek, V., Janssen, R. and Van Rompay, P., Z. Naturforsch, 37a, 39 (1982).
8. E.g.: Müller, S.C., Kai, S., and Ross, J., Science, 216, 635 (1982). Stern, K.H. Chem. Rev., 54, 79 (1954).
9. Möckel, P., Naturwiss., 64, 224 (1977).
10. Kagan, M., Levi, A. and Avnir, D., ibid., 69, 548 (1982).
11. Avnir, D., Kagan, M. and Levi, A., ibid., 70, 141 (1983).
12. (a) Gimenez, M. and Micheau, J.-C., ibid., 70, 90 (1983); (b) Micheau, J.-C., Gimenez, M., Brockmans, P. and Dewel, G., Nature, in press.
13. Zinkovskaya, O.V., Deposited Doc., VINITI, 1772-1782 (1982).
14. (a) Plesser, T., and Müller, S., this volume; (b) Silveston, P., Forsch. Ing-Wes., 24, 59 (1958).
15. Block, M.J., Nature, 178, 650 (1956).
16. Photoreduction of ferric chloride by oxalic acid in visible light. 0.2% oxalic acid/0.2% $K_3Fe(CN)_6$/0.1% $FeCl_3$; Fry, M.S. and Gerwe, E.G., Ind. Eng. Chem., 20, 1392 (1928).
17. Berg, J.C., Boudart, M., and Acrivos, A.J., Fluid Mech., 24, 721 (1966).
18. Ross, J. and Nitzan, A., J. Chem. Phys., 59, 241 (1973); Ross, J., Ber. Bunsen Phys. Chem., 80, 112 (1976).
19. (a) Avnir, D. and Kagan, M., Naturwiss., 70, 361 (1983); (b) Patterns formed at aqueous surfaces by air oxidations were also independently found by M. Orban (the system $O_2/I^-/H_2SO_4$, private communication) and by L. Weissenborn and R. Bausch, Naturwiss., 70, 307 (1983); (c) P. Möckel reported structure formation during H_2O and CO_2 exchange through liquid/air interface (Naturwiss., 66, 575 (1979)).
20. Avnir, D., and Kagan, M., submitted for publication; Kagan, M. and Avnir, D., to appear in the Proceedings of the Seventh International Conference on the Origins of Life, Mainz, July, 1983.
21. Spectropore 2 Dialysis Membrane, molecular cutoff 12,000.
22. Weinberger, H., unpublished manuscript.
23. Stern, M.E., J. Fluid Mech., 35, 209 (1969); Shirtcliffe, T.G.L. and Turner, J.S., J. Fluid Mech., 41, 707 (1970).
24. Dewel, G., Brockmans, P. and Walgraef, D., preprint.
25. Pearson, J.R.A., J. Fluid Mech., 4, 489 (1958).
26. Velarde, M.G., Lecture Notes in Physics, ed. Sorenson, T.S., 105, 260 (1979); Scriven, L.E. and Sternling, C.V., AIChEJ, 5, 514 (1959).
27. Segal, L., private communication; Saunders, D.A. and Schmidt, P.J., Proc. Roy. Soc. A. 165, 216 (1938).
28. Experimental evidence of a scaling law is presented by Gimenez et al. (ref.12)
29. Avnir, D., Peleg, S. and Kagan, M., manuscript in preparation.
30. E.g., Jurs. P.C., "Chemical Application of Pattern Recognition" (Wiley, N.Y. 1975), and references therein.
31. Peleg, S., and Rosenfeld, A., IEEE Trans. PAMI-3, 208 (1981).
32. Peleg, S., Naor, J., Hartley, R. and Avnir, D., submitted for publication.

DYNAMIC PATTERNS IN EXCITABLE MEDIA

James P. Keener
Department of Mathematics
University of Utah
Salt Lake City, UT 84112 USA

For normal individuals, the heartbeat is a time periodic pattern of electrical excitation which spreads in rapid wavelike fashion across the surface of the heart. In contrast, fibrillation, which occurs at death in many heart patients, corresponds to a much different pattern of excitation which is not capable of sustaining life. In this paper we will show one way in which the transition from a normal heartbeat to fibrillation may occur.

Contraction of heart muscle is triggered by electrical polarization of the cardiac tissue. Under normal conditions a wave of depolarization (or excitation) spreads from the sinoatrial (SA) node in the right atrial wall, throughout the left and right atria. At the base of the right atrium, this wave of excitation encounters the atrioventricular (AV) node, which is the only conducting link between atria and ventricles. The depolarizing wave continues from the AV node through specialized fibers (Purkinje fibers) to the base of the ventricles from where it spreads across the left and right ventricles.

This sequence of events may change in a patient with heart disease. A patient with a relatively normal heartbeat first experiences ventricular tachycardia (V-tach), a sequence of rapid beats, characterized by a broad electrocardiogram, signifying slowly propagating waves of excitation. Shortly thereafter the patient's heart begins to fibrillate. In fibrillation, the heart no longer beats in a coordinated fashion, and the electrocardiogram signal consists of irregular, rapid, small fluctuations around the base voltage. The transition to fibrillation is instantaneous, and unless the fibrillation is interrupted, the patient will die within minutes.

Although the precise sequence of electrical events in fibrillation is not known, it is certainly quite different from the normal situation. Among the theories to explain fibrillation, the most popular is that there are one or more reentrant (also called circus or spiral) waves which are self sustained on the ventricles without the need for external stimulation. There are numerous experiments [1,4,14,21,22] showing that self sustained waves can occur, but they do not conclusively show that these indeed occur in clinical situations, nor is it understood how such waves are initiated. There is also a significant mathematical literature [3,6,8,10,15,23,24] on the existence of spiral waves in excitable and oscillatory media. Again, in these works, the question of how the normal situation degenerates into a spiral wave is ignored.

Of previous work on the problem of sudden death, the work of Winfree [25] is most intriguing His theory depends upon two assumptions, first that the entire ventricular surface is self oscillatory, and second that the onset of fibrillation is caused by a spatially nonuniform phase resetting impulse. Unfortunately, the heart consists almost entirely of non-oscillatory cells, and to my knowledge, evidence for a phase resetting impulse in clinical situations is weak.

In this paper we give one explanation for the sudden transition from V-tach to fibrillation observed in the clinic. Our model is based upon the physiology of infarcted myocardium. When blood flow to the heart is restricted, the myocardial tissue loses its ability to respond to electrical stimuli. If such a region of infarction occurs on the ventricles, then ventricular conduction is around the infarcted area, and therefore, the conduction network is topologically equivalent to conduction on a one dimensional ring with an oscillatory input at one point. Conduction on this ring may occur in two ways. Waves originating at the oscillatory source will propagate in opposite directions from the source around the loop and terminate when they collide at the point opposite from the source. Self sustained waves which travel in one direction around the loop are also possible if the loop is long enough or the speed of conduction slow enough. The first scenario corresponds to normal conduction around an infarct, and the second to a reentrant wave of fibrillation around an infarct. Although both states may occur on the same medium, the transitions between these states are not apparent.

In addition to infarcted regions there are regions of less than total electrical failure called ischemic zones. Ischemic tissue recovers from excitation more slowly than normal tissue. Most important to our model is the observation that an ischemic zone may be a region of one way block, that is, a region where some or all waves travelling in one direction fail to propagate through the region while waves travelling in the opposite direction are not similarly hindered.

In the sections that follow we will show some of the implications of the above physiological observations. First, in Section 2 we introduce a mathematical model for conduction of electrical excitation (the Fitzhugh-Nagumo model) and show that when parameters are chosen to simulate ischemia and infarction, the model exhibits one way block. In Section 3, we develop and analyze a simple model of periodic stimulation of a conducting ring with a one way block to show how hysteretic transitions between normal and reentrant waves can occur.

2. ONE WAY BLOCK

Excitable tissue, such as nerve membrane, are able to maintain a potential difference across the membrane when at rest. This potential difference is due to an imbalance of ions such as sodium, potassium, calcium, etc. These ions move across the membrane through specialized channels in response to changes in transmembrane potential. A standard way to depict such a membrane is with the equivalent electrical circuit shown in Figure 1. Kirchhoffs laws give that

$$C\frac{dv}{dt} = I - i_{Na^+} + -i_{K^+} - i_0 \tag{2.1}$$

where C is the transmembrance capacitance, v is the transmembrane potential, and i_{Na^+}, i_{K^+}, i_0 are sodium, potassium, and miscellaneous other ionic currents, respectively.

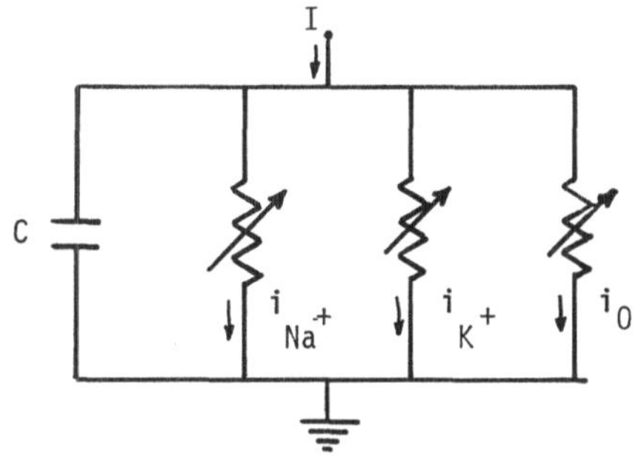

FIGURE 1

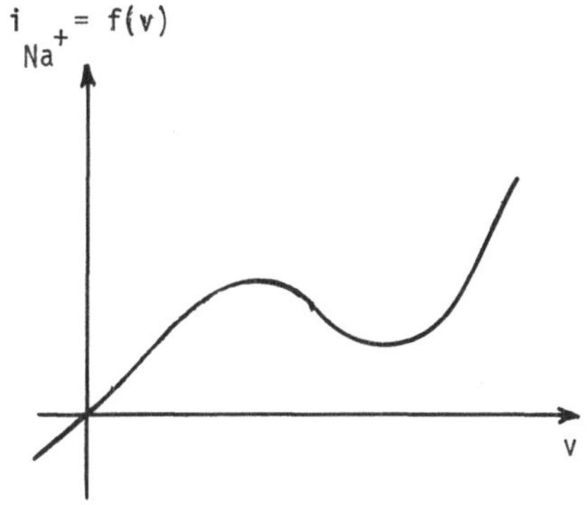

FIGURE 2

There are three well known ways to describe the behavior of the ionic currents. The Hodgkin-Huxley model [7] is based on careful measurements in squid axon. The McAllister-Nobel-Tsien [11] model is based on measurements on cardiac tissue, and the FitzHugh-Nagumo (FHN) [5,16] model is a simplified caricature of the Hodgkin-Huxley model. For the FitzHugh-Nagumo model the sodium current is taken to be a "fast" current with a nonlinear cubic-like current-voltage relationship as shown in Figure 2. The actual details of this curve seem not to be especially important [12]. All other currents are lumped together as a recovery variable i which responds linearly, but not instantaneously, to potential changes

$$L\frac{di}{dt} = v+v_0 - Ri, \quad i = i_{K^+} + i_0 \tag{2.2}$$

The parameters of this equation suggest an electrical circuit analogue using an inductor, and voltage source [9,16].

The equations (2.1), (2.2) are commonly called the space clamped equations as there is no spatial dependence. Spatial dependence is found by assuming that many such elements are connected by a resistive transmission line, so that in the continuum limit

$$I = \frac{\partial}{\partial x}\left(\frac{1}{R_2}\frac{\partial v}{\partial x}\right) \tag{2.3}$$

where R_2 is the resistance per unit length [20]. In terms of nondimensional parameter and variables, the system (2.1), (2.2), (2.3) is

$$\frac{\partial v}{\partial t} = \frac{\partial^2 v}{\partial x^2} - i - f(v)$$

(2.4)

$$\frac{\partial i}{\partial t} = \varepsilon(v+v_0-\rho i)$$

where $\varepsilon = \frac{V_0^2 C}{I_0^2 L}$ is a small dimensionless parameter, I_0, V_0 are typical values of current and voltage, respectively, and $f(v)$ is the cubic-shaped normalized, dimensionless function of Figure 2. We have assumed that the cable resistance R_2 is constant.

This model has been studied by numerous authors [5,8,12,16,18,19,24], and many of its properties are well known. The property we wish to discuss is that of one way block. In particular, we will indicate that with spatial variation of parameters suggested by infarction and ischemia, the equations (2.4) exhibit one way block.

Previous studies of conduction block [13,17,18,19] have indicated that sudden changes in cable diameter, such as when a cable branches, can lead to block. This can be modelled by spatial variations in the parameter R_2, the cable resistance. Ischemia and infarction are more reasonably modeled by spatial variations in the bias voltage v_0, since changes of this parameter can render the space clamped system self oscillatory and more or less excitable. In addition, there are values of v_0 for which no pulses propagate in the system (2.4). There may be other parameter variations which are relevant for this problem, but for this discussion we take $v_0(x)$ to be the only parameter that varies in space.

An action potential for the equation (2.4) moving into a medium at rest can be constructed using singular perturbation arguments [8] in the limit $0 < \varepsilon \ll 1$. To construct an action potential there are four component parts: the leading edge of the excitation, the excited state in which i and v change slowly, the trailing edge which repolarizes the medium, and the slow return of i and v to the rest state. We concentrate here on the leading edge, since if it fails to propagate throughout the entire medium, the action potential is blocked.

For $\varepsilon > 0$ sufficiently small, the leading edge of an action potential is governed approximately by the equation

(2.5)
$$\frac{\partial v}{\partial t} = \frac{\partial^2 v}{\partial x^2} - f(v) - i_0(x).$$

where $i_0(x)$ is the resting current which satisfies the equations

$$i_0(x) = -f(v(x))$$

(2.6) where

$$\rho f(v(x)) + v(x) + v_0(x) = 0.$$

As is standard, we assume that $\frac{1}{\rho} > -\min_v(f'(v))$ so that for each v_0 there is one and only one value v for (2.6). We take $f(v)$ to be a cubic-like function such as shown in Figure 2. There is an open interval $I=(i_-,i_+)$ so that for $i \in I$, $f(v)+i$ has three zeros. We denote by $v_\pm(i)$ the largest and smallest of these zeroes. There is a value of $i \in I$, say $i = i^*$, at which

$$\int_{v_-(i^*)}^{v_+(i^*)} (f(v)+i^*)dv = 0. \tag{2.7}$$

Our goal is to show that for certain nonconstant profiles $i_0(x)$ (equivalently $v_0(x)$) there is one way block. Our main tool is the use of comparison methods for parabolic equations, as established in [2]. For easy reference, we state the main theorems that we will use.

Lemma 2.1. Suppose $u(x,t)$, $v(x,t)$ satisfy the inequalities

$$\begin{gathered} u_t - u_{xx} + f(u) + i(x) \leqslant v_t - v_{xx} + f(v) + j(x) \quad \text{on} \quad (-\infty,\infty) \times [0,T] \\ \text{and} \\ u(x,0) \geqslant v(x,0),\ x \in (-\infty,\infty). \end{gathered} \tag{2.8}$$

Then

$$u(x,t) \geqslant v(x,t) \quad \text{on} \quad (-\infty,\infty) \times [0,T].$$

Furthermore, if $u(x,0) > v(x,0)$ on an open subinterval of $(-\infty,\infty)$ then $u(x,t) > v(x,t)$ on $(-\infty,\infty) \times (0,T]$.

Corollary 2.2. Suppose the functions $u(x)$, $w(x)$ satisfy

i) $u(x) > w(x)$
ii) $u'' - f(u) - i(x) \leqslant 0$
iii) $w'' - f(w) - i(x) \geqslant 0$.

Then there is a solution of $v'' - f(v) - i(x) = 0$ which satisfies $w(x) \leqslant v(x) \leqslant u(x)$.

From these two statements one can easily show

Lemma 2.3. Suppose $i_0(x) \in (i_-,i_+)$. Let $i_{min} = \inf_x(i_0(x))$ and $i_{max} = \sup_x(i_0(x))$. Then there are two functions $q_L(x)$, $q_u(x)$ which are both solutions of

$$q'' - f(q) - i_0(x) = 0$$

satisfying

$$v_-(i_{max}) \leqslant q_L(x) \leqslant v_-(i_{min})$$

$$v_+(i_{max}) \leqslant q_u(x) \leqslant v_+(i_{min})$$

PROOF: To find $q_L(x)$, note that $u(x) = v_-(i_{min})$ and $w(x) = v_-(i_{max})$ are upper and lower solutions, respectively. That is,

$$w'' - f(w) - i_0(x) = i_{max} - i_0(x) \geqslant 0$$

and

$$u'' - f(u) - i_0(x) = i_{min} - i_0(x) \leqslant 0,$$

and $u(x) > w(x)$, so that Corollary 3.2 applies. To find $q_u(x)$, we use $u(x) = v_+(i_{min})$ and $w(x) = v_+(i_{max})$ in a similar way.

Using $q_L(x)$, $q_u(x)$, we have other important properties of solutions of (2.5) [2].

Lemma 2.4: i) Suppose $q_L(x) \leqslant v(x,0) \leqslant q_u(x)$ for $x \in (-\infty,\infty)$. Then the solution $v(x,t)$ of (2.5) satisfies

$$q_L(x) \leqslant v(x,t) \leqslant q_u(x)$$

ii) Suppose $q^*(x)$ satisfies $q'' - f(v) - i_0(x) = 0$ on $x \in (a,b)$ and $q^*(x) > q_L(x)$ on $x \in (a,b)$

$$q^*(a) = q_L(a), \quad q^*(b) = q_L(b).$$

If $v(x,0) \geqslant \begin{cases} q^*(x) & x \in (a,b) \\ q_L(x) & x \notin (a,b) \end{cases}$ then $v(x,t)$ is a nondecreasing function of t for each x. Furthermore $\lim_{t\to\infty} v(x,t) = \tau(x)$ uniformly on bounded intervals where $\tau(x)$ is the smallest solution of $\tau'' - f(\tau) - i_0(x) = 0$ which satisfies

$$v(x,0) \leqslant \tau(x) \leqslant q_u(x).$$

The function $q_L(x)$ corresponds to the rest state of the medium and $q_u(x)$ is the excited state. If there are no other steady solutions of (2.5), then with a sufficient stimulus, as defined by $q^*(x)$ in Lemma (2.4), the entire medium will become excited. On the other hand, if there are other steady solutions of (2.4) between $q_L(x)$ and $q_u(x)$, and excitatory stimulus will be blocked from exciting the entire medium. Thus, a necessary and sufficient condition for blocking is the existence of a steady solution of (2.4) between $q_L(x)$ and $q_u(x)$. Such a solution we will call a blocking solution and the corresponding $i_0(x)$ will be called a blocking current profile.

Lemma 2.5. Suppose that for $i = i_0(x)$ there is a blocking solution $q(x)$ with $q_L(x) \leqslant q(x) \leqslant q_u(x)$. Then for every larger current profile $i = j_0(x) \geqslant i_0(x)$ there is also a blocking solution.

PROOF: If $q'' - f(q) - i_0(x) = 0$, then $q'' - f(q) - j_0(x) = i_0(x) - j_0(x) \leqslant 0$ so that $q(x)$ is an upper solution and $q_L(x)$ is a lower solution.

Now we consider a specific choice of $i_0(x)$,

$$(2.9) \qquad i_0(x) = \begin{cases} i_1 & -\infty < x < 0 \\ i_2 & 0 < x > d \\ i_3 & d < x < \infty \end{cases}$$

where $i_- \leqslant i_1$, $i_3 < i^* < i_2$. Using phase plane methods we can readily find blocking solutions for the current profile $i_0(x)$.

Lemma 2.6. If $i_- \leqslant i_1$, $i_3 < i^* < i_2$, then for all d sufficiently large there is a steady solution $q(x)$ of (2.4) which satisfies $q_L(x) \leqslant q(x) \leqslant q_u(x)$ and

$$\lim_{x\to-\infty} q(x) = v_+(i_1)$$

$$\lim_{x\to+\infty} q(x) = v_-(i_3).$$

PROOF: To find steady solutions of (2.4) we look at the phase plane trajectories

$$v' = w$$

$$w' = f(v) + i_k \qquad k = 1,2,3$$

First integrals for these trajectories are the curves

$$(2.11) \qquad \frac{w^2}{2} = F(v) + i_k v - K, \qquad F'(v) = f(v), \qquad k = 1, 2, 3.$$

We must choose pieces of these trajectories which connect and traverse from $v_+(i_1)$ to $v_-(i_3)$. In Figure 3 we show this phase portrait with the relevant trajectories labeled. The trajectory labeled A is the arc $\frac{w^2}{2} = F(v) - F(v_+(i_1)) + i_1(v-v_+(i_1))$ and approaches the critical point at $v = v_+(i_1)$ as $x \to -\infty$. The trajectories labeled C are those trajectories $\frac{w^2}{2} = F(v) - F(v_-(i_3)) + i_3(v-v_-(i_3))$ which approach the critical point at $v = v_-(i_3)$ as $x \to +\infty$. To get a connection between these two trajectories we must choose one from the family of trajectories, labeled B, which are arcs for $k = 2$. Since there is an entire family to choose from, there is a wide range of values of d, corresponding to the spatial length required to traverse a particular B arc. Note that d has a lower bound, but no upper bound. Take d^* to be the smallest permissible distance d. Then for any $d \geqslant d^*$, there is a blocking solution.

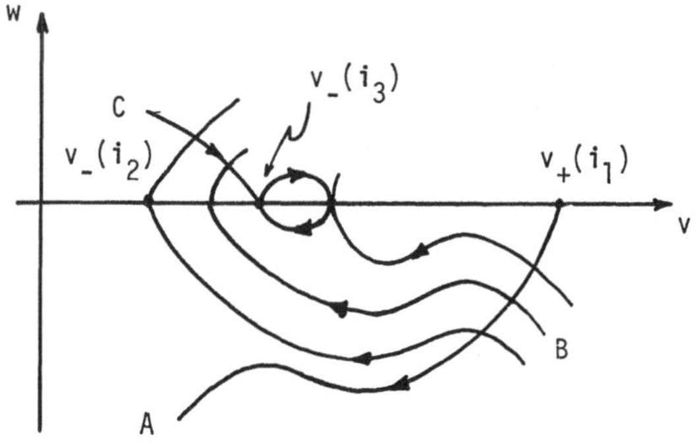

FIGURE 3

For the piecewise constant current profile $i_0(x)$, in (2.9), there is a smallest distance $d = d^*(i_1,i_2,i_3) > 0$ for which blocking solutions exist. If $d^*(i_1,i_2,i_3) \neq d^*(i_3,i_2,i_1)$ then the profile $i_0(x)$ has one way block for every value of d between $d^*(i_1,i_2,i_3)$ and $d^*(i_3,i_2,i_1)$.

We will show that $d^*(i_1,i_2,i_3)$ is not symmetric about $i_1 = i_3$. To do so, consider an arc connectin? $v_+(i_1)$ to $v_-(i_3)$ via a B arc with spatial length d. Keeping the B arc fixed, we vary slightly the values of i_1 and/or i_3 and note that the spatial length d must change by the amount $dx = \frac{dv^*}{w^*}$ where dv^* is amount of change of v of the point (v^*,w^*) of intersection of A and B arcs or B and C arcs. The intersection of A with B arcs occurs at

$$v_+^* = \frac{K-i_1v_+(i_1)-F(v_+(i_1))}{i_2 - i_1}$$

and the intersection of B with C arcs occurs at

$$v_-^* = \frac{K-i_3v_-(i_3)-F(v_-(i_3)}{i_2 - i_3}$$

where K is fixed by the B arc. Clearly,

$$\frac{dv_+^*}{di_1} = \frac{v_+^*-v_+(i_1)}{i_2 - i_1} \quad \text{and} \quad \frac{dv_-^*}{di_3} = \frac{v_-^*-v_-(i_3)}{i_2 - i_3} ,$$

We claim the numbers $\frac{dx}{di_1}$ and $\frac{dx}{di_3}$ are different along the curve $i_1 = i_3$, at least for some values of i_1. To see that this is so, note that in the limit $i_2 \to i^*$, the family of admissible B arcs shrinks to the one arc with $K = 0$ (provided $F(v_+(i^*)) = 0$). On this arc,

$$\frac{dx_\pm}{di_k} = \frac{v_\pm^*-v_\pm(i_k)}{-i_k\sqrt{2F(v_\pm^*)}} = \frac{f(v_\pm(i_k))}{-i_k^2\sqrt{2F(v_\pm^*(i_k))}} \qquad k = 1,\ 3.$$

A direct calculation shows that for i near i^*,

$$\frac{F(v_\pm)}{\sqrt{F(v_\pm^*)}} \sim \frac{i^* - i}{\sqrt{f'(v_\pm(i^*))}}\left(1 + \frac{5}{12}\frac{f''(v_\pm(i^*))}{f'^2(v_\pm(i^*))}(i-i^*) + \ldots\right).$$

If $f(v)$ has either $f'(v_+(i^*)) \neq f'(v_-(i^*))$ or $f''(v_+(i^*)) \neq f''(v_-(i^*))$ then on the arc $K = 0$, $i_2 = i^*$, and for all nearby B arcs

$$\frac{dx}{di_1} \neq \frac{dx}{di_3} \quad \text{at} \quad i_1 = i_3.$$

In other words, for i_2 near i^*

$$\left.\frac{\partial d^*(i_1,i_2,i_3)}{\partial i_1}\right|_{i_1=i_3} \neq \left.\frac{\partial d^*(i_1,i_2,i_3)}{\partial i_3}\right|_{i_1=i_3}$$

which establishes the assymetry of d^*.

Finally, since we are concerned with closed loops, we would like to have one way block on regions whose limits as $x \to \pm\infty$ are the same. To this end we have

Lemma 2.7: Suppose the current profile $i_0(x)$ given by (2.9) has one way block. Then the current profile

$$j_0(x) = \begin{cases} i_1 & -\infty < x < -d \\ i_2 & -d < x < 0 \\ i_3 & 0 < x < d_1 \\ i_1 & d_1 < x < \infty \end{cases}$$

also has one way block for all d_1 sufficiently large. We leave the proof of this statement to the reader.

3. A SIMPLIFIED MODEL

The geometry of infarction can have a dramatic inpact on how waves of excitation travel, and may lead to one way block. We now use this observation to motivate a simple model for reentrant waves.

Suppose we view the conducting pathway of the ventricles as a closed loop (with infarcted interior) with one way block at some point of the loop, stimulated by a periodic input at some point on the loop near the block. A rendering of this idealization is shown in Figure 4.

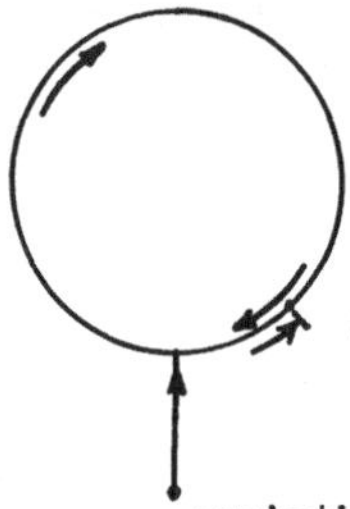

Figure 4

Suppose the oscillatory cell fires (becomes excited) at time $t = t_n$. If nothing else interferes, it will recover and fire again at time $t = t_{n+1} = t_n + T$ where T is the natural period of oscillation. However, because of wave propagation around the loop, the impulse will travel around the loop and try to restimulate the oscillatory cell. If the loop is too small, or the speed of propagation too fast, the excitation will return while the oscillatory cell is still refractory and there will be no effect. If the excitatory wave returns after the refractory period ends and before the next natural oscillation, the cell will be fired prematurely and a self sustained oscillation is established.

We can express this as

$$t_{n+1} = \min \begin{cases} t_n + T \\ t_n + \Delta \quad \text{provided} \quad \Delta > R \end{cases}$$

where Δ is the propagation delay and R is the minimum recovery time of the oscillatory cell.

The delay Δ depends upon the period of stimulation. It is known [18] that for periodic travelling waves in FitzHugh Nagumo equations, there is a relationship between frequency of oscillation and speed of propagation $c = c(\omega)$ along the fiber. In the usual situation the function $c = c(\omega)$ is a decreasing function of frequency ω in the range $0 < \omega < \omega_0$. Using this dispersion relation we calculate the delay time for propagation of one pulse of a periodic wave train

$$\Delta = \frac{L}{c(\frac{1}{\Delta T})} = L\delta(\Delta T) \tag{3.1}$$

where L is the distance of propagation and $\Delta T = \frac{1}{\omega}$ is the period of oscillation. The function $\delta(\Delta T)$ is (typically) a monotone decreasing function of ΔT on the semifinite interval $\frac{1}{\omega_0} \leqslant \Delta T < \infty$. Using this dispersion relation in a kinematic way [19] we find

$$\Delta t_{n+1} = t_{n+1} - t_n = \min\begin{cases} T \\ L\delta(\Delta t_n) \quad \text{provided} \quad L\delta(\Delta t_n) > R. \end{cases} \tag{3.2}$$

Equation (3.2) is a map of $\Delta t_n \to \Delta t_{n+1}$ which we can study graphically as a function of the parameters T, L.

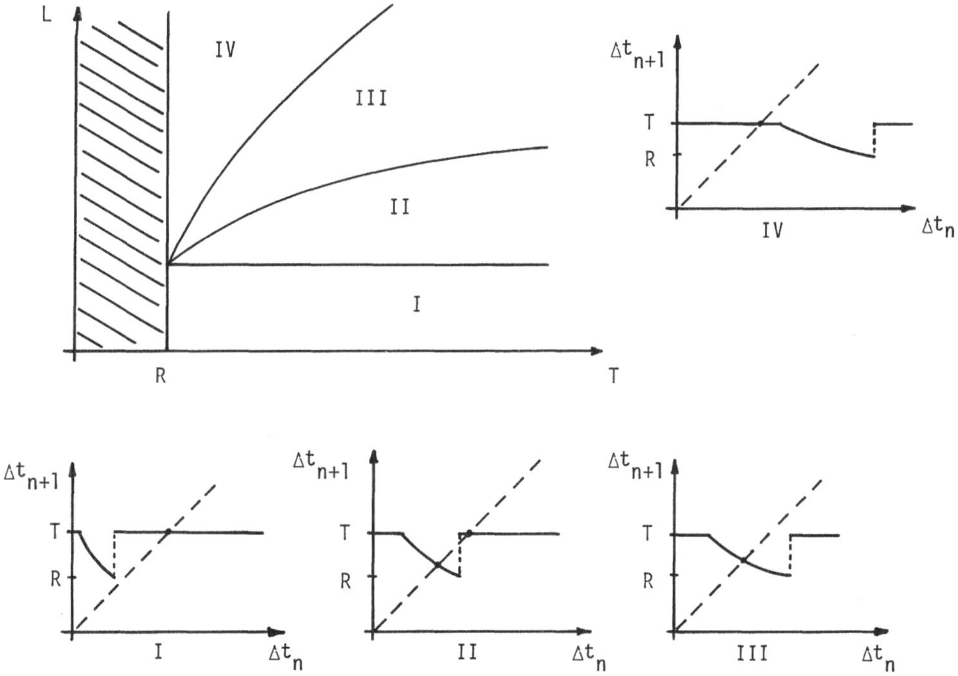

FIGURE 5

In Figure 5 we show the L-T parameter space in the case $R = \frac{1}{\omega_0}$. There are four regions of parameter space to consider, and the map $\Delta t_n \to \Delta t_{n+1}$ is shown for each region. In region I, there is a unique stable steady state corresponding to periodic stimulation at period T. In this region the heart functions properly. In region II, there are two steady states, the first corresponding to periodic excitation at period T and the second corresponding to a self sustained wave travelling continuously around the loop with significantly smaller period. In region III, the circulating wave is the only steady state solution.

With this parameter space in mind consider the following scenario. Suppose an individual acquires a small infarction. As the infarction grows, its circumference increases, and, in all likelihood, the conduction speed of the surrounding tissue decreases, increasing the effective length of the conduction loop L. As L increases, the patient moves from regionI into region II. This transition from region I to region II produces no noticeable change as the steady solution remains unchanged. Once in region II, a gradual increase in heartrate, due to exercise or deteriorating efficiency of the heart, may move the patient from region II into region III. The transition from region II to region III produces a dramatic change as the steady state corresponding to normal heartbeat suddenly disappears, and the only steady solution is the circulating (fibrillating) solution. Furthermore, efforts to restore the normal heartbeat by increasing T have no effect, since there is a hysteresis. That is, even with an increase in T, the circulating steady solution persists. To eliminate the circulating solution one must move back into region II and then do something dramatic (like defibrillation) to move back to the steady solution corresponding to normal heartbeat.

In conclusion, according to this model, certain infarcted conditions can lead to a strong hysteresis between regular propagation of waves of excitation and fatal, circulating waves.

REFERENCES

1. M. A. Allessie, F. I. M. Burke and F. J. G. Schopman, Circus movement in rabbit atrial muscle as a mechanism of tachycardia, Circ. Res. 41, 9-18 (1977).

2. D. G. Aronson and H. F. Weinberger, Nonlinear diffusion in population genetics, combustion and nerve propagation, Lecture Notes in Mathematics, 446, 5-49 (1975) Springer.

3. D. S. Cohen, J. C. Neu, and R. R. Rosales, Rotating spiral wave solutions of reaction-diffusion equations, SIAM J. Appl. Math., 35, 536-547 (1978).

4. P. F. Cranefield, The Cardiac Impulse, Futura, New York, 1975.

5. R. FitzHugh, Impulses and physiological states in theoretical models of nerve membrane, Biophys. J. 1, 445-466 (1961).

6. J. M. Greenberg and S. P. Hastings, Spatial patterns for discrete models of diffusion in excitable media, SIAM J. Appl. Math., 34, 515-523 (1978).

7. A. L. Hodgkin and A. F. Huxley, A quantitative description of membrane current and its application to conduction and excitation in nerve, J. Physiol. 117, 500-544 (1952).

8. J. P. Keener, Waves in Excitable Media, Stud. Appl. Math., 39, 528-548, (1980).

9. J. P. Keener, Analogue Circuitry for the van der Pol and FitzHugh Nagumo Equation, IEEE Trans. Systems, Man. and Cybernetics, to appear, 1983.

10. V. I. Krinsky, Mathematical models of cardiac arrhythmias (spiral waves), J. Pharm. Theur. Ser. B, 3, 539-555 (1978).

11. R. E. McAllister, D. Noble and R. W. Tsien, Reconstruction of the electrical activity of cardiac Purkinje fibers, J. Physiol. (London) 251, 1-59 (1975).

12. H. P. McKean, Nagumo's Equation, Adv. Math., 4, 209-223 (1970).

13. R. N. Miller, A simple model of delay, Block and one way conduction in Purkinje fibers, J. Math. Biol., 7, 385-398 (1979).

14. G. R. Mines, On circulating excitations on heart muscles and their possible relation to tachycardia and fibrillation, Trans. R. Soc. Can., 4, 43-53 (1914).

15. G. K. Moe, W. C. Rheinboldt and J. A. Abildskov, A computer model of atrial fibrillation, Am. Heart J., 67, 200-220 (1964).

16. J. Nagumo, S. Arimoto and S. Yoshizawa, An active pulse transmission line simulating nerve axon, Proc. IRE, 50, 2061-2070 (1962).

17. J. Pauwelussen, One way traffic of pulses in a neuron, J. Math. Bio., 15, 151-171 (1982).

18. J. Rinzel, Repetitive nerve impulse propagation: Numerical results and methods, in Nonlinear Diffusion (ed. W. E. Fitzgibbon and H. F. Walker), Pitman, 1977.

19. J. Rinzel, Models in Neurobiology, Lect. Appl. Math., 19 281-297 (1981), AMS.

20. A. C. Scott, Active and nonlinear propagation in Electronics, Wiley-Interscience, NY (1970).

21. M. Shibata and J. Bures, Reverberation of critical spreading depression along closed-loop pathways in rat cerebral cortex, J. Neurophy., 35, 381-388 (1973).

22. G. R. Stibitz and D. A. Rytand, On the path of the excitation wave in atrial flutter, Circulation, 37, 75-81 (1968).

23. N. Wiener and A. Rosenblueth, The mathematical formulation of the problem of conduction of impulses in a network of connected excitable elements, specifically in cardiac muscle, Arch. Inst. Cardio. Mexico, 16, 206-265 (1946).

24. A. T. Winfree, Rotating solutions to reaction-diffusion equations in simply connected media, SIAM-AMS Proc., 8, 13-31 (1974).

25. A. T. Winfree, Sudden Cardiac Death, A Problem in Topology, Sci. Am., 248, 5, 144-161 (1983).

HEAD REGENERATION IN HYDRA : BIOLOGICAL STUDIES AND A MODEL

Wolfgang Kemmner

Universitaet Heidelberg,
Institute of Zoophysiology,
Heidelberg, FRG.

ABSTRACT

The study of head regeneration in hydra should lead to insights into the more general mechanisms of pattern formation. From the tissue of hydra, two morphogenetic substances have been isolated, the head activator and the head inhibitor, which seem to have high significance for the regulation of such processes. Studies of the changes of the concentrations of head inhibitor and head activator during hydra head regeneration have shown that free head inhibitor blocks its own release from sources and that of head activator as well. On the basis of this feedback mechanism a system of differential equations was formulated which describes the changes of the free and bound substances during regeneration in a computer simulation.

Address for correspondence:
Universitaet Heidelberg, Institut fuer Zoologie II,
Im Neuenheimer Feld 230, D-6900 Heidelberg, FRG.

INTRODUCTION

One of the most interesting aspects of ontogenesis in higher organisms is the generation of form starting from an apparently formless origin. Out of one cell, the egg, a multitude of different structures develops. Two mechanisms seem to be of fundamental significance for this development: the expression of the genetic information on the molecular level which leads to molecular differentiation of the cells on one hand and the formation of a spatial pattern of the arrangement of the cells. When disturbed, this development of spatial patterns is subject to considerable regulation which requires intercellular communication about the position of the cell within the whole organism.

A biological system that shows all these basic self-regulating properties in a relatively simple fashion is the small fresh-water polyp hydra, which has long been known for its striking regenerative capacities (Trembley,1744). Hydra consists essentially of a cylindrical tube, 3-4 mm long, with a hypostome and tentacles at the apical end and a sticky foot at the basal end. The cells of the gastric region are not terminally differentiated, but are able to regenerate head or foot-like structures after removal of the respective end. Any tissue, cut out of the body column, regenerates a new animal with head and foot, preserving the original polarity, and it does so without excessive growth. Grafting and transplantation experiments which were done with the intention to elucidate the underlying mechanisms led to the formulation of a concept of pattern formation in terms of gradients of activation and inhibition (Webster& Wolpert,1966; MacWilliams,1983a,b). As the ability of tissue pieces to induce a head after transplantation to a host animal is highest in tissue close to the head, the existence of a gradient of head activation within the animal is assumed (Figure 1). Head formation in the host, on the other hand, is lower at sites close to an existing head, which is interpreted as resulting from an underlying gradient of inhibition extending from the head of the host.

That gradients are involved in pattern formation has been discussed since long (Child,1941), but its material basis remained undisclosed. The most simple case may be a gradient of morphogenetic substances giving a one-to-one correlation between substance concentration and position to specify positional information (Wolpert,1969) within the tissue. Wolpert´s theory states that cells in a developmental system may have their position specified with respect to the ends of the system. This positional information determines the way of molecular differentiation of the respective cell. Thus there are clearly defined positional values which determine expression of genetic information leading for example to head formation. In a model of hydra regeneration Wolpert (1974) defines two gradients, one consisting of an inhibitor S which is diffusible over the range of the system and a second one which should be a stable cellular parameter P. Following removal of the head the concentration of S falls by diffusion until it is at a critical concentration below the value of P, which in turn leads to new synthesis of P until its value reaches the original positional value of the head end.

The model of lateral inhibition by Gierer&Meinhardt (1972) emphasises the significance of mechanisms of auto- and crosscatalysis interactions of morphogenetic substances for pattern formation. Two diffusible substances are postulated, a short-range activator and a long-range inhibitor with mutual interaction on their respective rates of production from sources. Due to auto- and crosscatalytic properties of these substances small variances of a homogeneous distribution get enhanced which process results in the formation of gradient-like distributions of the concentrations of both substances. Removal of the head leads to the formation of an activated zone close to the cut surface which determines site and size of the new head. Recently MacWilliams (1982) used a proportion-regulating version of this model for simulations of hydra head regeneration, which underlined the high relevance of injury effects for this process.

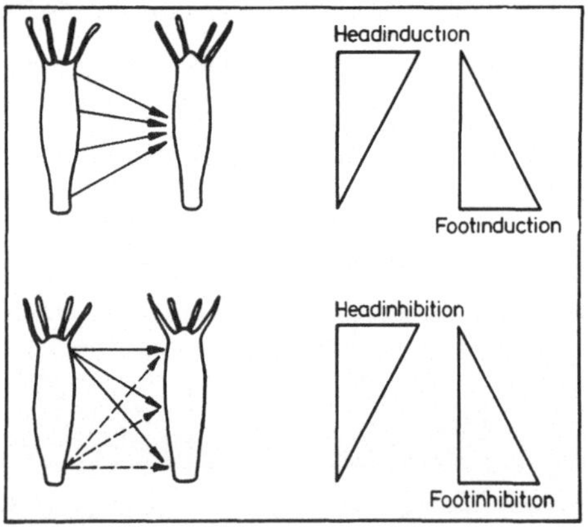

Figure 1

Schematic representation of some transplantations experiments. In the experiment depicted at the top the position in the donor from which the implant was taken is varied, at the bottom the position in the host is varied. (taken from Grimmelikhuijzen&Schaller,1976)

Biochemical work now has revealed the existence of four substances in tissue of hydra which act upon head and foot regeneration in very low concentrations. These are the head activator (Schaller&Bodenmueller,1981) a hydrophobic peptide (1124 d) and the head inhibitor (Schaller et al,1979), a small hydrophilic molecule (< 500 d), which is the major component of Berking's inhibitor (Berking,1974,1977; Schaller,1984). Foot regeneration, which is not further discussed in this contribution, is thought to be controlled by the foot activator (Grimmelikhuijzen&Schaller,1977) and the foot inhibitor (Schmidt&Schaller,1980). Within the animal the head substances are stored in a structure-bound form, probably in nerve cells (Schaller& Gierer,1973; Berking,1977), from which they are released e.g. during regeneration (Kemmner&Schaller,1983). These investigations raised the question whether it would be possible to construct a model of hydra head regeneration based on the properties of these "real" substances. First our experimental knowledge concerning the role of the head activator and head inhibitor during regeneration will be summarised, and then I will describe a model (Kemmner,1983) which tries to fit such data.

Morphogen	Molecular Weight	Nature	Purification (X-Fold)	Active Concentration	Gradient
Head Activator	1142	Peptide	10^9	10^{-13} M	
Head Inhibitor	<500	Non-Peptide	10^5	$< 10^{-9}$ M	
Foot Activator	~1000	Peptide	10^5	$< 10^{-9}$ M	
Foot Inhibitor	<500	Non-Peptide	10^4	$< 10^{-8}$ M	

Table 1

Properties of morphogenetic substances from hydra controlling head and foot formation.

MORPHOLOGICAL STUDIES OF HEAD REGENERATION

Removal of a hydra´s head leads to a sequence of events, which finally results in the regeneration of a fully restored head 48 hours later. At the cellular level, following closure of the wound within 20 minutes (Ham&Eakin,1958) after removal of the hypostome and the tentacles, there is a rapid decrease in the mitotic rate of the interstitial cells. After a certain recuperation period the mitotic index starts to increase again (Park et al,1970). Determination of interstitial stem cells to become nerve cells of the future head is initiated 2-5 hours after cutting (Schaller,1976b).

During this early period the morphogenetic properties of the head regenerating tissue changes too. That activation increases has shown by transplantation experiments. MacWilliams (1983a) finds that a piece of head regenerating tissue is able to induce a secondary head in the gastric region of a host animal if it is transplanted 4-6 hours after removal of the donor´s head. The same piece fails to induce a head if it is transplanted at time 0. A second result states (MacWilliams,1983b) that the head itself inhibits formation of a secondary head. Removal of the head leads to a drastical reduction of head inhibition within the gastric tissue after 6 hours. These results indicate that both at the cellular and at the morphogenetic level decisive changes occur during the very early hours after cutting, which lead to the reprogramming of the gastric tissue to become a head.

The first visible changes of regeneration appear 24 hours after cutting with the outgrowth of tentacles. Head regeneration is fully accomplished after 48 hours.

THE ROLE OF MORPHOGENETIC SUBSTANCES FOR REGENERATION

The head activator as well as the head inhibitor act at concentrations below 10^{-9} molar (Table 1). Their effects are measured as activation or inhibition of head formation as expressed by an increase or decrease in the average tentacle number of animals regenerated in medium with added substance. A comparison of the low concentrations needed for action with a hydra´s content in these substances, shows that under normal conditions only a very small portion (nearly 1:10000) of both substances is in the free and active form, but that they are stored almost totally in a structure-bound and inactive form. Both substances occur in nerve cells whose density is high in the head and decreases basally. From these sources head activator and head inhibitor are released under certain conditions e.g. after feeding, osmotic shock or during regeneration. Initiation of head regeneration is characterised by extensive release of both substances from their sources (Kemmner& Schaller,1983). Already 30 min after removal of the head the content of bound head inhibitor within the animal has decreased to about 50% of its original value (Figure 2). The decrease in the mitotic rate of interstitial cells at the same time indicates a high level of free inhibitor in the tissue of the regenerating animal. These results suggest extensive release of head inhibitor after removal of the head.

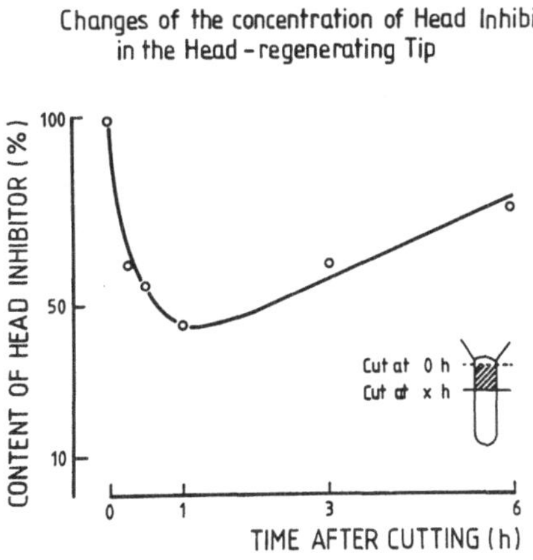

Figure 2
Changes of the content of head inhibitor in the upper part (apical third) of animals regenerating a head. The amount of head inhibitor in the upper part of regenerates at time 0 was taken as 100%.

The amount of head activator is also changing during head regeneration (Schaller,1976a), but more slowly. Three hours after removal of the head the content of head activator decreases to about 50% of its original value. Medium collected between 1 and 4 hours contains only head activator. The later appearance of head activator in the medium may be due to slower diffusion and degradation or to a somewhat delayed release.

We thus find that removal of the head provokes an extensive release of head activator and head inhibitor from the sources close to the cut surface. If such a process is essential for the initiation of regeneration, control mechanisms are necessary. A possible candidate for such control functions is the inhibitor. Incubation of regenerating animals in head inhibitor prevents release of head activator and of head inhibitor from the sources (Kemmner&Schaller,1983). From these results we postulate an inhibitory effect of the head inhibitor on its own release and that of the head activator as well. The fact that more head inhibitor is required to block its own release than for a blockage of head activator release may explain why head inhibitor is released faster than head activator.

Incubation of head regenerating animals in head activator shows that head activator has no effect on the release of head inhibitor (Kemmner& Schaller,1983), a finding which excludes a possible crosscatalytic action of head activator on the release of head inhibitor.

At the cellular level the head activator functions as one of the substances which control determination of stem cells (Schaller,1976b). In the presence of head activator determination of interstitial cells to nerve cells is stimulated, determination to nematocytes suppressed. Maximal increase in new nerve cells is obtained if regenerating animals are incubated in head activator between 0 to 5 hours after removal of the head (Schaller,1976b). A second property of the head activator is its action as growth hormone, which induces proliferating cell types to divide (Schaller,1976c). The head inhibitor acts antagonistically by suppressing the determination to nerves and inhibiting mitosis (Berking,1979; Schaller,1976a and unpublished). From these results one may assume that the determination of nerve cells depends on the local ratio of free head activator over free head inhibitor.

Thus it appears that hydra head regeneration depends on two processes which are differently regulated (Figure 3). A fast one is the release of head activator and head inhibitor after removal of the head, which is controlled by self- and cross-inhibitory actions of the free head inhibitor. A slow one is the restoration of the head region as indicated by the determination and differentiation of new nerve cells, which in turn are sources of head activator and head inhibitor. Source restoration depends on the presence of head activator which in this way acts autocatalytically on the production of its own producer cells.

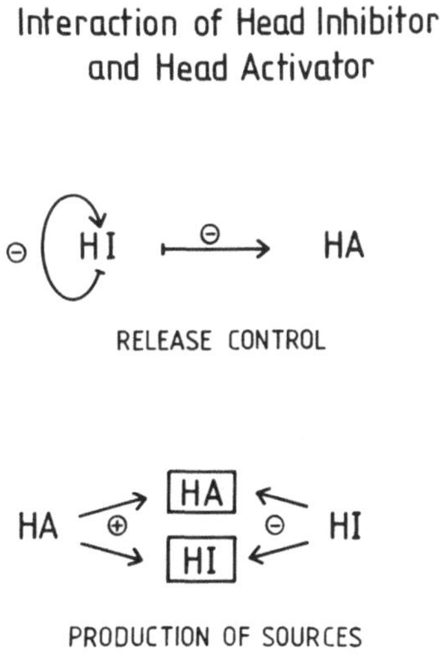

Figure 3

FORMULATION OF A MODEL

In this chapter a model will be presented which has been developed along the described properties of the head activator and the head inhibitor (Kemmner,1983). In accordance with the experimental data a basic feature of the model is its distinction between temporal changes of the structure-bound and free portions of the substances, expressed by two sets of equations.

The concentrations of the free head inhibitor I and the free head activator A depend on their release from the sources S_I, S_A respectively. The extent of the release is controlled by the concentration of free head inhibitor, with some time delay T. In reference to work from Nordstroem & Bartfai (1982), Bartfai (1982), about presynaptic inhibition via auto-receptors, the assumption is made that it takes some time until a change of the signal - the concentration of free head inhibitor around receptor sites - may cause a change in the extent of the release. This transmission of the signal may involve changes in the activity of enzymes along a cascade, which includes adenylate cyclase, protein kinase and protein phosphatase. By activation of a molecule of adenylate cyclase due to binding to the autoreceptor, generation of many molecules of cAMP is initiated, which may also cause a large amplification of the signal (Koshland et.al,1982).

Release is limited by the amount of stored substances within the sources. For amplification of small changes of the concentration of the free head inhibitor a power function k is used. The drop of I under a certain threshold should lead to release. This was taken in account by using the formula $1 / (1 + I^k)$, $k>1$, for release. Activator and inhibitor are removed according to first order kinetics with constants b and d, e.g. by enzyme degradation, and diffuse with constants D_I, D_A.

$$\frac{\partial I}{\partial t} = \frac{S_I}{1+I(t-T)^k} - bI + D_I \frac{\partial^2 I}{\partial x^2} \qquad (1a)$$

$$\frac{\partial A}{\partial t} = \frac{S_A}{1+I(t-t)^k} - dA + D_A \frac{\partial^2 A}{\partial x^2} \qquad (1b)$$

The ends of system (1) are assumed to be nonpermeable for the diffusing substances, except after removal of head or foot, or after incisions elsewhere. Then diffusion is permitted from the respective end for a certain time by using Dirichlet boundary conditions.

Sources are depleted upon release and degradation v, and restored depending on the ratio of free head activator to free head inhibitor. The production of new sources as determined by the local ratio of A to I, which can be enhanced using the power function $j \leq 1$, needs some maturation time W. Therefore the changes of the sources can be written as:

$$\frac{\partial S_I}{\partial t} = z \frac{A(t-W)^j}{I(t-W)^j} - \frac{S_I}{1+I^k} - vS_I \qquad (2a)$$

$$\frac{\partial S_A}{\partial t} = z \frac{A(t-W)^j}{I(t-W)^j} - \frac{S_I}{1+I^k} - vS_A \qquad (2b)$$

SIMULATIONS OF THE MODEL

a) Selection of parameter values

The choice of parameter values is subject to several constraints. Highly relevant is the ratio of free to bound substances. Under normal conditions only a very small portion of the inhibitor occurs in the free form. Thirty minutes after decapitation most of the sources of the head inhibitor close to the cut surface are depleted (Kemmner& Schaller,1983). That means that the concentration of free head inhibitor increases dramatically (nearly 10000 fold) during the first minutes of head regeneration within the region near the cutsurface. Such an explosive release cannot be triggered on the molecular level alone e.g. by a bimolecular reaction, which would be expressed with values of $k=2$ in equation 1. Instead of this, neural processes which lead to release have to be assumed. A possible mechanism is the drop of the concentration of the free inhibitor under a certain threshold which triggers fusion and subsequent release of a large number of vesicles containing inhibitor and activator. Such threshold mechanisms are abundant in neural systems. High values of the power function k ($k \geq 12$) of the concentration of free inhibitor yield the low level of free substances under steady state conditions as well as the extent of the release which we expect from experimental data. Selection of high values for k does not change the qualitative properties of the model but changes its quantitative predictions. Instead of such high values of k the use of an exponential function seems to be more appropriate (in preparation). Thus a threshold mechanism is postulated which allows a different extent of release of the substances according to the concentration of free inhibitor.

Selection of diffusion and decay parameters was done by evaluation of experimental data. No direct measurements of the propagation of the head activator or the head inhibitor within the animal are available. A diffusion constant of 10^{-5} cm^2/sec for the head inhibitor (such as for amines) is assumed, and for the head activator a diffusion constant which is some orders of magnitudes lower. These selections are supported by data about the chemical properties of the molecules. The head inhibitor is a small hydrophilic molecule and therefore should be more diffusible than the hydrophobic head activator, which absorbs easily to all types of structures including membranes.

The total amount of head inhibitor decreases rapidly during the first hours after decapitation, suggesting that the amount of free head inhibitor must decrease at least as rapidly. Results of Schaller (1976a) show a slower decrease of the amount of head activator. One interpretation of these results states that the decay of free head inhibitor is faster than the decay of free head activator. Therefore in the simulations a median decay of 30 minutes for the head inhibitor and of 90 minutes for the head activator is used.

b) Course of the simulation

The first step in the simulation is the calculation of steady state values of the free substances on the basis of the assumed values of the sources. According to the gradient of sources release of both substances is maximal in the head. Due to its smaller diffusion range the slope of free head activator is steeper than the slope of the free head inhibitor, which leads to determination of new sources preferentially in the head.

Removal of the head is simulated by eliminating the first segments (Figure 4a-c). Cutting provokes leakage of the free substances from the tissue of the animal. Due to the decreasing concentration of free head inhibitor in the region close to the cut surface the blockage of release is eliminated and extensive release of head activator and head inhibitor occurs. The extent of the release depends on the concentration of the free head inhibitor and the local source values. Diffusion of inhibitor from adjacent segments can partially compensate the loss by leakage and therefore diminish release.

Different diffusion ranges of both substances lead to a spreading of the newly released head over the entire tissue, and an accumulation of the head activator near the cut surface. The distribution of the free inhibitor reaches a new steady state 3 hours after cutting. Then the ratio of free activator to free inhibitor is high in the region close to the cut surface but low in regions more distant to the head. Therefore determination of new sources occurs mainly in the region of the presumptive head. In this way regeneration of the gradient of sources is assured. Removal of both head and foot results in release of the substances at both cut surfaces. Different local ratios of the free substances due to the different concentrations of the activator in the apical and basal part and the spreading of the inhibitor over the tissue, leads to restoration of the original polarity.

DISCUSSION

The model predicts the release of head activator and head inhibitor after any cut as locally restricted to the region near the cut surface which is in accordance with the experimental data. A minor depletion of sources due to release occurs after each cut but its extent is higher after removal of the head - 70 % versus 98 % after removal of the foot. Therefore the specificity of regeneration after regionally different cuts depends not only on the specificity of the release, expressed as an all or none mechanism, but on a different extent of release and on regulation of head determination on the source level. With the model it is possible to simulate depletion and subsequent restoration of the gradient of sources after various types of amputations as expected from experimental data.

It is assumed that the gradient of the sources is restored and maintained by an autocatalytical effect of the head activator on the production of cells which in turn produce head activator and head inhibitor. This point is similar to the autocatalytical term in the model of lateral inhibition of Gierer&Meinhardt, insuring that whereever a head has formed more head-specific substances will be produced. Since the primary event of pattern formation is thought to happen on the level of release control, in the present model the amplification of sources is of minor importance for the pattern formation process during regeneration.

The main function of the sources seems to be to act as a memory of the original polarity, depending on the slope of the gradients of sources. A consequence of this hypothesis concerns mutants with altered morphogenetic behaviour (Sugiyama & Fujisawa,1977;Sugiyama,1982). Slope and values of the sources should be changed in such mutants. Because source gradients for several mutants have been determined (e.g. Kemmner& Schaller,1981), simulations of the mutants´ properties should be easily accessible.

The basic feature of the model is the inhibition of the release of two morphogenetic substances from sources by one of them, i.e autoinhibition. This seems to be in sharp contrast to other models of pattern formation like that mentioned above which postulates self enhancement via autocatalysis. On a second look principal similarities are apparent. The drop of the inhibitor below a certain threshold eliminates blockage of release resulting in an explosive release and corresponding increase of the inhibitor´s concentration shortly after decapitation. Thus self enhancement is a necessary prerequisite for this model too. As an outcome, our studies therefore suggest that self enhancement may be realized in hydra on the basis of an autoinhibition of release.

ACKNOWLEDGEMENT

The work was carried out in the laboratory of HC Schaller and supported by the Deutsche Forschungsgemeinschaft (SCHA 253/8). I am very indebted to Dr. Schaller for close co-operation while writing this contribution.

Figure 4

The figures are designed along the following scheme: in the vertical axis the concentration of the respective substance is depicted (logarithmic scale for the free substances), in the horizontal axis the position within the animal. The different graphs within each figure represent distributions of the substances after several time periods. In figure 4 simulations of head activator and head inhibitor concentrations during regeneration of the head are depicted.

4a) The concentration of free head activator in the most apical segment increases from a steady-state value (graph 1, broken line) of 8 units to 7000 units (graph 3), and decreases again due to decay three hours later (graph 7). The smaller increase between six hours (graph 8) and 28 hours (last graph) depends on the restoration of the sources.

4b) The concentration of free head inhibitor rises from 2 units (graph 1) to 220 units after removal of the head (graph 3). Free head inhibitor gets leveled all over the whole animal after 48 minutes (graph 6). After decay of the newly released head inhibitor its concentration reaches a new stable state at three hours (graph 7).

4c) The sources of head activator and of head inhibitor are depleted due to release in tissue close to the cut surface (graph 3). Restoration of the sources starts nine hours later (graph 9), and is complete at twelve hours (graph 10).

For the simulations of source restoration the square of the ratio, j=2, was used. Simulation parameters were b=0.0001, d=0.0004, D_I=0.1, D_A=0.0001, $z=5*10^{-5}$, $v=10^{-6}$.

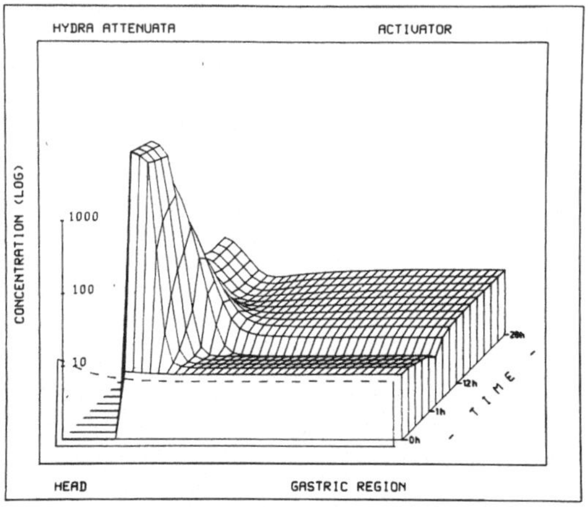

Figure 4a

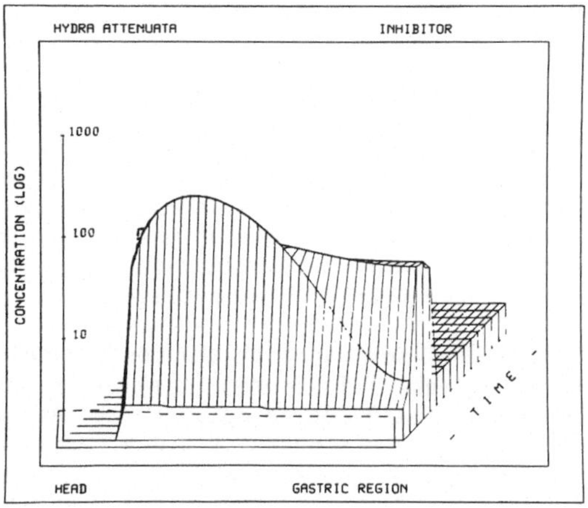

Figure 4b

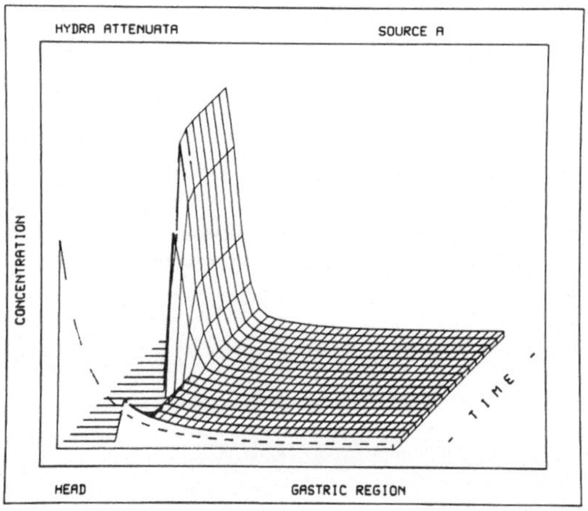

Figure 4c

REFERENCES

Bartfai T (1982) Cyclic nucleotides and the nervous system. TIPS 8 338-340

Berking S (1974) Nachweis eines morphogenetisch aktiven Hemmstoffs in Hydra attenuata und Untersuchung seiner Eigenschaften und Wirkungen. Ph.D. Thesis, Eberhard Karls Universitaet, Tuebingen.

Berking S (1977) Bud formation in hydra: Inhibition by an endogenous morphogen. Wilhelm Roux's Archives 181,215-225

Berking S (1979) Control of nerve cell formation from multipotent stem cells in hydra. J.Cell Sci. 40,193-205

Child,C.M (1941) Patterns and problems of development. Chicago: Chicago University Press

Gierer A, Meinhardt M (1972) A theory of biological pattern formation. Kybernetik 12,30-39

Grimmelikhuijzen CJP, Schaller HC (1977) Isolation of a substance activating foot formation in hydra. Cell Differ. 6,297-305

Ham RG, Eakin RE (1958) Time sequence of certain physiological events during regeneration in hydra. J.exp.Zool. 139,33-53

Kemmner W (1983) A model of head regeneration in hydra. Differentiation, in press

Kemmner W, Schaller HC (1981) Analysis of morphogenetic mutants of hydra IV. Reg-16, a mutant deficient in head regeneration Wilhelm Roux's Archives 190,191-196

Kemmner W, Schaller HC (1983) Actions of head activator and head inhibitor during regeneration of hydra. Differentiation, in press

Koshland DE, Goldbeter A, Stock JB (1982) Amplification and adaptation in regulatory and sensory systems. Science 217,220-225

MacWilliams HK (1982) Numerical simulations of hydra head regeneration using a proportion-regulation version of the Gierer-Meinhardt model. J.theor.Biol. 99,681-703

MacWilliams HK (1983 a) Hydra transplantation phenomena and the mechanism of head regeneration. I. Properties of the host. Dev.Biol. 96,217-238

MacWilliams HK (1983 b) Hydra transplantation phenomena and the mechanism of head regeneration. II. Properties of the donor. Dev.Biol. 96,239-257

Nordstroem O, Bartfai T (1982) Mechanisms of action of the muscarinic autoreceptors in the rat hippocampus. Proceedings of the 4th Meeting of the European Society for Neurochemistry, Catania.

Schaller HC (1973) Isolation and characterisation of a low molecular weight substance activating head and bud formation in hydra. J.Embryol.exp.Morphol. 29,27-38

Schaller HC, Gierer A (1973) Distribution of the head activating substance in hydra and its localisation in membranous particles in nerve cells. J.Embryol.exp.Morphol. 29,39-52

Schaller HC (1976 a) Head regeneration is initiated by the release of head activator and head inhibitor. Wilhelm Roux's Archives 180,287-295

Schaller HC (1976 b) Action of the head activator on the determination of interstitial cells in hydra Cell Differ. 5,13-20

Schaller HC (1976 c) Action of the head activator as a growth hormone in hydra. Cell Differ. 5,1-11

Schaller HC, Bodenmueller H (1981) Isolation and amino acid sequence of a morphogenetic peptide from hydra. Proc.Natl.Acad.Sci.78,7000-7004

Schaller HC (1984) The head and foot inhibitor from hydra are no Dowex-artefacts. Wilhelm Roux's Archives, in press

Schmidt T, Schaller HC (1980) Properties of the foot inhibitor from hydra. Wilhelm Roux's Archives 188,133-139

Sugiyama T (1982) Roles of head-activation and head-inhibition potentials in pattern formation of hydra: Analysis of a multiheaded mutant strain. Am.Zool. 22,27-34

Sugiyama T, Fujisawa T. (1977) Genetic analysis of developmental mechanisms in hydra. III.Characterisation of a regeneration deficient strain
J.Embryol.exp.Morphol. 42,65-77

Trembley,A. (1744) Mémoires pour servir à l'histoire d'un genre de polypes d'eau douce, a bras en forme de cornes.
J & M. Verbeek, Leyden.

Webster G, Wolpert L (1966) Studies on pattern regulation in hydra. I. Regional differences in time required for hypostome differentiation.
J.Embryol.exp.Morph. 16,91-104

Wolpert L (1969) Positional information and the spatial pattern of cellular differentiation.
J.theor.Biol. 25,1-47

Wolpert L, Hornbruch A, Clarke MRB (1974) Positional information and positional signalling in hydra.
Am.Zool. 14,647-663

PATTERN FORMATION IN ASPECT

Yehoshua Keshet* and Lee A. Segel**
Department of Applied Mathematics
The Weizmann Institute of Science
76100 Rehovot, Israel

INTRODUCTION

This volume documents many recent advances in the theory of pattern formation in physical space. It is the purpose of the present communication to continue the effort started by Levin and Segel (1982) to demonstrate how ideas about physical pattern formation can also find important application in spaces of descriptive but somewhat unconventional independent variables. Such variables have been associated with the term "aspect".

We extend here the Levin and Segel (1982) illustrative analysis of an ecological interaction where aspect is connected with the appearance of the prey. This appearance is generally of multi-dimensional character and could include facets such as color, spot size and distribution, overall shape, beak size, etc. We restrict ourselves to a single aspect variable z. For definiteness let us think of z as associated with color; thus z could be the dominant frequency of reflected light, or its logarithm.

There is evidence that predators such as birds temporarily tend to concentrate on capturing prey of a certain color. It seems that the "search image" in their brain of a particular prey type helps them locate their victims. As the prey of a given aspect become scarce, the probability grows that the predator will switch to a different type; e.g. when they have eaten most of the yellow butterflies, the birds are likely to switch to butterflies of a different color.

In the presence of predators with search image, there is a tendency for prey to evolve with a variety of aspects, for a type with a rare aspect will be relatively unmolested. The starting point of our investigation thus will be a situation in which all prey aspects are equally represented in the population. We shall show that under suitable conditions the uniform state can become unstable, resulting in a new prey distribution in which aspect clumps prevail. That is, the medley for which all colors are equally likely undergoes a transition to a state where a few dominant colors stand out.

Following Levin and Segel (1982) we shall start our characterization of the predator-prey interaction with an extremely simple situation, building by stages toward the full complexity that we wish to consider.

---

* Present address: Dept. of Geol. Sci., Brown Univ., Providence, R.I. 02912, U.S.A.
** Also at Dept. of Math. Sci., RPI, Troy, N.Y. 12181, U.S.A.

DISCRETE MODELS

An elementary model that incorporates some features of search image behavior is the following:

$$dE/dt = c - E(\nu-\mu V) \ , \quad dV/dt = rV - kVE \ . \qquad (1)$$

Here we assume that the predators or exploiters, of density $E(t)$ at time t, have a constant probability per unit time c of switching from a pool of possible prey or victims to the particular victim population under consideration [density $V(t)$]. Switching back to the pool is at a rate $\nu-\mu V$ that decreases with increasing victim population density. (We can restrict ourselves to cases where $\nu-\mu V>0$.) In the victim equation there are standard terms for the consumption of victims by exploiters (kVE), and for victim birth (rV). Analysis of steady states and their stability leads to predictions of qualitative behavior shown in Fig. 1. This behavior is expressed in terms of two dimensionless parameters, the predator pressure $P=ck/r^2$ and the switching parameter $S=\nu/r$. Relatively high values of the former lead to extinction of the victims; coexistence occurs otherwise.

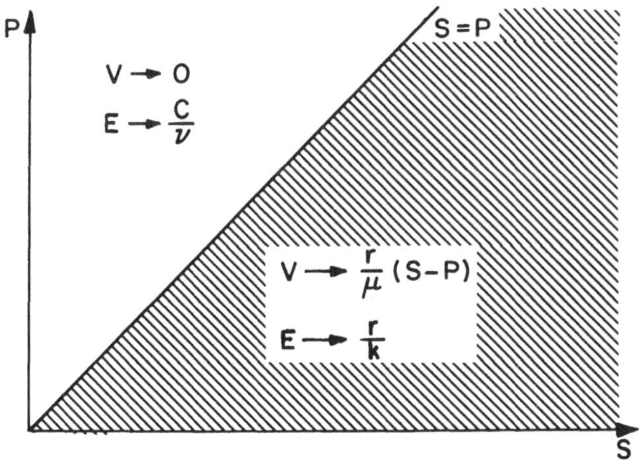

Fig. 1. Qualitative behavior of (1) as a function of predator pressure $P=ck/r^2$ and switching parameter $S=\nu/r$. Coexistence is possible in the shaded domain.

If dominant types are preyed upon preferentially, we have noted that there is an advantage in diversity of appearance. To study this matter more closely, let us postulate the existence of two types of victims, with densities $V_1(t)$ and $V_2(t)$. Let $E_i(t)$ denote the density at time t of exploiters currently searching for and eating type-i victims, $i=1,2$. Exploiters will be assumed to behave as postulated earlier, but the victim equation will be extended to incorporate new features related to mating.

Let us assume that the number of males and females is equal. If mating is random, then the probabilities that a type 1 female will mate respectively with types 1 and 2 males are in the ratio V_1/V_2. We shall assume that this ratio is $\alpha V_1/V_2$, α a positive constant, thereby taking into account the possibility of preferential or so-called assortative mating. The results of an m-n mating will be described by p^i_{mn}, the probability that such a mating will result in an i-offspring. To fix ideas, let us consider variation within a species. Then one may assume that

$$p^i_{ii} = 1 - \varepsilon , \quad p^k_{ij} = \frac{1}{2} , \quad p^j_{ii} = \varepsilon ; \quad i,j,k = 1,2 ; \quad i \neq j . \tag{2}$$

We expect that ε will be small so that offspring of like types are usually identical to their parents.

With the assumptions just described, keeping a constant reproduction rate r (average number of offspring/mating) the governing equations become the following [as in Levin and Segel (1982)]:

$$\frac{dV_i}{dt} = -kV_iE_i + rV_i + \frac{r(V_i-V_j)}{(\alpha V_1+V_2)(\alpha V_2+V_1)}\left[\frac{\alpha-1}{2} V_1V_2-\varepsilon\alpha^2V_1V_2-\varepsilon\alpha(V_1^2+V_1V_2+V_2^2)\right] ;$$

$$\frac{dE_i}{dt} = c - E_i(\nu-\mu V_i) ; \quad i,j = 1,2 ; \quad i \neq j . \tag{3}$$

We assume that $dV_i/dt=0$ if $V_1=V_2=0$.

The system (3) reduces to (1) if $V_1=V_2=V$ and $E_1=E_2=E$. In particular, (3) possesses homogeneous steady-state solutions (with equal populations of exploiters and victims) that correspond to the steady-state solutions of (1):

$$V_1 = V_2 = 0 , \quad E_1 = E_2 = c\nu^{-1}, \tag{4a}$$

and, if $S > P$,

$$V_1 = V_2 = r\mu^{-1}(S-P) , \quad E_1 = E_2 = c\nu^{-1} . \tag{4b}$$

As is illustrated in Fig. 2, the second homogeneous steady state (coexistence) is stable for a restricted domain of parameters when it is regarded as a solution of (3). This domain is described in terms of

$$L \equiv \alpha(\alpha+3)(1-2\varepsilon)/(\alpha+1)^2 . \tag{5}$$

Levin and Segel (1982) show that there are also nonhomogeneous steady states of (3), which are stable in the domain labelled B in Fig. 2 and may be stable in the shaded domain. Keshet (1981) investigated various solutions numerically. He found, as one would expect, that oscillatory instabilities of the coexistence state (Hopf bifurcations) result in temporally periodic solutions, while monotonic instabilities lead to inhomogeneous steady states. Two new parameters enter (see Fig. 3):

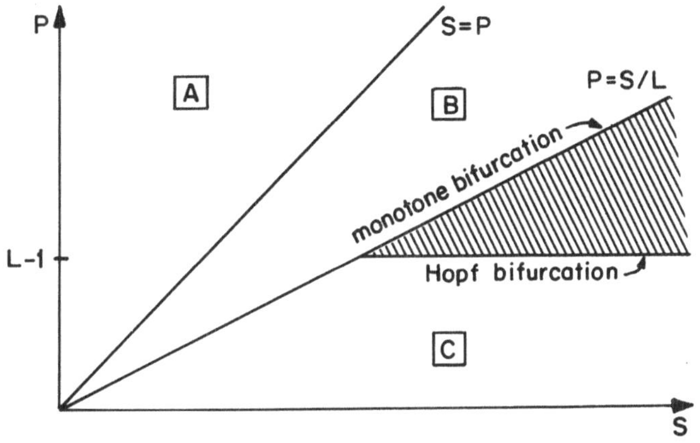

Fig. 2. Stability of the coexistence state (4b), according to the model (3), when parameter L of (5) is greater than unity. [L>1 for sufficiently great "positive assortativity" ($\alpha-1$ positive and sufficiently large) combined with sufficient reproductive fidelity (ε small enough).] State (4b) is stable in the shaded region (compare Fig. 1). Stable non-homogeneous steady states exist in B and may exist in the shaded domain. In region A the victims become extinct. [The diagram also applies to the system (9), (10) if L is given by (12).]

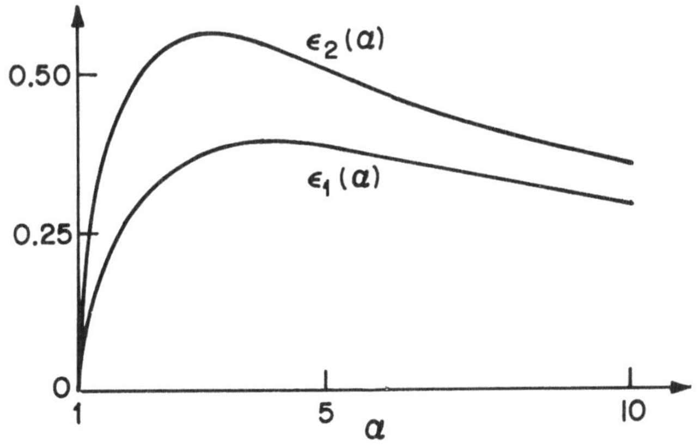

Fig. 3. Graphs of the parameters ε_1 and ε_2 of (6) as functions of the assortativity parameter α. Here $\alpha=1$ corresponds to random mating, and $\alpha>1$ to a situation where mating between like types is more probable.

$$\varepsilon_1 \equiv \frac{1}{2} - \frac{\alpha+1}{2\alpha(\alpha+3)}\left[(\alpha+1)^2+4\,\frac{\alpha-1}{\alpha+3}\right]^{\frac{1}{2}}\ , \qquad \varepsilon_2 \equiv \frac{1}{2} - \frac{(\alpha+1)^2}{2\alpha(\alpha+3)} > \varepsilon_1\ . \tag{6}$$

The results for positive assortativity ($\alpha>1$) are summarized in Fig. 4. There is a rich variety of behavior, with the possibility of multiple steady states. In particular, although the two victim types do not differ in their intrinsic properties (only in their aspect) there are parameter domains where the final state is strongly inhomogeneous, with one of the victim types predominating. (Which predominates is determined by the accident of initial conditions.)

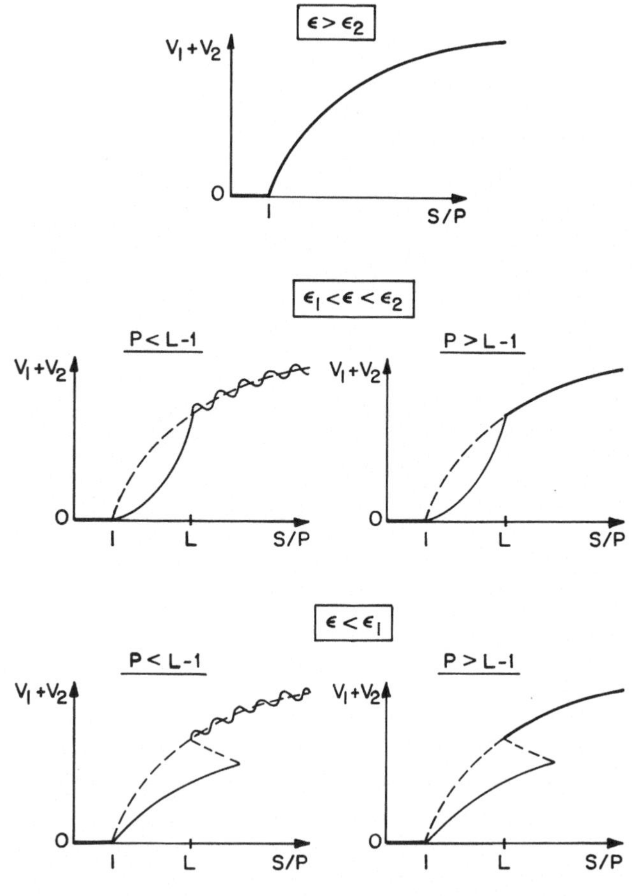

Fig. 4. Schematic presentation of possible limiting solutions of (3). ━━━ : Stable homogeneous ($E_1=E_2$, $V_1=V_2$) steady solution. ─── : Stable inhomogeneous steady solution. ---- : Unstable steady solution. ∿∿∿ : Stable periodic solution.

We remark that the stability of the steady state that yields victim extinction is difficult to investigate, for the system (3) cannot be linearized in the neighborhood of this state. Only when $S<P$ does it seem important from the ecological point of view to pursue this matter further, for otherwise no attracting final state has been revealed. When $S<P$ simulations show that initial conditions with a moderate number of victims quickly are succeeded by very low victim levels. Stochastic effects, not considered in our model, would then bring about victim extinction. For this reason, in Fig. 4 we have indicated as stable the state corresponding to (4a).

A CONTINUOUS MODEL

Following Levin and Segel (1982) we now consider situations with an arbitrarily large number of types. To this end we postulate the probability that a particular j-female will mate with a k-male to be $\alpha_{jk}V_k/\sum_p \alpha_{jp}V_p$, and we denote by $\phi_{jk\ell}$ the probability of a type ℓ offspring from a mating of this kind.

When many different victim types may occur, it is often convenient to pass to a continuous model. To this end, we assume that aspects can be linearly ordered (as may be the case for position, dominant color, mean spot size). We imagine the aspect line to be infinite, and to be composed of subintervals of uniform length. Let E_i and V_i denote the numbers of exploiters and victims in the ith compartment, where the central aspect is z_i. We define victim and exploiter densities in aspect space by

$$V(z_i,t)\Delta z = V_i(t) \quad , \quad E(z_i,t)\Delta z = E_i(t) \quad ; \quad \Delta z \equiv z_{i+1} - z_i \ . \tag{7}$$

We also define an assortativity function α and a redistribution function ϕ by

$$\alpha(z_i,z_j) = \alpha_{ij} \quad , \quad \phi(z_i,z_j,z_k) = \phi_{ijk} \ . \tag{8}$$

For simplicity we shall continue to assume that the reproduction rate r is constant, in which case passage to the limit gives the following as governing equations of the continuum model:

$$\frac{\partial E(z,t)}{\partial t} = c - E(z,t)[\nu-\mu V(z,t)] \ , \tag{9a}$$

$$\frac{\partial V(z,t)}{\partial t} = -kV(z,t)E(z,t) + r\iint V(\eta,t)V(\xi,t)\,\frac{\alpha(\eta,\xi)}{W(\eta,t)}\,\phi(\eta,\xi,z)\,d\eta d\xi \ . \tag{9b}$$

$$W(\eta,t) \equiv \int \alpha(\eta,s)V(s,t)ds \ . \tag{9c}$$

Here $-\infty<z<\infty$; all integrals in (9) and in what follows range from $-\infty$ to $+\infty$. For definiteness and simplicity, it is assumed that the frequency of encounter between organisms is a gaussian function of the "aspect distance" between them and that porgeny are gaussianly distributed about the parental mean:

$$\alpha(\eta,\xi) = (2\pi\sigma_\alpha^2)^{-\frac{1}{2}}\exp[-(\eta-\xi)^2/2\sigma_\alpha^2] , \quad (10a)$$

$$\phi(\eta,\xi,z) = (2\pi\sigma_\phi^2)^{-\frac{1}{2}}\exp[-(\frac{\eta+\xi}{2}-z)^2/2\sigma_\phi^2] . \quad (10b)$$

BEHAVIOR OF THE MODEL (9), (10)

A linear stability analysis of the uniform coexistence state (for $S>P$) in which

$$V = r\mu^{-1}(S-P) , \quad E = rk^{-1} , \quad (11)$$

continues to provide much information concerning the location of different behavioral domains. The results (Levin and Segel, 1982) for the continuous model are very similar to those for the four-compartment discrete model (3). Indeed, Fig. 2 can still be used in this case when $L>1$, but now

$$L = (2-\tilde{\alpha})\tilde{\alpha}^{\beta+\frac{1}{4}} , \quad \tilde{\alpha} \equiv \exp(-\frac{1}{2}\sigma_\alpha^2 q^2) , \quad (12)$$

where q is the assumed wave number of the disturbance.

Assortativity in the discrete model was incorporated in the parameter α; it now is measured by σ_α, the standard deviation of (10a) that characterizes how disparate in aspect mates are likely to be. Similarly, fidelity of reproduction was measured by ε in the discrete model; this quality now is reflected in the parameter σ_ϕ of (10b), a measure of how tightly the offspring are clustered about the parental mean aspect. Thus the ratio

$$\beta \equiv \sigma_\phi^2/\sigma_\alpha^2 \quad (13)$$

appears in the current expression (12) for L, instead of the quantitities α and ε that appeared in its counterpart (5).

Levin and Segel (1982) show that if

$$L_{cr} \equiv \min(P+1,S/P) < 1.27 \quad (14)$$

then when $\beta<\beta_{cr}$ there is a band of wavenumbers corresponding to growing disturbances to (11) where β_{cr} is given implicitly by

$$\left[\frac{2}{\beta_{cr}+5/4}\right]^{\beta_{cr}+\frac{5}{4}} (\beta_{cr}+\frac{1}{4})^{\beta_{cr}+\frac{1}{4}} = L_{cr} . \quad (15)$$

We used the linear stability results to guide the selection of parameters for a numerical analysis of the full nonlinear equations. One can debate as to the most appropriate numerical scheme for the integrodifferential equation of (9b), but it must be borne in mind that this equation was introduced as an approximation to a system of discrete equations governing a large number of species. For the numerical analysis, we therefore chose to return to the original discrete system. We integrated 220 ordinary differential equations, corresponding to 110 evenly spaced values of the aspect variable. Periodic boundary conditions were used, with a wavelength equal to that predicted to be "most dangerous" by linear stability theory.

One major calculation was for a parameter set where oscillatory instability of the uniform steady state (11) is predicted by linear theory. A temporally periodic solution was found, with a period within a few percent of the prediction of linear theory. [Details of this and other matters can be found in Keshet (1981).] Fig. 5 shows the results of our second major calculation, employing a parameter set for which linear theory predicts monotonic instability of (11). Here the solution settled down to a new periodic nonuniform steady state where spikes of concentrated victims and exploiters are separated by segments of aspect that are virtually devoid of victims.

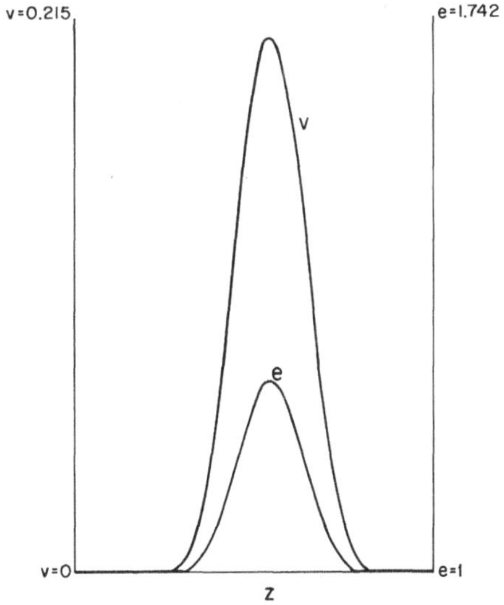

Fig. 5. Inhomogeneous steady state (Keshet, 1981) of discrete version of (9), (10) when $P=4.25$, $S/P=1.177$, $\beta=0.08$, $\sigma_\alpha=5.0$ so $\lambda_{cr}=2\pi/0.229$. The graph depicts dimensionless exploiter and victim population levels $e\equiv\nu E/c$ and $v=\mu V/\nu$.

Many questions remain open, both mathematical and ecological, but our principal purpose has been served -- to illustrate pattern formation in aspect space. Note also the strong similarities of the results we have obtained in a system with nonlocal interactions, governed by an integrodifferential equation, and more conventional pattern formation in systems of differential equations.

AN ECONOMIC EXAMPLE

One more example will be given, a fanciful one, but one that perhaps well emphasizes the possibilities of pattern in aspect. The equation to be derived is very similar to equations that have already been used by Levin and Segel (1982) in discussing how organisms might switch their location in physical or aspect space.

Let $n(z,t)$ be the number of individuals at time t that are buying toothpaste of aspect z -- where z denotes some linearly orderable property such as color or flavor. Suppose that in the absence of a tendency to switch brands, $n(z,t)$ tends to a uniform value K -- all brands are equally patronized. Assume that individuals abandon their present brand at a rate S that increases when there are fewer buyers of similar brands. (One wants to "keep up with the Joneses", and if it appears that one's normal brand is relatively unpopular then one abandons it for something different.) Let a new brand be selected according to a normal probability distribution centered on the location just abandoned. This situation can be described by the equation

$$\frac{\partial n(z,t)}{\partial t} = rn(z,t)\{1-\frac{n(z,t)}{K}\} - S[N(z,t)]\cdot n(z,t) + \int\psi(z-\xi)S[N(\xi,t)]\cdot n(\xi,t)d\xi \ , \tag{14}$$

where

$$N(z,t) = \int\theta(z-\eta)n(\eta,t)d\eta \ ,$$

and where ψ and θ are gaussians with standard deviations σ_ψ and σ_θ respectively.

By an analysis of the type needed to treat (9) it can be shown that the uniform state can become unstable if $S'<0$, that is if switching is more likely when one's present toothpaste and similar brands are relatively unpopular. By analogy with the previous example, one again expects a situation similar to that of Fig. 5. Thus the Jones effect, keeping up with the Joneses, can lead to the breakup of a market into a collection of rather isolated products -- each with its loyal following.

We hope that our two examples have convinced the reader to keep his or her eyes open for the possibility of pattern in aspect.

ACKNOWLEDGEMENTS

We thank Simon Levin for many helpful discussions. LAS was partially supported by the National Science Foundation. Much of the detailed analysis formed a portion of YK's M.Sc. thesis (Keshet, 1981) and the entire work benefitted from the support of the US-Israel Binational Fund.

REFERENCES

Keshet, Y. (1981) Numerical studies of effect of predation on aspect diversity in prey. M.Sc. Thesis, Dept. of Applied Math., Weizmann Inst., Rehovot, Israel.

Levin, S.A. and Segel, L.A. (1982) Models of the influence of predation on aspect diversity in prey populations. J. Math. Biol. **14**, 253-285.

CHEMOTAXIS AND CELL AGGREGATION

Douglas A. Lauffenburger
Department of Chemical Engineering
University of Pennsylvania
Philadelphia, PA USA 19104

Chemotaxis is a phenomenon exhibited by motile cells, in which the direction of cell movement is influenced by the concentration gradient of a chemical stimulus. Chemotactic movement is usually toward the direction of higher stimulus concentration, so that the stimulus is termed a chemoattractant. This phenomenon is exhibited by cells of both procaryotic and eucaryotic types, ranging from bacteria and protozoa to white blood cells and tumor cells.

Because chemotaxis is not a simple dispersive movement process, it has provided a natural explanation to invoke in circumstances where motile cells aggregate. One of the earliest applications of their now well-known mathematical model for chemotactic movement, by Keller and Segel (1970) was, in fact, to aggregation of the slime mold amoebae Dictyostelium discoideum. These cells normally feed on smaller bacteria, but when the environment of a population doing so becomes depleted of this nutrient source, the amoebae form dense aggregates which then develop into multicellular slugs. A fruiting body arises from each slug, releasing spore cells which then can germinate in favorable environments. This life cycle continues to repeat, depending on the environmental conditions. For more biological details, the interested reader is referred to the monograph by Bonner (1967). The interest of Keller and Segel was focused on the aggregation step, as they desired to provide an explanation for the evolution of cell aggregates from a dispersed cell population. They considered the possibility that the cells, which can migrate across surfaces, could exhibit a chemotactic response to a diffusible chemical attractant, with the cell population flux governed by the expression:

$$\vec{J}_c = -\mu \nabla c + \chi c \nabla a \qquad (1)$$

where $\vec{J}_c$ is the cell population flux, c is the cell density, and a is the attractant concentration. The two phenomenological parameters characterizing cell movement behavior are the random motility coefficient,

μ, and the chemotaxis coefficient, χ. Assuming then that the attractant is produced by the cells and is degraded in the extracellular environment, the conservation equations for the cell density and attractant concentration take the form:

$$\frac{\delta c}{\delta t} = \nabla \cdot [\mu \nabla c - \chi c \nabla a] \tag{2}$$

$$\frac{\delta a}{\delta t} = D\nabla^2 a + f(a)c - k(a)a \tag{3}$$

where D is the attractant diffusivity, f(a) is the rate of production of attractant per cell, and k(a) is the rate of degradation of attractant. Assuming no-flux boundary conditions on either a finite or an infinite spatial domain, these equations possess a uniform steady-state solution

$$c = c_o \qquad\qquad a = a_o$$

provided that c_o and a_o satisfy the equation

$$a_o k(a_o) = c_o f(a_o) \tag{4}$$

This uniform state is presumed to represent the situation at the time of nutrient depletion; the cells are neither growing nor dying, and are distributed uniformly throughout the spatial domain. Then, Keller and Segel reasoned, if this steady state were unstable with respect to nonuniform fluctuations, the uniform distribution would not be maintained and the system might evolve into a nonuniform distribution of cell density, representing aggregates. Using a standard linear stability analysis of the model equations at the uniform steady state, they derived a criterion necessary and sufficient for instability of the uniform state:

$$\frac{\chi a_o}{\mu} > 1 - \frac{k'}{k} + \frac{f'}{f} \tag{5}$$

From this inequality, a possible role for chemotaxis in aggregation is apparent: chemotactic movement can provide a destabilizing effect on uniform cell distributions. What is required for aggregation to occur is that the ratio of the chemotaxis coefficient, χ, to the random motility coefficient, μ, be large enough. This result is intuitively satisfying, because chemotaxis is not a simple dispersive phenomenon, while random motility is.

In the ensuing years, this problem has become a favorite of applied mathematicians and theoretical biologists, with a great number of papers directed to analysis of these model equations or variations on them. Among such papers, three major results can be highlighted. First, Childress and Percus (1981) showed that the instability does, in fact, evolve into a nonuniform steady state, thus bearing out the original speculation of Keller and Segel. Second, Cohen and Hagen (1981) showed that spiral wave patterns can evolve as well under certain circumstances. Such phenomena are indeed observed for some related slime mold amoebae species. Third, Schaaf (this volume) has developed the nonlinear bifurcation structure for this system, so that the solution character is known for all parameter combinations. In all of these results, the conclusion is that chemotaxis serves as an aggregative force, as large values of the ratio χ/μ lead to spatial patterns in cell density.

Because this problem and conclusion has so dominated discussion of the role of chemotaxis in cell aggregation, the purpose of this paper is to point out that other, more subtle roles are possible as well. Two examples of recent work will be presented: one is an investigation of the possible role of chemotaxis in swarming movement of myxobacteria, and the other is a speculation on a possible role of chemotaxis in localized lesions in the inflammatory response to bacterial infection. In both of these instances, chemotaxis appears to be involved in cell aggregation phenomena, but not in the obvious way that might be expected from the results for the classical slime mold amoebae problem. Thus, I hope to show that the relationship of chemotaxis to cell aggregation is much less obvious than the currently predominant notion.

MYXOBACTERIA SWARMING

Myxobacteria are a class of procaryotic microorganisms characterized by gliding motility on surfaces and the capacity to degrade other microbial cells in order to obtain nutrients for growth. These bacteria have a life cycle somewhat similar to that of the slime mold amoebae (for a recent review see Kaiser et al, 1979). When there is sufficient nutrient available, the cells grow vegetatively, dividing by binary fission. Under starvation conditions, they aggregate and construct a fruiting body, in which resting spore cells are held until being released for a new cycle of germination and growth. When the resting cells are released from fruiting body cysts, they congregate in swarms that move across the substrate surface, retaining a coherent swarm integrity. The purpose of these swarms is apparently to provide a "wolfpack" feeding effect, since the high cell density in the swarms allows high concentrations of the enzymes released by the cells for purposes of degrading microbial cells for nutrient (Rosenberg et al, 1977). Individual cells are about 6 μm long and move with a speed of about 12 μm/min. The swarms have diameters of roughly 150 to 200 μm. The swarm cell density is very great, except for a fringe of about 30 to 50 μm wide of decreasing cell density at the swarm edge. Cells in the fringe make frequent brief sorties out from the edge, only to return to the main body. Each swarm moves around as a coherent body on the substrate surface, maintaining its integrity.

Motion picture studies by Reichenbach (1974) suggest that a swarm can sense the presence of another at a distance, and that the two will move toward each other. At the edges where the two swarms are closest the fringe cells gradually move farther out from the main body and the rest of the swarm follows. In this manner the two swarms approach each other with a rate that increases from about 6 μm/hr when the initial center-to-center distance separating the swarms is about 600 μm, to approximately 36 μm/hr just before the fringes meet. Forming larger swarms clearly makes sense when the "wolfpack" feeding phenomenon is considered, so the major questions are: How are coherent swarms maintained? And how do two swarms sense and approach each other?

One simple hypothesis which provides an answer for both of these questions is that the cells continuously secrete a single diffusable chemical attractant, toward which the cells exhibit a chemotactic response. However, since the attractant must serve the dual purposes of maintaining swarm coherence and allowing swarm approach, the role of chemotaxis must be more subtle than in the classical slime mold amoebae aggregation problem. This role can be best elucidated through a mathematical modeling approach. We assume that two circular swarms with radius R are separated by a center-to-center distance d on a surface (see Figure 1). Within these swarms there are spatially uniform cell densities, n, continuously secreting a diffusible chemoattractant at specific rate q. At the edges of the swarms, we assume the existence of fringes of width h in which the cell density decreases gradually from n to 0.

Since chemical diffusion should be rapid compared to the gliding cell movement, then for any given swarm separation distance we can assume that the attractant concentration throughout the substrate (which is three-dimensional, although the cells can move only on the two-dimensional surface) is at steady state. This concentration profile can be deter-

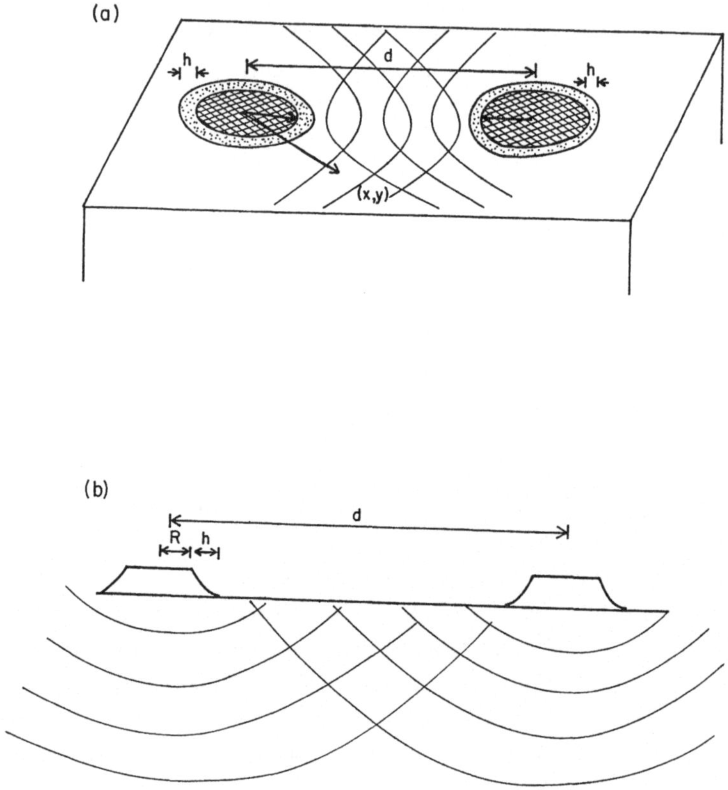

Figure 1. Illustration of model for myxobacteria swarms. (a) top view, and (b) side view, of swarms on surface of three-dimensional substratum.

mined by solving the steady-state diffusion equation in three dimensions, with attractant generation in two disks on one surface. Details of this calculation can be found elsewhere (Lauffenburger et al, in press). Figure 2 shows a plot of attractant concentration on the substrate surface (which is, of course, what is of interest to the cells) for an isolated single swarm, $d \to \infty$.

Given the attractant concentration profile, the movement behavior of the cells within a swarm fringe can be determined, assuming that cell movement is governed by the Keller-Segel flux expression, Equation (1). It should be evident that the random motility term will cause net cell movement away from the main body of the swarm, while the chemotaxis term will cause net cell movement back toward the main body. This explains the frequent brief sorties of the fringe cell. The fringe edge will remain in place, though, when these two effects are in balance. When they are not in balance, the cells will show net movement either away from or toward the main body, depending on which effect is greater.

In order to determine swarm movement, we assume that the details of cell movement within the dense main body are unimportant, and focus on the fringe cells. The main body is assumed to serve primarily as a reservoir, providing a source of more cells when the fringe widens. Thus, if we know the rate and direction of fringe movement at all points around the swarm, we will postulate that the swarm follows. Within the fringe of width h, the cell density, c, is governed by the conservation equation

$$\frac{\delta c}{\delta t} = \frac{1}{r}\cdot\frac{\delta}{\delta r}\left(r\mu\frac{\delta c}{\delta r}\right) - \frac{1}{r}\cdot\frac{\delta}{\delta r}\left(rc\chi\frac{\delta a}{\delta r}\right) \tag{6}$$

where r is the distance from the center of the swarm. We have already calculated the attractant concentration profile within the fringe, for any swarm separation distance d. If we now approximate the cell density in the fringe as a second-order polynomial, decreasing from $c = n$ at $r = R$ to $c = 0$ at $r = R + h$, we find that

$$c(r) = n\left(1 - \frac{2}{h}(r - R) + \frac{1}{h}(r - R)^2\right) \tag{7}$$

gives the cell density profile in the fringe. Substituting this profile into Equation (6), we obtain an expression for the rate of movement of the fringe edge:

$$\frac{dh}{dt} = \frac{6\mu}{h} + 3\chi\frac{da}{dr} \tag{8}$$

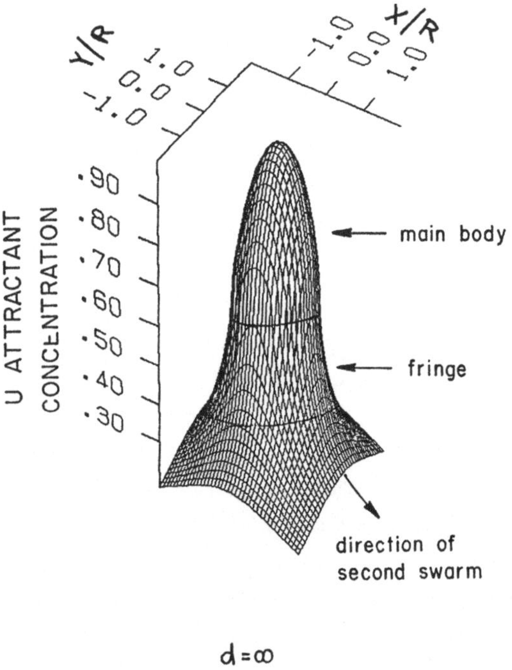

Figure 2. Attractant concentration profile for an isolated swarm (d = ∞). U is a scaled concentration.

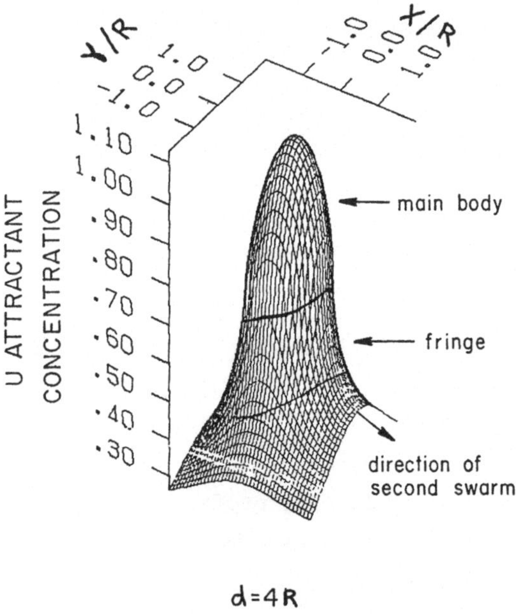

Figure 3. Attractant concentration profile of a swarm when another swarm is at a distance of 4 swarm radii. U is a scaled concentration.

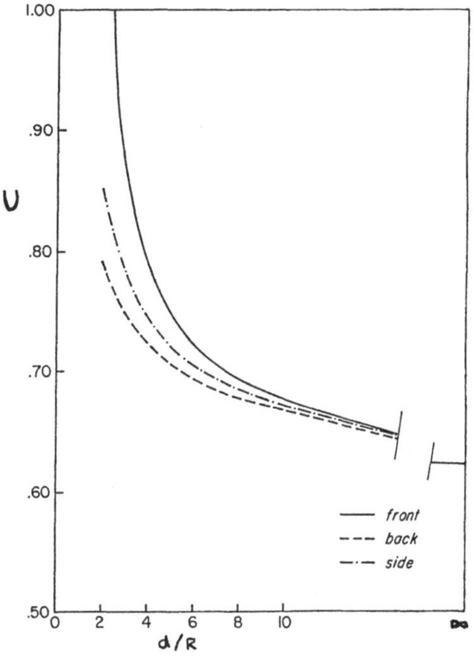

Figure 4. Effect of swarm separation distance on attractant concentration at the fringe, $r = R$. U is a scaled concentration.

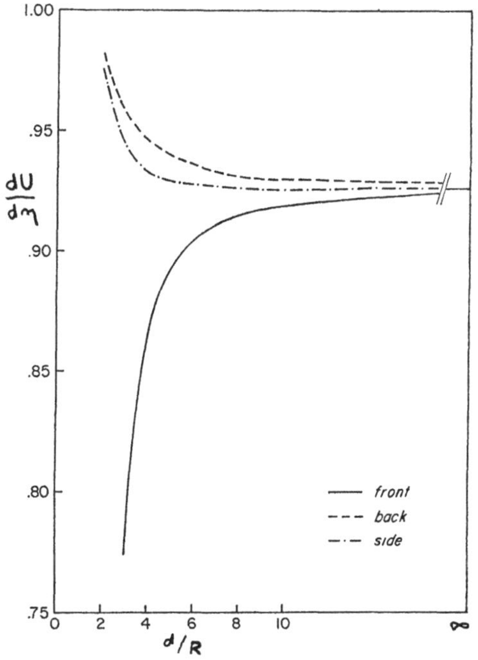

Figure 5. Effect of swarm separation distance on attractant gradient at the fringe, $r = R$. U is a scaled concentration, and η is a scaled distance.

Again, details are given elsewhere (Lauffenburger et al, in press).

For a single isolated swarm, we can set dh/dt = 0, since the swarm holds its form fairly constant. This yields a fringe width

$$h_\infty = -\frac{2\mu}{\chi \frac{da}{dr}\Big|_{r=\infty}} \tag{9}$$

Now, the fringe width will tend to increase if the ratio of chemotactic movement to random movement is decreased from the isolated swarm state. This is precisely what happens when a second swarm appears in the neighborhood of the first. Figure 3 shows the alteration of the attractant concentration distribution around one swarm that occurs when the swarms are, for example, a distance of four swarm radii apart. The concentration increases everywhere, but the largest increase is in the direction of the second swarm. The radial slope at the front edge decreases and the radial slopes at the back and side edges increase as the swarm separation distance decreases. These trends are illustrated in Figures 4 and 5. The decrease in the gradient at the front edge diminishes the inward chemotactic movement, permitting the outward random movement to dominate. This results in net movement of the front fringe toward the other swarm, and the main body follows. The mechanism of swarm approach is thus actually random cell movement as the constraint of chemotactic movement is diminished. However, the swarm approach appears to be directed because the random movement is least constrained by chemotaxis in the particular direction in which the attractant gradient is most diminished.

Equation (8) can also be used to predict the rate of swarm approach and the maximum separation distance at which two swarms can "perceive" each other, as well as other interesting quantities. These predictions are discussed elsewhere (Lauffenburger et al, in press), and are consistent with the motion picture observations. For purposes of this volume, the central point of this work is that the cell aggregation phenomena involved in myxobacteria swarm approach is actually due to random movement. Cell aggregation occurs in the direction in which chemotaxis is diminished to the greatest degree, and net cell movement during the approach is down the attractant concentration gradient. This is a much more subtle role for chemotaxis than might have been expected at first, given the slime mold amoebae conclusions.

LOCALIZED INFLAMMATORY LESIONS

Inflammation is the initial nonspecific host defense response to tissue invasion by foreign bodies such as bacteria. The phagocytic white blood cells -- the polymorphonuclear leukocytes and the macrophages -- are primarily responsible for elimination of bacteria from this tissue. These cells enter the tissue from the bloodstream in response to chemical stimuli caused by the presence of bacteria in the tissue. Once in the tissue, the leukocytes and macrophages move about, phagocytosing and killing the bacteria. For more details on the inflammatory response to bacterial infection, the interested reader is referred to another source, such as the monograph by Mims (1982). From a dynamical perspective, however, this system can be viewed in simple terms: the defense response will be adequate if the phagocytes can destroy the bacteria more rapidly than the latter can grow or produce toxins. We have previously published some attempts to mathematically model the inflammatory response as a dynamical system, in order to elucidate how the effectiveness of the defense response depends upon key parameters representing various cell processes (Lauffenburger & Keller, 1979; Lauffenburger & Kennedy, 1981, 1983).

In the context of this volume, one aspect of the inflammatory response is extremely relevant. In most cases, the inflammatory response to bacterial infection results in a fairly diffuse lesion, in which bacteria and phagocytes are distributed reasonably uniformly throughout the immediate tissue region. However, in a number of infectious diseases, the inflammatory lesions consist of many foci of high densities of bacterial and phagocytic cells. One major example of this latter situation is chronic granulomatous inflammation, with tuberculosis being an important particular disease (Adams, 1976). Thus, in some types of inflammatory lesions there is a clear demonstration of cell aggregation, while in others there seems to be no such aggregation. Since phagocyte chemotaxis is a key component of the inflammatory response (see, for example, Wilkinson, 1982), again chemotaxis seems to be a ready explanation for cell aggregation. As in the previous example, though, its role is not as clear as might be expected from the classical slime mold amoebae problem.

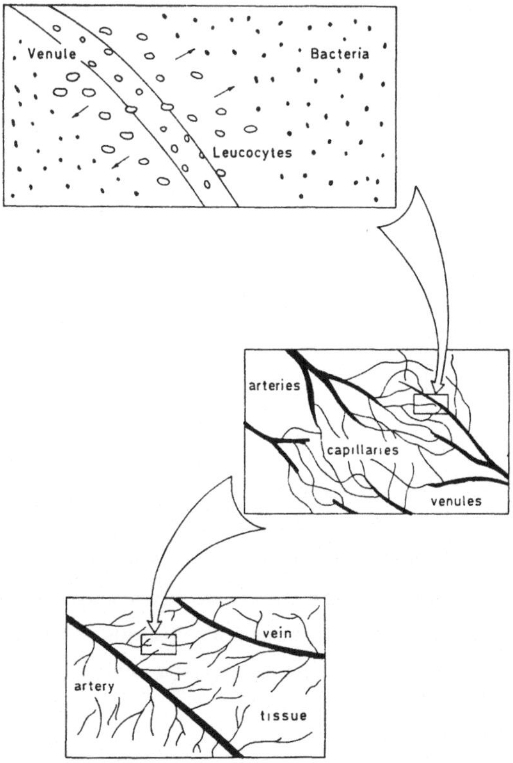

Figure 6. Illustration of tissue inflammatory response to bacterial infection, showing different scales. The model presented here uses the macroscopic scale of the bottom picture.

A mathematical model intended to bring understanding to this situation can be formulated in the following manner (Lauffenburger & Kennedy, 1983). Figure 6 illustrates the system under consideration. Phagocytes are shown emigrating from the local bloodstream, by diapedesis, through the walls of post-capillary venules. Emigration takes place at a small base rate under normal conditions when no bacteria are present, and the rate increases when bacteria appear in the tissue. This increase is presumably due to vasoactive and chemotactic chemical mediators produced either by the bacteria directly or by activation of complement by the bacteria. Once in the tissue, the phagocytes move about and killing the bacteria until they themselves die. The bacteria multiply, utilizing nutrients available in the tissue. Phagocytosis occurs only after contact between phagocytes and bacteria. This contact may result from random movement of both cell types, possibly enhanced by chemotactic movement of the phagocytes up concentration gradients of attractant produced by the bacteria, again either directly or indirectly. In vitro studies have suggested that phagocytes may "home in" on individual bacteria by means of chemotaxis, but in vivo observations are less clear. Often, the phagocytes appear to move within an inflammatory lesion randomly, and only show a propensity for directed movement at the periphery of the lesion (Wilkinson, 1982).

For purposes of our model, we assume that the tissue can be treated as a homogeneous medium, using a macroscopic perspective in which micro-anatomical structure can be neglected (see Figure 6), so that leukocyte emigration from the venules is considered to occur continuously throughout the tissue. Assuming saturation of the rates of both growth and phagocytosis of bacteria at high bacterial densities, and that the concentration of chemoattractant is proportional to bacterial density, b, and the phagocyte density, c:

$$\frac{\delta b}{\delta t} = \mu_b \frac{\delta^2 b}{\delta x^2} + \frac{k_g b}{1 + b/K_i} - \frac{k_d bc}{K_b + b} \tag{10}$$

$$\frac{\delta c}{\delta t} = \mu_c \frac{\delta^2 c}{\delta x^2} - \chi \frac{\delta}{\delta x}\left(c \frac{\delta b}{\delta x}\right) + (h_o + h_1 b)c_b - gc \tag{11}$$

where μ_b is the bacterial random motility coefficient, k_g is the specific bacterial growth rate constant, k_d is the specific phagocytosis rate constant, and K_i and K_b are the saturation constants for growth and phagocytosis, respectively. μ_c is the phagocyte random motility coefficient; χ is the phagocyte chemotaxis coefficient; h_o is the normal emigration rate constant; h_1 is the enhanced emigration rate constant, c_b is the phagocyte density in the bloodstream, and g is the specific phagocyte death rate constant. We also assume no-flux boundary conditions on the region, of either finite or infinite extent. The details of the following analysis and results can be found elsewhere (Lauffenburger & Kennedy, 1983).

This system possesses two types of uniform steady-state solutions:

1. "elimination"

$$b_o = 0 \qquad c_o = \frac{h_o c_o}{g}$$

2. "compromise":

$$b_o > 0 \qquad c_o = \frac{h_o c_b}{g}\left(1 + \frac{h_i k_i}{h_o}\right) b_o$$

The first is termed elimination since the bacteria are gone, and the phagocyte density is at its normal background level. This should be the state of healthy tissue. The second type is called a compromise state, since a nonzero bacterial density persists and the phagocyte density is increased, which may represent a state of chronic inflammation.

Using a linear stability analysis just as in the classical slime mold amoebae problem, it can be shown that a compromise steady state becomes unstable with respect to nonuniform fluctuations when the following two conditions are violated:

$$\frac{\mu_b}{\mu_c} < \rho_c < 1 \tag{12}$$

$$\frac{\chi}{\mu_c} < \delta_c \tag{13}$$

where ρ_c and δ_c are quantities which depend upon the other model parameters.

The first condition means that a large ratio of phagocyte random motility to bacterial random motility can lead to instability of a uniform steady state. The second means that a small ratio of phagocyte chemotaxis to phagocyte random motility can lead to instability of a uniform steady state. If we speculate that when a uniform steady state becomes unstable to nonuniform fluctuations a nonuniform steady state might evolve, then we see that the nonuniform steady states should result from small values of the ratio of chemotaxis to random motility. This is exactly the opposite of the result from the classical problem. Upon reflection, this is reasonable, because the chemotactic population in the classical problem serves as an "activating" species, producing the chemical aggregation mediator, while in our inflammation problem, the chemotactic population serves as a "inhibiting" species, destroying the source of the aggregation mediator.

Figures 7 and 8 show two example numerical solutions of the full transient equations, Equations (10) and (11), for values of the phagocyte chemotaxis coefficient smaller and larger than the critical value for instability, respectively. It can be seen that a large value of the chemotaxis coefficient leads to uniform densities, while it is a small

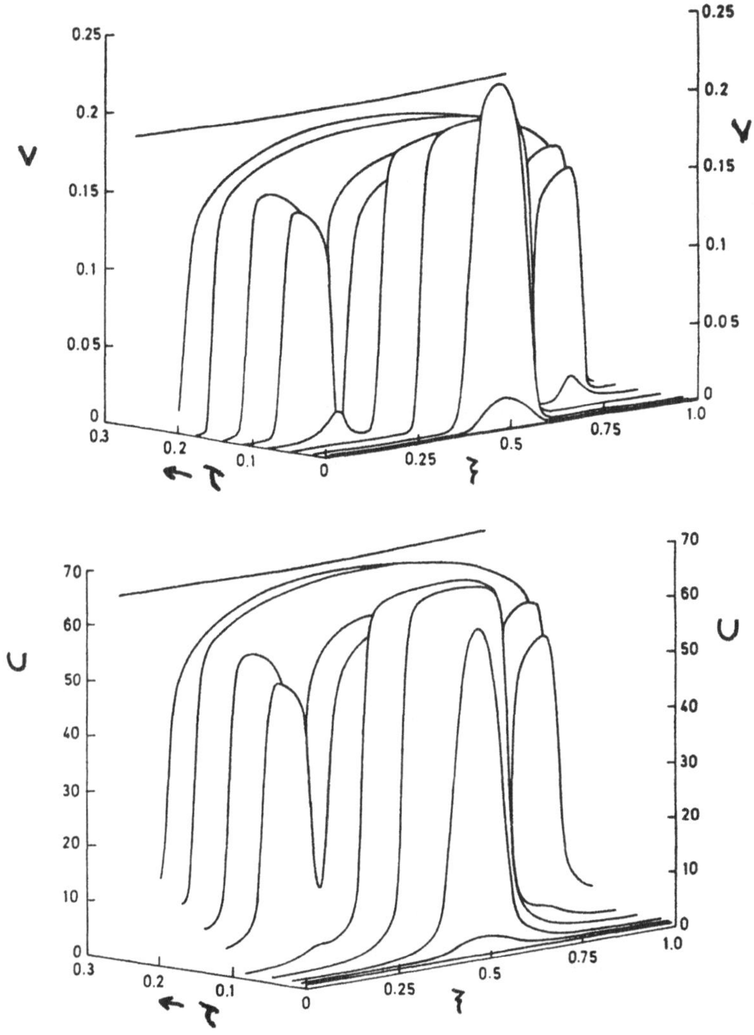

Figure 7. Example transient cell density profiles for bacteria (top) and phagocytes (bottom), for parameter values such that inequalities (12) and (13) are satisfied. V and U are scaled bacteria and phagocyte densities, and τ and ξ are scaled time and position variables, respectively.

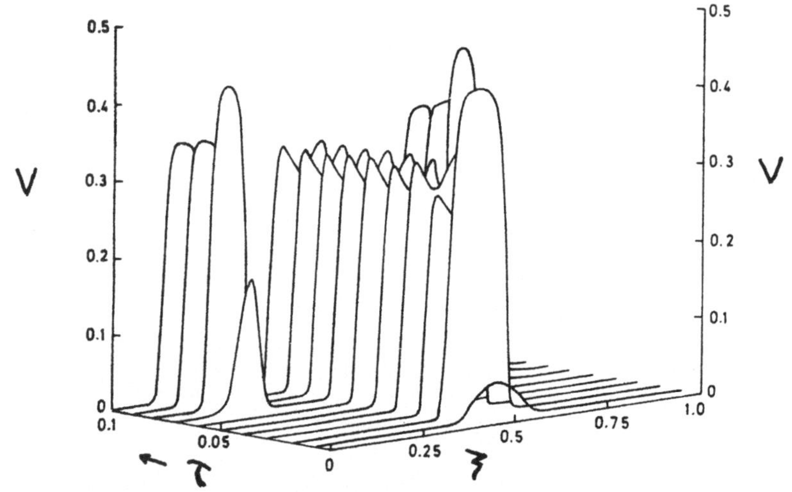

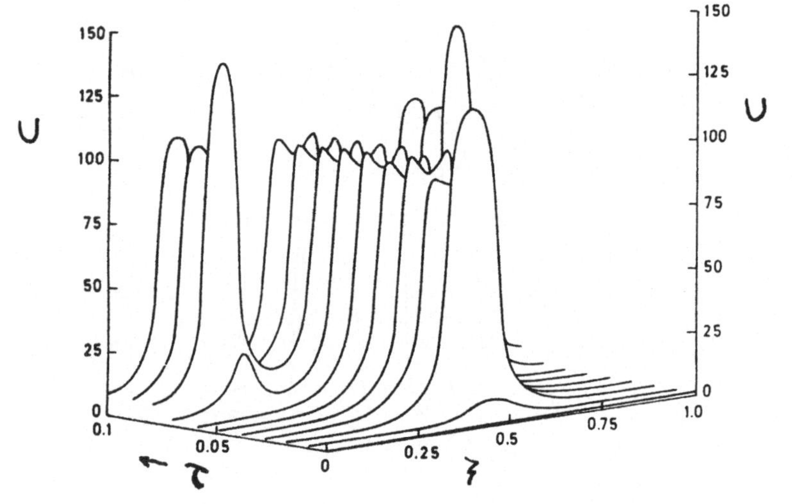

Figure 8. Example transient cell density profiles for bacteria (top) and phagocytes (bottom), for parameter values such that inequalities (12) and (13) are violated. V and U are scaled bacteria and phagocyte densities, and τ and ζ are scaled time and position variables, respectively.

value of the chemotaxis coefficient that leads to cell aggregation and spatial pattern. One interesting implication of these results is that the lack of observation of noticeable directed phagocyte movement in in vivo inflammation may not necessarily indicate that a chemotactic response plays no part in host defense. Rather, chemotaxis might be of importance by preventing dispersion of phagocytes away from the affected lesion.

SUMMARY

Through the study of two particular examples, it is evident that the role of chemotaxis in cell aggregation can be much more subtle and interesting than the straightforward role found for it in the classical slime mold amoebae problem. I hope that this gives encouragement for study of individual cell aggregation problems on their own merits, with the possiblity that further variations might be developed for the role of this quite exquisite cell behavioral phenomenon.

REFERENCES

1. Adams, D. O. Am. J. Path. 84, 164 (1976).
2. Bonner, J. T. The Cellular Slime Molds, Princeton Univ. Press (1967).
3. Childress, S. & J. K. Percus, Math. Biosci. 56, 217 (1981).
4. Cohen, M. S. & P. S. Hagan, J. Theor. Biol. 93, 881 (1981).
5. Kaiser, D., C. Manoil, & M. Dworkin, Ann. Rev. Microbiol. 33, 595 (1979).
6. Keller, E. F. & L. A. Segel, J. Theor. Biol. 26, 399 (1970).
7. Lauffenburger, D. A., M. Grady, & K. H. Keller, J. Theor. Biol. (in press).
8. Lauffenburger, D. A. & K. H. Keller, J. Theor. Biol. 81, 475 (1979).
9. Lauffenburger, D. A. & C. R. Kennedy, Math. Biosci. 53, 189 (1981).
10. Lauffenburger, D. A. & C. R. Kennedy, J. Math. Biol. 16, 141 (1983).
11. Mims, C. A. The Pathogenesis of Infectious Disease, Academic Press (1982).
12. Reichenbach, H. (1974). Encyclop. Cinematogr. Film E779/1965, ed. G. Wolf, Göttingen: Inst. Wiss. Film, p. 3.
13. Rosenberg, E., K. H. Keller, & M. Dworkin, J. Bacteriol. 129, 770 (1977).
14. Wilkinson, P. C. Chemotaxis and Inflammation, Churchill-Livingston (1982).

ACKNOWLEDGEMENTS

I would like to thank the Amoco Foundation for its generous financial support, and Catherine Barnes for preparing this manuscript.

TURING STRUCTURES, PERIODIC AND CHAOTIC REGIMES IN COUPLED CELLS

M. Marek
Prague Institute of Chemical Technology
166 28 Prague 6

Coupled cells with reactions and mutual mass exchange are classic model systems used for description of processes in living cell aggregates and tissues. They also form the basis of compartmental models, and serve as discretized description of distributed systems with reactions combined with diffusive transport (e.g. of various types of chemical reactors). If the reactions are nonlinear and the reaction rate dependences on concentrations have proper nonlinear forms (describing, e.g., autocatalysis or inhibition), then dissipative structures - concentration profiles non-monotonous in space and/or time - can occur[1-16]. An accumulation of experimental knowledge on observations of both multiple steady states and oscillations in single continuous flow stirred cell reactor enables now to study these structures experimentally (most of the quoted studies[1-16] were based either on approximate analytical description or on computer simulation). Here we first report on multiple stationary states in system consisting of up to seven cells, with stability dependent on the intensity of interaction between cells (Turing structures[1,7,8]). Then various periodic and aperiodic (chaotic) nonstationary regimes observed in two coupled cells are described. One modelling result is then discussed to stress complexity of the regimes possible in these systems.

Experimental Apparatus

Experimental system consisting of two coupled reaction cells with mass exchange through common walls was used earlier for the study of synchronization of chemical oscillators[17]. Several types of systems consisting of two or three cells were constructed later on until versatile experimental system of seven cells (cf. Fig. 1a, b) was developed[18]. The apparatus consists of seven hexagonal stirred cells (volume 100 ml) (cf. Fig. 1a) with controlled mutual mass exchange through common walls. The walls between the cells contain apertures of adjustable dimensions. Each cell is independently stirred, contains temperature control (heating coil acts as a flow distributor), inlet and outlet of reactants and products and measurement of concentrations. Pt and calomel electrodes are used in the case of the Belousov-Zhabotinski reaction system[19]. The system can be operated as batch, semibatch or continuous flow system with an inlet (by multichannel peristaltic pump) directed into any of the seven reactors (or combination

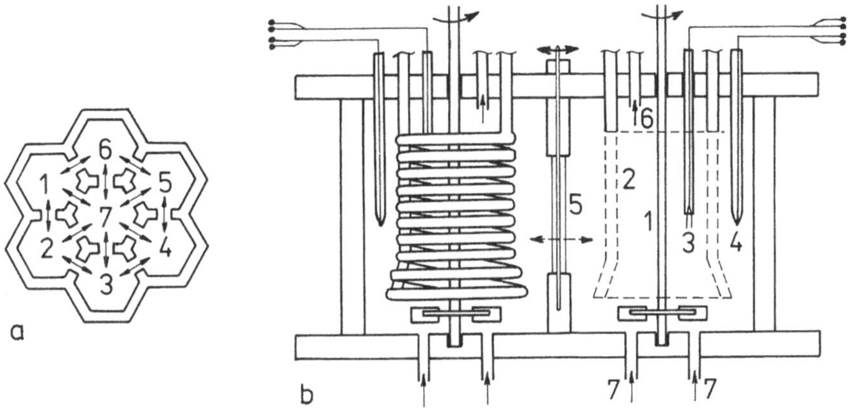

Fig. 1a Experimental set-up of seven flow-through hexagonal stirred cells

Fig. 1b Details of experimental arrangement of two neighboring cells; 1 - stirrer; 2 - flow distributor; 3,4 - Pt and calomel electrodes; 5 - aperture between the cells; 6 - outlet of products; 7 - inlet of reactants

of them) and an outlet taken also from one of the cells or from their chosen combination. Thermistors are used for the measurement of the temperature, conductivity cells located either inside the cells or at the cell outlets are used in the tracer experiments.

Coefficients of Mutual Mass Exchange between Cells

The pattern of liquid passage through the system can be varied by opening or closing the apertures in the common walls. The intensity of turbulence (micromixing) in the reactors can be controlled by varying stirring intensity and impeller type (the rate of mass exchange between the reactors also depends on the intensity of stirring). The intensity of mutual mass exchange between the cells was determined by independent tracer experiments using conductivity probes[20]. The course of tracer concentrations C(i) in the i-th cell can be described as that of a system of well mixed flow-through stirred cells with mutual mass exchange by convection and diffusion (cf. Fig. 2).

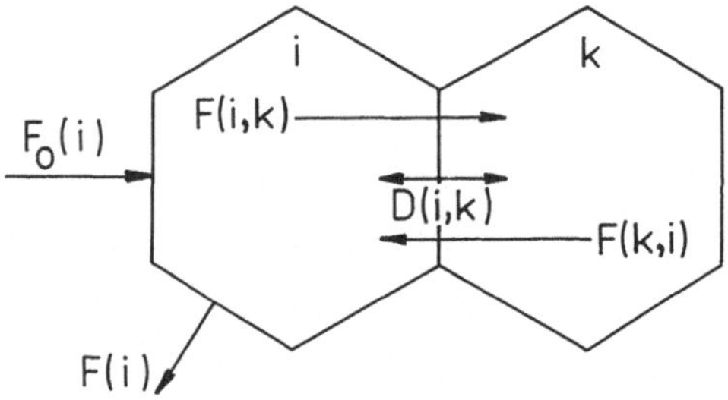

Fig. 2 Two neighboring cells schematically - tracer balance

$$V(i)\ \frac{dC(i)}{dt} = F_o(i)C_o(i) - F(i)C(i) + \sum_{k=1,k\neq i}^{M} \Big\{F(k,i)C(k) - F(i,k)C(i) + D(i,k)\left[C(k) - C(i)\right]\Big\} \qquad (1)$$

Here V(i) is cell volume, $F_o(i)$, F(i) are external inlet and outlet flow rates, $C_o(i)$ tracer concentration in the external inlet stream, F(k,i) convection flux between k-th and i-th reactor and D(i,k) dispersion (backmixing) coefficient. The convection and dispersion terms can be summed into one exchange coefficient D constant for all cells under conditions where there is one inlet and one outlet of the entire system and the apertures between the cells and the rates of stirring are the same[18]. Then, for example, the tracer concentrations in two cells are described by Eqs. (2) ($C(i) = C_i$).

$$\frac{dC_1}{d(t/\tau)} = C_{o1} - C_1 + D(C_2 - C_1)$$

$$\frac{dC_2}{d(t/\tau)} = C_{o2} - C_2 + D(C_2 - C_1) \qquad (2)$$

Here τ is the mean residence time and the value of D evaluated from tracer experiments for various intensities of stirring and sizes of apertures by the least squares fitting[21,18] ranges from 0.01 to 0.08. We shall denote relative intensities of mass exchange by numbers D = 1, 2,..., 8.

Turing Structures

An extensive experimental study of various types of dissipative structures used the Belousov-Zhabotinski reaction as a model reaction system. In this reaction malonic acid is oxidated by bromate in the presence of sulfuric acid and ceric/cerous redox catalyst[19]. Two stable steady states can be observed in a single continuous stirred cell[22]. Two coupled reactors with mutual mass exchange were used for the study of the coexistence of the two different steady states in dependence on the intensity of mutual mass exchange (D) and the value of malonic acid concentration, c_M. The courses of the reaction between natrium bromate ($2x10^{-2}$M), cerous ions ($1.2x10^{-3}$M), and malonic acid in 1.5 M sulfuric acid in the two cells were recorded as potential changes of Pt wire electrodes coupled with calomel electrodes via sulfate salt bridges. In all experiments the same inlet solutions were fed into the cells. First, the cells were operated independently with closed apertures 5. Two different states were established in the cells applying perturbations by Br^- and Ce^{4+} ions (Br^- supports transition to the lower and Ce^{4+} to the upper stationary state). Then the apertures between the cells were opened to reach required intensity of mass exchange (D). Next, value of D was increased, until one of the stationary states became dominant in both cells, cf. Fig. 3a for record of the potential changes and Figs. 3b, c for schematic picture of experiments where upper stationary state (Fig. 3b, $c_M = 1x10^{-2}$M) and lower stationary state (Fig. 3c, $c_M = 1.6x10^{-2}$M) became dominant. Hence, dependent on the values of D and c_M, either inhomogeneous steady state or one of the homogeneous steady states are stable. Turing's original work[1] is primarily concerned with the stability analysis of the homogeneous stationary solutions. The main point of biological interest, however, is whether stable spatial structures may be generated by a proper perturbation[8]. In our case both homogeneous and inhomogeneous stationary states can be stable and values of other parameters (i.e. intensity of mass exchange, D, inlet concentration of malonic acid, c_M) determine the regions of attraction of individual stationary states.

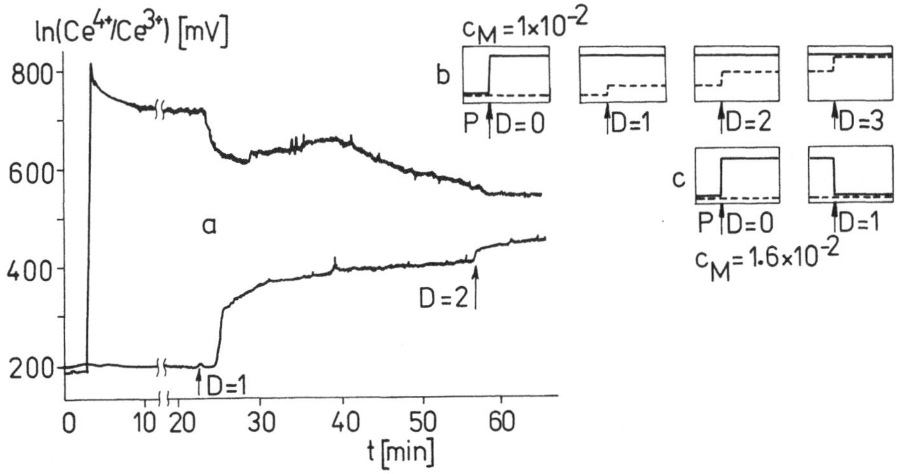

Fig. 3a The time course of redox potential in two coupled cells $c_M = 1x10^{-2}$ M

b upper steady state dominant

c lower steady state dominant

Evolution of inhomogeneous stationary states after instability of periodic oscillations in two coupled cells with mutual mass exchange was reported earlier[25].

Evolution of inhomogeneous stationary structures in three coupled cells as a result of perturbations is shown in Fig. 4a (part of the redox potential recordings) and Fig. 4b (schematic pictures of several transitions). If the steady state with high value of the redox potential is denoted by 1 and the one with low value by 0, then all six possible structures 000, 001, 010, 011, 101, 111 were observed in the system[26]. The examples in Fig. 4 were recorded for relatively low value of D = 2. We can observe in Figs. 4a,b that Ce^{4+} perturbation applied in the first cell of the stationary state 000 causes transition to the state 100. However, this state spontaneously changes into the state 111 after several minutes of its existence. Thus this particular perturbation cannot be used for creation of the state 100 (perturbation of lower intensity is necessary). It was found that the regions of attractions of individual structures are intertwined and character of perturbation will determine which of the observable

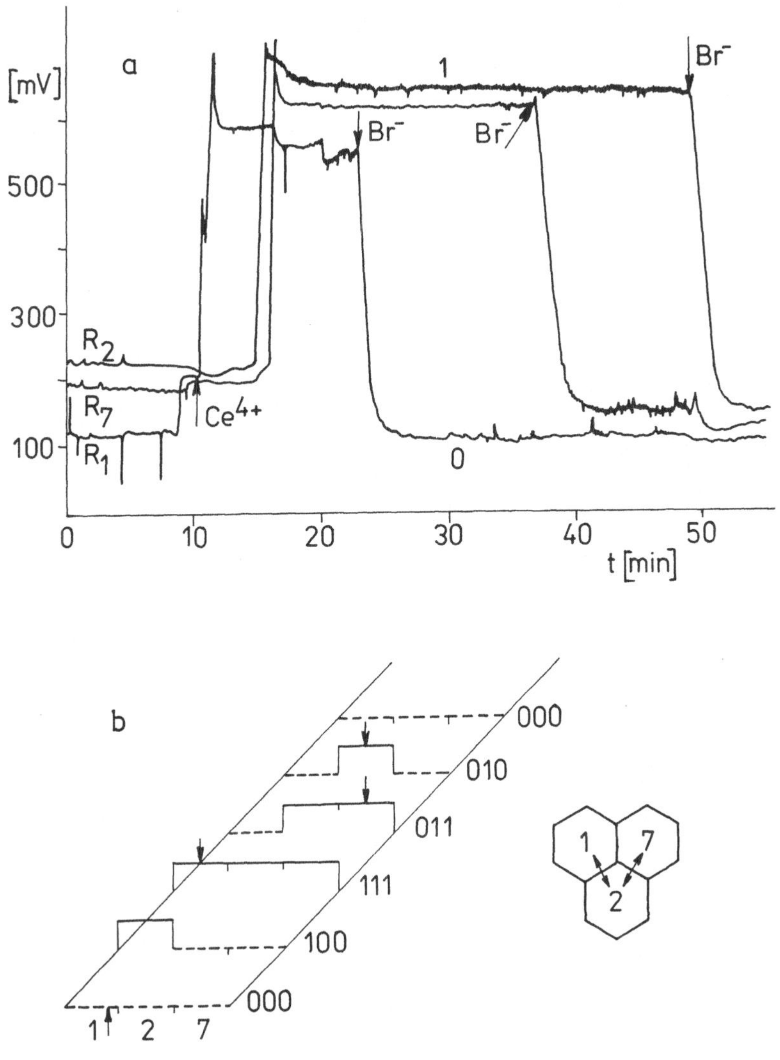

Fig. 4a The time course of redox potential in three coupled cells. Concentrations: $NaBrO_3$ - $2x10^{-3}$M; Ce^{3+} - $1.6x10^{-3}$M; c_M - $1x10^{-2}$M. T = 313.7 K, rate of stirring 600 rpm, residence time τ = 4.25 min.

b Transition between inhomogeneous stationary states, resulting from perturbations, schematically.

↑ Ce^{4+} perturbation; ↓ Br^- perturbation

structures will be actually in the system. When D = 5, then not all of the structures realized for D = 2 are stable. For example, the structure 010 becomes unstable when the value of D is increased. This is ilustrated in Fig. 5. We observe that for D$\geqslant$4 is this structure unstable and stable structure 011 arises.

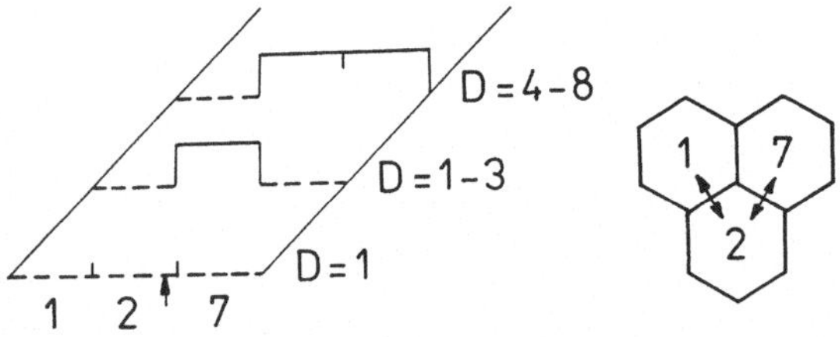

Fig. 5 Effects of intensity of mass exchange on stability of inhomogeneous stationary states in three coupled cells

Several stationary structures realized for seven coupled cells and D = 2 is shown in Fig. 6. We can observe formation of inhomogeneous structures 1010101, 1011101, 1000110. The structure 1011101 is relatively very stable (it has large region of attraction).

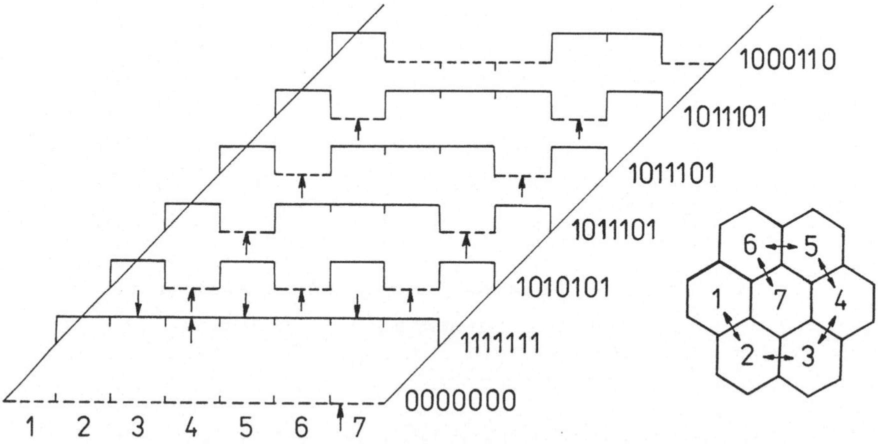

Fig. 6 Transition between inhomogeneous steady states in seven coupled cells, schematically

Periodic and Chaotic Regimes

Limit cycle type oscillations occur in the Belousov-Zhabotinski reaction in single flow-through stirred cell in a wide range of inlet components concentrations[19].,Chaotic oscillations were also observed in a narrow range of flow-rates. When oscillations occur in two mutually coupled cells, periodic, quasiperiodic or chaotic regimes can be observed. Synchronization of oscillations on the common frequency or on subharmonics have been reported earlier[17]. Here we shall describe several examples of chaotic regimes.

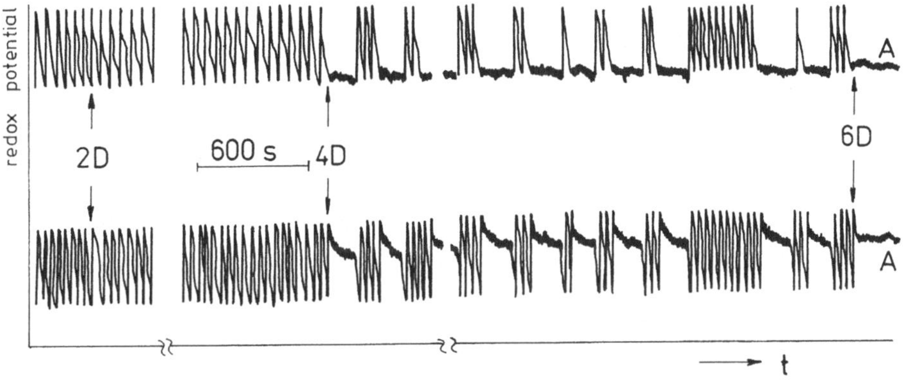

Fig. 7 Time course of redox potential oscillations in two coupled cells.
Temperature 308.1 K; initial periods in the left and right cell, T_{Pl} = 47.5 s; T_{Pr} = 37.8 s, respectively; residence time τ = 13.8 min; inlet concentrations : malonic acid 0.05 M, natrium bromate 0.05 M, sulfuric acid 1.5 M, Ce^{4+} - $1x10^{-3}$M; A - asymmetric steady state

Two cells have been operated under the same inlet conditions. The experiments were started with two cells separated and the periods of oscillations in the cells have been controlled by the temperature or the intensity of mixing in each cell. Then the aperture between the cells was opened and chosen value of D was adjusted. The time course of oscillations of the redox potential in the two cells is shown in

Fig. 7. We can observe that at the intensity of interaction equal to 4 D an aperiodic regime, which is an irregular combination of oscillations with short and long periods occur. Similar chaotic regime was reported in the experiment on chemical chaos in a single flow-through reaction cell[28]. When the intensity of interaction is increased to 6 D, oscillations in both cells disappear and an asymmetric steady state with different levels of the redox potential in the two cells (simple Turing structure) sets in.

Next two figures will be reported in the form of periodograms, i.e. in the form of the dependence of the "periods" of oscillations (time intervals between two sharp rises of the redox potential, cf. Fig. 7) on time. Only sections from the data series will be shown in Figures (each series contained several hundred periods). An example of the periodogram illustrating the dependence of periods of oscillations in two cells on time for various levels of the intensity of mass exchange is shown in Fig. 8. This sequence of regimes is typical for a number of data series obtained at various initial ratios of T_{Pl}/T_{Pr}.

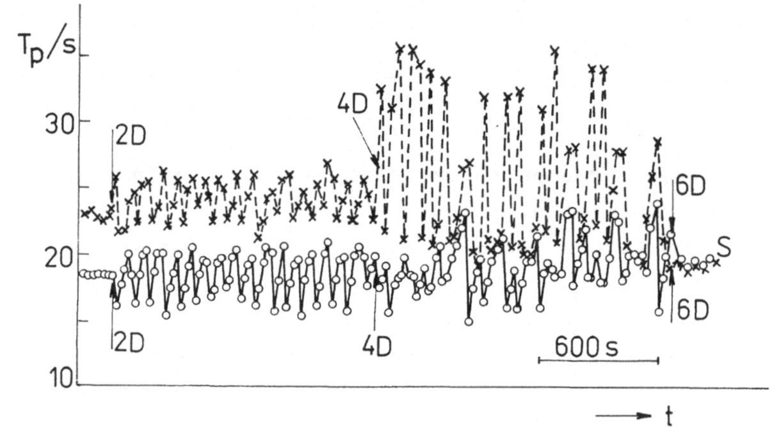

Fig. 8 Time course of "periods" (periodogram); temperature and concentrations cf. Fig. 7; initial periods, T_{Pl} = 22.8 s, T_{Pr} = 18.4 s; S - Synchronization of oscillations; $T_P = T_{Pr}$ = 19.6 s

At the intensity of interaction equal to 2 D a nearly quasiperiodic regime sets in - we can observe that the time course of periods is again periodic, particularly in the faster oscillator. Then at the higher intensity of interaction equal to 4 D a chaotic course of periods sets in. When the intensity of interaction is even higher (equal to 6 D), a regime of periodic oscillations synchronized (S) at the value of the period close to the initial period of faster oscillator sets in.

If the ratio of the initial periods T_{Pl}/T_{Pr} is close to the ratio of 1 : 2, then at the weak mutual interaction are both oscillations synchronized on the same ratio and the resulting regime is periodic. At medium intensity of interaction the behavior is chaotic (cf. Fig. 9a) and when the intensity of interaction is even higher, the behavior is

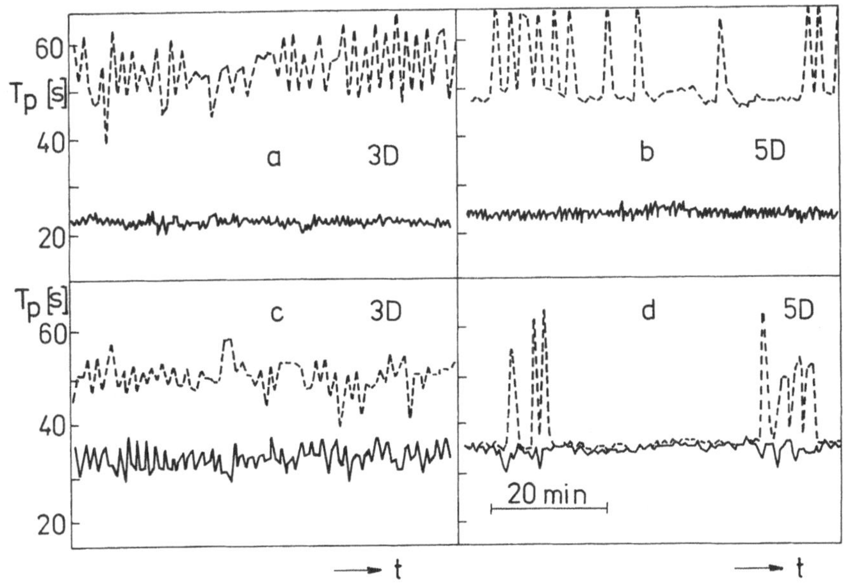

Fig. 9 Time course of "periods". Initial periods T_{Pl}= 24.1 s, T_{Pr} = 49.4 s in Fig. 9a,b and T_{Pl} = T_{Pr} = 50 s in Fig. 9c,d.
a) chaotic oscillations, coupling intensity 3 D
b) intermittency, coupling intensity 5 D
c) chaotic oscillations, coupling intensity 3 D
d) intermittency, coupling intensity 5 D

typically intermittent, cf. Fig. 9b. Intermittent route to chaos have been reported for a single flow-through stirred tank reactor by Pomeau and coworkers[29]. We can observe that the time course of periods is nearly periodic in the cell with faster oscillations but it is clearly chaotic in the cell with slower oscillations (this behavior is typical for a number of experiments). The interaction between the cells is here actually uni-directional, even if the original experimental arrangement is symmetric. If the ratio of the initial periods is close to 5 : 7, then the behavior of periods is chaotic in both cells; cf. Fig. 9c (medium intensity of interaction). We can again observe intermittency at higher intensities of interaction (cf. Fig. 9d). At very strong intensities of interaction (e.g. 8 D) the resulting oscillatory regime is periodic, and the oscillations are synchronized at the ratio of periods 1 : 2 (the case of Fig. 9a, b) or 1 : 2 (the case shown in Fig. 9c, d).

Comparison with Computations

System of identical coupled cells with reaction can be described by the set of relations

$$\frac{d\underline{c}^k}{dt} = \underline{R}(\underline{c}^k) + \underline{D}\,\Delta^k \underline{c}^k , \qquad k = 1,2,\ldots,N \tag{3}$$

Here $\underline{c}^k = (c_1{}^k, \ldots, c_n{}^k)$ is a vector of concentrations of n components reacting in the k-th cell, $\underline{R}(\underline{c}^k)$ is a vector function describing reaction kinetics and mass exchange with the environment, $\underline{D}$ is $n \times n$ matrix of mass exchange coefficients and $\Delta^k \underline{c}^k = \sum_m (\underline{c}^m - \underline{c}^k)$ (summation is made over m cells which are in contact) is a difference operator describing the structure of linear coupling (mass exchange) between the cells. For example, in the case of linear array we have $\Delta^k \underline{c}^k = (\underline{c}^{k+1} - 2\underline{c}^k + \underline{c}^{k-1})$.

The Belousov-Zhabotinski reaction is complex and not fully elucidated. The Noyes-Field-Körös model[19], which is now considered to describe best the mechanism of the reaction, contains set of kinetic equations which cause stiffness of the system (3). Together with relatively high dimension of the entire set of equations it prevents its full investigation by numerical methods. However, similar studies were performed with simple model kinetic systems. One of them is a Brusellator.

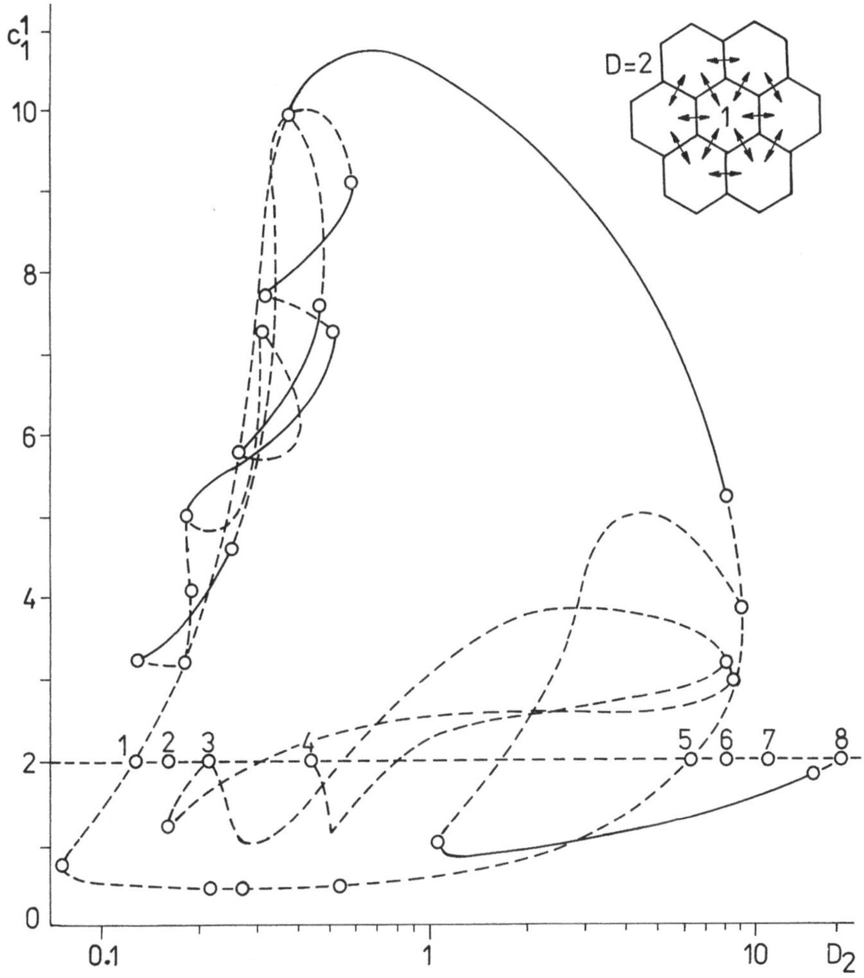

Fig. 10 Dependence of some of the stationary solutions of (3) (concentrations in the middle cell c_1^1) on the parameter D_2. $A = 2$, $B = 6$, $D_1/D_2 = 0.1$,
—— stable solution, - - - - unstable solution,
o bifurcation point, 1, 2,..., 8 bifurcations from the homogeneous periodic solution

Here the kinetic functions are in the form (n = 2)

$$\underline{R} = \begin{pmatrix} R_1 \\ R_2 \end{pmatrix}, \qquad \begin{aligned} R_1 &= A - (B+1)c_1 + c_1^2 c_2 \\ R_2 &= Bc_1 - c_1^2 c_2 \end{aligned} \tag{4}$$

and the mass exchange coefficients are D_1 (component c_1) and D_2 (component c_2). The dependence of the solutions of the system (3) with hexagonal structure (N = 7) on the parameter D_2 (it is assumed equal for all cells) is shown in Fig. 10. Numerical continuation techniques[30] have been used to compute this Figure. We can observe that number of coexisting inhomogeneous stable stationary states can be high and their mutual relations complex, even if not all solutions are contained in the Figure. If interacting periodic solutions are studied for the same kinetics in two coupled cells, both periodic and chaotic states are found[31].

References

1 A.M. Turing, Philos. Trans. R. Soc. London Ser. B237, 37 (1952).
2 W.R. Loewenstein and Y. Kanno, Nature (London) 209, 1248 (1966).
3 W.R. Loewenstein, Dev. Biol. Supp. 2, 151 (1968).
4 I. Prigogine and R. Lefever, J. Chem. Phys. 48, 1659 (1967).
5 J.I. Gmitro and L.E. Scriven, in Intracellular Transport, ed. by K.B. Warren; Academia Press, N.Y., 1966.
6 L. Wolpert, J. Theor. Biol. 25, 1 (1969).
7 H. Martinez, J. Theor. Biol. 36, 479 (1977).
8 H. Martinez and R.M. Baer, Bull. Math. Biol. 35, 87 (1973).
9 M.G. Othmer and L.E. Scriven, J. Theor. Biol. 32, 507 (1971).
10 T. Pavlidis, Biological Oscillators: Their Mathematical Analysis; Academia Press, N.Y., 1973.
11 B. Bunow and C.K. Colton, Biosystems 7, 160 (1975).
12 V. Torre, Biol. Cyber. 17, 137 (1975).
13 S. Smale, in Lectures on Mathematics in the Life Sciences 7, ed. J. Cowan, Amer. Math. Soc., Providence, 1975.
14 M. Ashkenazi and M.G. Othmer, J. Math. Biol. 5, 305 (1978).
15 A. Shapiro and F.J.M. Horn, Math. Biosci. 44, 19 (1979).
16 L. Lapidus and N.R. Amundson, editors: Chemical Reactor Theory, A Review, Prentice Hall, Inc., Englewood Cliffs, N.Y. 1977.
17 M. Marek and I. Stuchl, Biophys. Chem. 3, 137 (1975).
18 M. Marek, J. Havlíček and J. Vlček, "Network of Coupled CSTR's - Model Experimental System", Paper J4, Proc. Fourth Europ. Conf. on Mixing, April 1982, BHRA Fluid Engineering, Cranfield, Bedford, England.
19 J.J. Tyson, The Belousov-Zhabotinski Reaction, Lecture Notes in Biomathematics, Vol. 10, Springer, Berlin, 1976.
20 C.W. Sheppard, "Basic Principles of the Tracer Method", J. Wiley, New York, 1962.
21 M.L. Michelsen, The Chem. Eng. J. 4, 171 (1972).

22 W. Geiseler and H.M. Föllner, Biophys. Chem. 6, 107 (1977).
23 I. Stuchl and M. Marek, J. Chem. Phys. 77, 1607 (1982).
24 T. Erneux, J. Hiernaux and G. Nicolis, Bull. Math. Biol. 40, 771 (1978).
25 M. Marek, in Synergetics, Far from Equilibrium, edited by A. Pacault and C. Vidal, Springer, New York, 1979.
26 I. Stuchl and M. Marek, J. Chem. Phys. 77, 2956 (1982).
27 J.L. Hudson, J.C. Mankin and D. Marinko, J. Chem. Phys. 74, 6171 (1981).
28 M. Orbán and I. Epstein, J. Phys. Chem. 86, 3907 (1982).
29 Y. Pomeau, J.C. Roux, A. Rossi, S. Bachelart and C. Vidal, J.Phys. Lett. 42, 271 (1981).
30 M. Kubíček and M. Marek, Computational Methods in Bifurcation Theory and Dissipative Structures, Springer, New York, 1983.
31 I. Schreiber and M. Marek, Physica 5D, 258 (1982); Phys. Lett. 91A, 263 (1982).

DIGITS, SEGMENTS, SOMITES - THE SUPERPOSITION OF SEQUENTIAL AND PERIODIC STRUCTURES

Hans Meinhardt
Max-Planck-Institut für Virusforschung
7400 Tübingen, West Germany

Segmented structures are among the most important classes of structures formed during development of higher organisms. They consist of a repetition of similar but not identical subunits. Examples are the segments of insects and arthropodes, the digits of vertebrates as well as the somites. The latter give rise to the vertebrae and to sequentially arranged muscles and nerves.

For both the segments of insects and the digits of vertebrates it has been shown that many experimental results can be accounted for by assuming that these structures arise under the control of a graded distribution of a morphogenetic substance (for details and references see Meinhardt, 1982a). For instance, in insects, a low concentration of

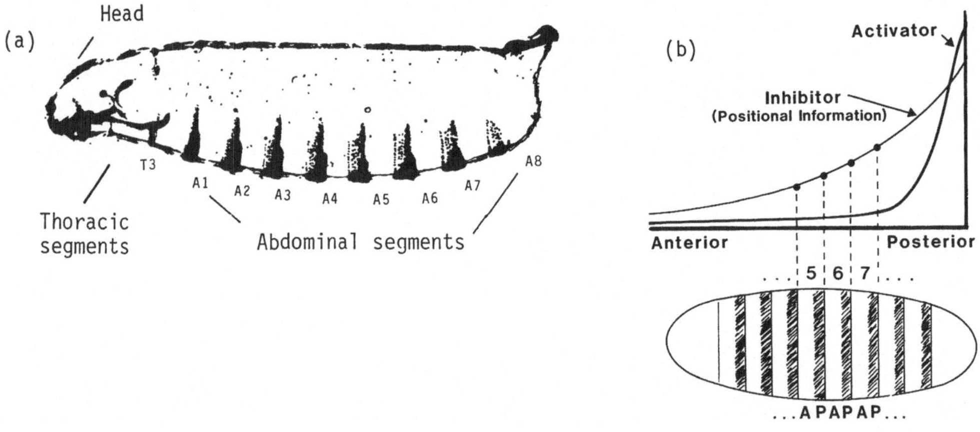

Fig.1 (a) An example for a sequential/periodic structure: a larvae of Drosophila. (b) The generation of the sequence of segments can be explained under the assumption of a controlling morphogen gradient. Such positional information can be generated by an activator-inhibitor mechanism. The smoothly graded inhibitor distribution supplies the positional information, determining in this way which segment (1,2,3....) is to be formed (Meinhardt, 1977). Each segment consists of at least two (or, as advocated for in the present paper, of three) subunits. The formation of this periodic structure (APAPAP) is assumed to be the primary event.

this (still hypothetical) morphogen would provide the positional information to form a head, a high concentration to form the abdominal segments (Fig. 1). In insects, basic structures which are repeated within each segment are also known: the compartments (Garcia-Bellido et al., 1973). At least the thoracic segments and the first abdominal segment become subdivided into an "anterior" (A) and a "posterior" (P) compartment. The formation of the A-P boundaries within a segment occurs almost simultaneously with the formation of the borders which separate the segments.

After a brief survey of how to generate the morphogen gradient I will discuss a model for the generation of sequential/periodic structures. The main stipulation of this model is that the formation of the periodic structure is the primary event and that under the influence of these periodic alternations between two (or probably three) states, a sequential activation of new control genes becomes possible which assures that each of these "segments" is different from the others.

PATTERN FORMATION BY AUTOCATALYSIS AND LATERAL INHIBITION

We have shown that many basic phenomena in biological pattern formation can be understood under the assumption that a local autocatalysis is coupled to a long-ranging inhibition (Gierer and Meinhardt, 1972; Gierer, 1981; Meinhardt, 1982a). It can be easily seen that this principle is also inherent to the pattern formation mechanism proposed by Turing (1952) in his pioneering paper. Turing (1952, p. 42) has exemplified his mechanism with the following pair of coupled differential equations:

$$\dot{x} = 5x - 6y + 1 \qquad (1a)$$

$$\dot{y} = 6x - 7y + 1 \text{ (+ diffusion)} \qquad (1b)$$

Both equations look very similar and it is not immediately obvious why this interaction leads to a patterned distribution of the substances x and y. However, a brief calculation reveals this feature. Both substances are in a steady state at $x = 1$ and $y = 1$. Let us examine a very local increase of x over this steady state concentration. The y-production will increase according to Eq.1b also but, due to the averaging effect resulting from the rapid y-diffusion, this has very little effect on the y-concentrations. For the initial phase after the perturbation, we can regard y as constant. Thus, we can simplify Eq.1a into

$$\dot{x} = 5x - 5$$

Thus, a local elevation of x above the steady state x = 1 will grow further since $\dot{x}$ is positive due to an autocatalytic feedback of x on its own production.

However, after a sufficient increase of x, the change of y-concentration can no longer be neglected. Assuming a rapid equilibration of y to a given x-concentration, we can express the y-concentration in a steady state as function of x and get from Eq. 1b.

$$7y = 6x + 1$$

Inserting this into Eq. 1a leads to

$$\dot{x} = 1/7\ (1 - x)$$

i.e. the inclusion of the rapidly diffusing substance y leads to an overall stabilization. Thus, also in a Turing system a local x-perturbation grows due to local autocatalysis and the long-ranging substance assures that this does not lead to a general explosion. A x-increase in a particular area leads, due to the antagonistic effect of the long-ranging substance, to a decrease of x in the surrounding area, i.e. to a stabilization of the emerging pattern.

A drawback of the Turing equations is that they are molecularly unreasonable. Eq.1a means that the number of molecules that disappear per time unit is independent of the number of x-molecules present. Thus, x-molecules can even disappear if no x-molecules are present. This could lead to negative concentrations and to stability problems of the emerging pattern.

The understanding that the basic mechanism of pattern formation is local autocatalysis and long-ranging inhibition has enabled us to propose equations which avoid these problems by assuming decay rates proportional to the number of molecules present. For instance, the interaction between an autocatalytic activator a(x) (x is in the following the space coordinate) and a long ranging inhibitor h(x) has pattern forming capabilities (Gierer and Meinhardt, 1972; Gierer, 1981;

Meinhardt, 1982a):

$$\frac{\partial a}{\partial x} = c\frac{a^2}{h} - \mu a + D_a\frac{\partial^2 a}{\partial x^2} \qquad (2a)$$

$$\frac{\partial h}{\partial t} = ca^2 - \nu h + D_h\frac{\partial^2 h}{\partial x^2} \qquad (2b)$$

Since the decay term has to be proportional to a, the autocatalysis must be overproportional. This will occur if two activator molecules have to cooperate for the autocatalytic production of a third activator molecule. This leads to the square-term in Eq.2a,b. A similar consideration as given above for the Turing model shows the local instability of the activator distribution and the overall stabilizing effect of the rapidly diffusing inhibitor.

The spread of an incipient activator maximum also can be limited by the depletion of a long-ranging substance which is required for the autocatalysis of the activator:

$$\frac{\partial a}{\partial t} = ca^2 s - \mu a + D_a\frac{\partial^2 a}{\partial x^2} \qquad (3a)$$

$$\frac{\partial s}{\partial t} = c_o - ca^2 s + D_s\frac{\partial^2 s}{\partial x^2} \qquad (3b)$$

Both mechanisms have slightly different properties. An analysis of biological regulation phenomena indicates that in most cases a real inhibition spreads out from an activator maximum (see Meinhardt, 1982a). For insects I have shown that such an activator maximum must be located at the posterior (rear) pole of the egg and that the smoothly graded inhibitor is appropriate to provide positional information for the cells (Fig. 1; Meinhardt, 1977).

GENERATION OF POSITIONAL INFORMATION FOR ARMS, LEGS AND WINGS

Substructures such as legs and wings arise at precise positions within the developing organism. They have a handedness and the axes of a limb are reproducibly aligned with respect to the main body axes of the embryo. How can this be achieved during normal development? One possibility would be that particular regions become specified to form a limb by a primary subdivision of the anteroposterior axis (head to tail) into several distinct bands (as shown in Fig. 1) and a second subdivision perpendicular to the first (dorsoventral, back to belly) and further, that the pattern formation within these "limb fields" proceeds by the same reaction-diffusion mechanism as discussed above. However, regulatory phenomena in limb formation have suggested a different mechanism: that such secondary structures are formed around the borders which result from the primary embryonic pattern formation. Let us assume that the primary embryonic pattern formation leads to, among others, two adjacent regions to be called A (anterior) and P (posterior) and further that the A and P tissues collaborate to produce a morphogen (m). The m-production will be restricted to the A-P

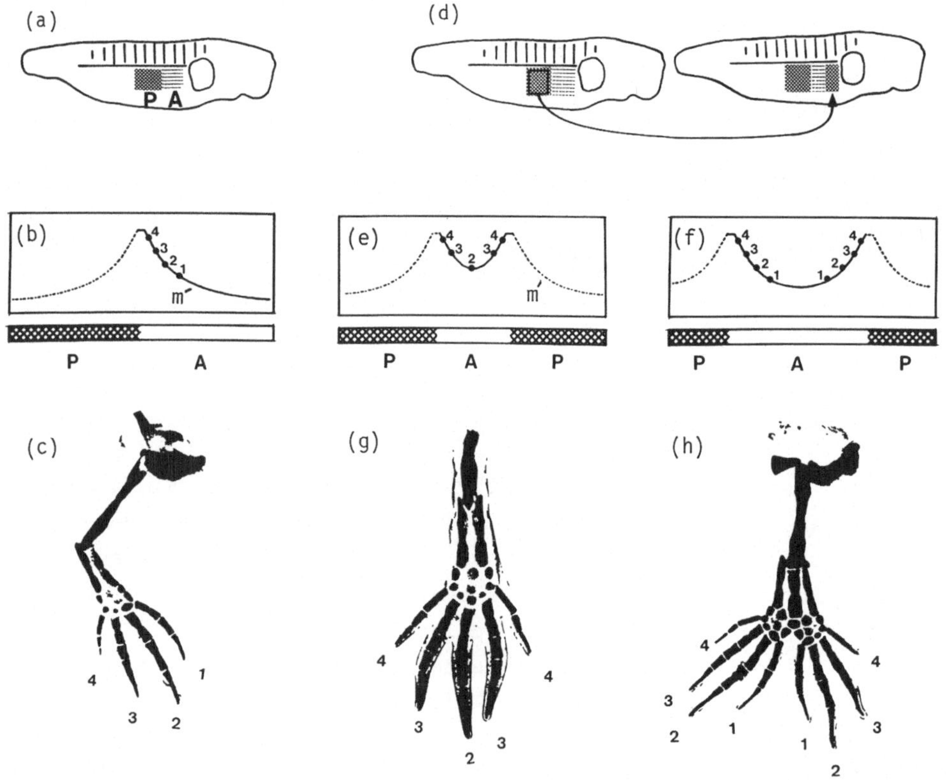

Fig.2 Induction of a vertebrate limb by the juxtaposition of a polarizing (P) and a competent (A) zone (Slack, 1976). It is assumed that both tissues (formed under the influence of a primary gradient as shown in Fig. 1) have to cooperate to produce a morphogen (m). The m-synthesis is possible only at the common border. (b) Only the A-cells can respond to this morphogen. The local concentration determines which digit will be formed. (c) A normal forelimb of an axolotl. (d) After grafting polarizing tissue (P) in front of competent (A) tissue, a symmetrical limb results (g,h). (e,f) Model: by this operation, a second AP border is formed; the competent zone becomes exposed to an U-shaped gradient. Depending on the distance between the two P-regions, the lowest concentration and thus the anterior-most digits may be missing (e,g). (Fig.c,g,h after photographs kindly supplied by J. Slack. For a more complete model of vertebrate limb formation, see Meinhardt 1982a, 1983b).

boundary region since only here can the collaboration take place (Fig. 2). The result would be a symmetrical m-distribution centered over the A-P boundary. In vertebrates, only the A-tissue seems to be able to respond to this morphogen gradient. The A-cells are thus exposed to an exponential m-distribution which leads to the polar sequence of digits (Fig. 2).

However, an A-P boundary is insufficient for limb determination. The A-P boundary would surround the embryo in a belt-like manner. It has a long extension and the precise location of a limb outgrowth remains to be determined. This can be achieved by an additional global dorso-ventral organisation of the developing embryo. The condition to form a limb would be that three particular tissues are close to each other, i.e. an intersection of two particular borders. Similar to three countries that border only at a particular location, such three tissues are in touch with each other only at a particular point defining in this way the position of a limb outgrowth. Also the handedness is determined by the clockwise or counterclockwise arrangement of the three areas. In insects, the legs are formed around the intersection of two boundaries. Our model accounts for the determination of the leg segments in concentric rings in a straightforward manner (Fig. 3). This boundary model for limb formation correctly explains many experimental observations (Meinhardt, 1982a, 1983a).

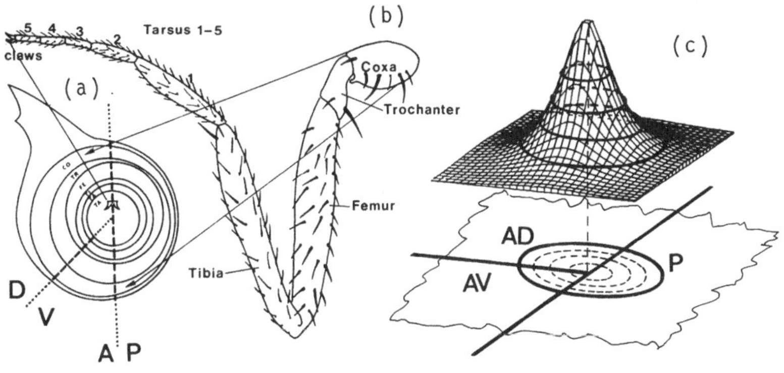

Fig.3 (a) The segments of an insect leg are layed down within the so-called imaginal discs in concentric rings. During metamorphosis, the central part of the disc evaginates and form in this way the adult leg (b). (c) Model: Cooperation of three different "compartments" (AD,AV,P) is required to produce the morphogen. A cone-shaped morphogen profile, centered over the intersection of the two borders result. The concentric contour lines determine the arrangement of the future leg segments (Meinhardt, 1982a, 1983a).

Our boundary model for secondary embryonic structures provides a rational explanation for why embryonic development is such a reproducible process: by a primary generation of positional information (via autocatalysis and lateral inhibition) a series of differently determined cells result which are separated by sharp boundaries. By cooperation of two pairs (one pair for each capital axis of the embryo), positional information is generated for the next finer subdivision. Since these substructures are organized from the primarily formed borders, they have necessarily the correct position, handedness and orientation with respect to the already determined parts of the embryo.

SEGMENTATION: THE PERIODIC SUBDIVISION IS THE PRIMARY EVENT

Let us go back to segmented structures. We have seen that they consist of a superposition of a sequential structure (1,2,3...) and of a periodic structure (APAPAP or /SAP/SAP/; see below). Both types of structures are precisely in register. That means that one process must be the master process and the other must be under its control. It may appear reasonable that the sequential structure is formed first and that in a secondary process, each particular area becomes subsequently subdivided, e.g. area 2 into 2A and 2P. However, Drosophila flies carrying a mutation in the Bithorax gene complex (Lewis, 1978) tell us that the periodic subdivision is the primary event (Fig. 4). This makes sense also from an evolutionary point of view: The segments in lower animals (Arthropodes and Annelides) seems to be much more similar, indicating that the invention of the repetition of structures was the primary event and that only later in evolution, these repetitive elements have become different from each other.

The periodic structures in insects are the alternating compartments (Garcia-Bellido et al., 1973). These compartments are formed in a cell layer underneath the egg shell, the blastoderm. The compartments are thus long belt-like stripes surrounding the insect egg. They have a very narrow extension in the anteroposterior dimension: a width of only one or two cell diameters.

HOW TO GENERATE STRIPES

Thus, in insects the periodic structure has the geometry of narrow stripes and since the formation of the periodic structure is the primary event, we have to explain their formation first. The mechanism

Superposition of sequential and periodic structure: what is the primary event?

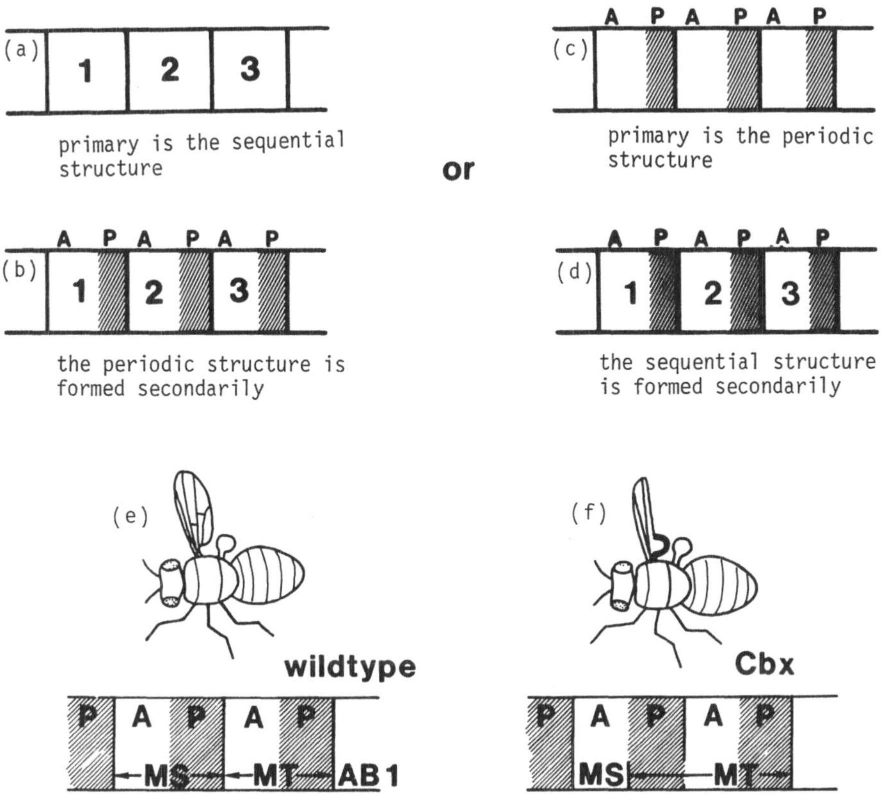

Fig.4 If the segmental and the periodic patterns are so precisely in register, one pattern must be the primary one, the other must arise under the control of the first. (a,b) Either the sequential pattern is formed first and, subsequently, each region of segmental specificity (1,2,3..) becomes subdivided into an A and a P region. Or, it is the other way round (c,d) and each primarily formed AP pair obtains a particular specification. (e,f) Drosophila flies carrying mutations in the bithorax gene complex (Lewis, 1978) indicate that the latter is correct. (e) Wildtype; (f) a fly carrying a Cbx mutation; the APAP.. pattern is unchanged but the metathoracic specificity (MT) extends into the Mesothorax (MS), indicating that it is not that the MT region becomes subdivided into an A and a P region.

of autocatalysis and lateral inhibition is unable to account for the generation of a pattern with a long extension in one dimension and a small extension in the other. However a stripe-like arrangement of cells in two different states (to be called A and P) can emerge if these two states activate each other mutually on long range but exclude each other locally (Fig. 5-7; Meinhardt and Gierer, 1980).

Then, both cell types need each other in a close neighborhood in a symbiotic manner. Stripes are the most stable state since in this geometry the long common boundaries between both cell types enable an efficient mutual stabilization.

The following interactions can lead to the formation of stripes.

$$\frac{\partial a}{\partial t} = \frac{\mu s_p a^2}{r} - \mu a + D_a \nabla^2 a \tag{4a}$$

$$\frac{\partial p}{\partial t} = \frac{\mu s_a p^2}{r} - \mu p + D_p \nabla^2 p \tag{4b}$$

$$\frac{\partial r}{\partial t} = \mu s_p{}^2 a^2 + \mu s_a p^2 - \beta r \tag{4c}$$

$$\frac{\partial s_a}{\partial t} = \nu(a - s_a) + D_s \nabla^2 s_a \tag{4d}$$

$$\frac{\partial s_p}{\partial t} = \nu(p - s_p) + D_s \nabla^2 s_p \tag{4e}$$

Such a system look complicated at the first glance. Its mode of operation is, however, easily understood (Fig. 5). According to

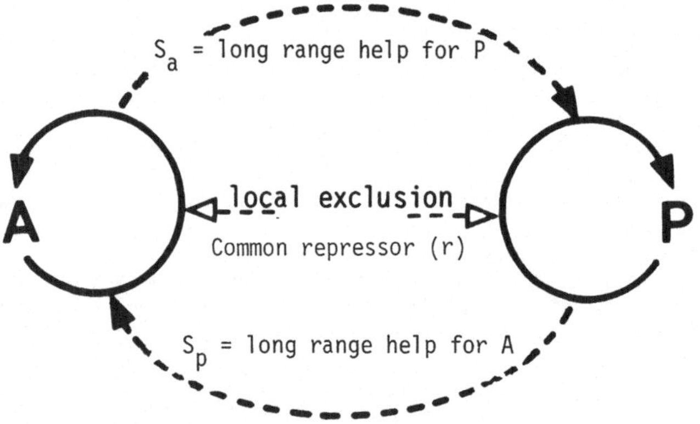

Fig.5 A molecular interaction which enables the generation of stripes. Both, the A and the P molecules feed back on their own production but both feedback loops compete via a common repressor R. This creates an instable situation and, in a particular cell, either the A loops become fully activated and the P loop completely suppressed or vice versa. If, on long range, both loops support each other (via S_a or S_p, see Eq.4) both loops need each other in a symbiotic way. Stripes of high A and high P concentration are especially stable since the long common boundaries between A and P allow an efficient stabilization.

Eq.4a,b, two autocatalytic feedback loops are assumed (a and p). Both these loops compete via a common non-diffusible repressor r which is produced by and which acts on both loops. This leads in a particular cell to a full activation of one loop and the suppression of the other (Meinhardt, 1978). Both loops help each other via s_a and s_p respectively. So, if in a particular group of cell the a-loop is turned on, the p-loop will be activated in neighboring cells (Fig. 6).

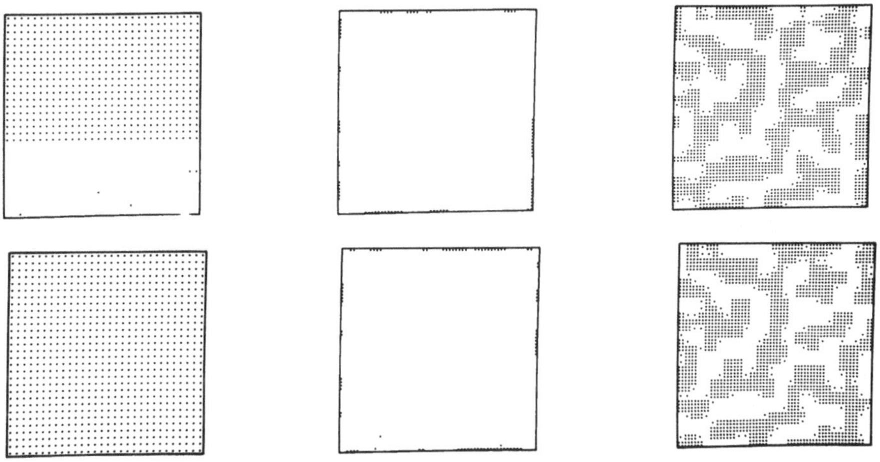

Fig.6 Stages in the formation of a stripe-like pattern. The emerging A- (top row) and P-pattern are complementary to each other. Due to the initiation by random fluctuations, the pattern is somewhat irregular but the regions of high A and high P concentrations have a much longer extension in one dimensions than in that perpendicular to the first.

A simpler system which enables stripe formation on the basis of the same principle is given in Eq.5a-c:

$$\frac{\partial a}{\partial t} = \mu\left(\frac{1}{s^2(\kappa+p^2)} - a\right) + D_a\nabla^2 a \tag{5a}$$

$$\frac{\partial p}{\partial t} = \mu\left(\frac{s}{\kappa+a^2} - p\right) + D_p\nabla^2 p \tag{5b}$$

$$\frac{\partial s}{\partial t} = \nu(a - ps) + D_s\nabla^2 s \tag{5c}$$

(κ limits a (and p) production at low p (and a) concentrations. To obtain a stable pattern, s has to equilibrate more rapidly than a and p, i.e. $\nu>\mu$.)

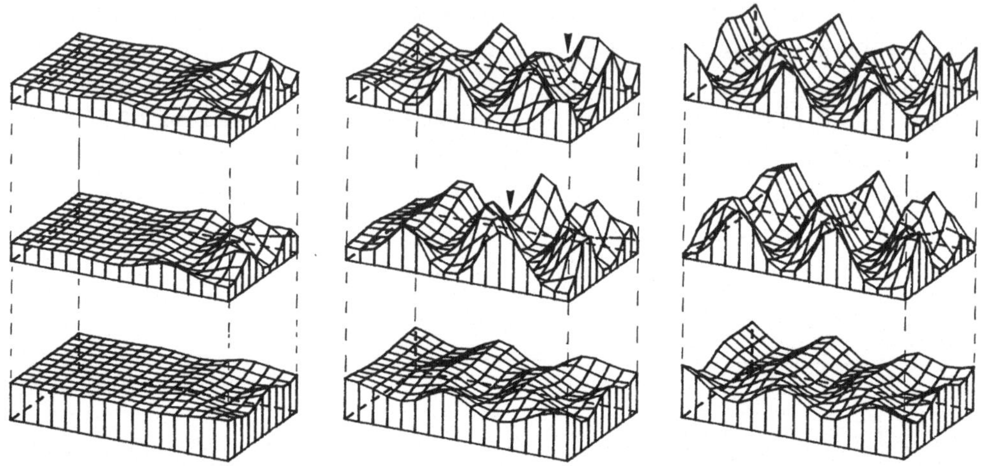

Fig.7 Formation of a stripe-like A-P pattern initiated by a local perturbation. Calculated with Eq.5a-c with $\mu=.02$, $\nu=.03$, $D_a=.004$, $D_s=.2$; plotted after 600,1200 and 3600 iterations. The stripe-like character improves in the course of time. Initial peaks and valleys (▼) on the ridges are smoothed out.

Here, the autocatalysis results from an inhibition of an inhibition. Imagine that a and p would be in a steady state. A small a-increase leads to a p-decrease which, in turn, leads to a further a-increase. Eq. 5a,b accounts for autocatalysis and for the mutual exclusion. The mutual help is mediated by the long-ranging substance s; s is produced by a and inhibits a. This selfinhibition provides a help for the competiting feedback loop p. On the other hand, cells with high p-concentration remove s and help in this way the a-production. Fig. 7 is calculated with this set of equations. Stripe formation is also possible if the s-removal is independent of the p-level ($-\nu s$ instead of $-\nu ps$) but then the system is more sensitive on the selection of parameters. Important is that the long-ranging substance acts on both feedback loops in an antagonistic manner. In contrast, if, for instance, the a-production has no selfinhibitory component (Eq.6), the pattern formation capability remains but the characteristics of the pattern formation is altered from the lateral activation type (stripes) to the lateral inhibition type (bristle-like pattern):

$$\frac{\partial a}{\partial t} = \mu\left(\frac{1}{\kappa+p^2} - a\right) + D_a\nabla^2 a \qquad (6a)$$

$$\frac{\partial p}{\partial t} = \mu\left(\frac{s}{\kappa+a} - p\right) + D_p\nabla^2 p \qquad (6b)$$

$$\frac{\partial s}{\partial t} = \nu(a - s) + D_s\nabla^2 s \qquad (6c)$$

Fig. 8 shows a simulation with Eq.6 and a comparison with Fig. 7 shows the different types of patterns produced by the two sets of equations (an analytical treatment which shows why the one type of equations form stripes (Eq.4,5) and others not (Eq.2,3,6) would be highly appreciated).

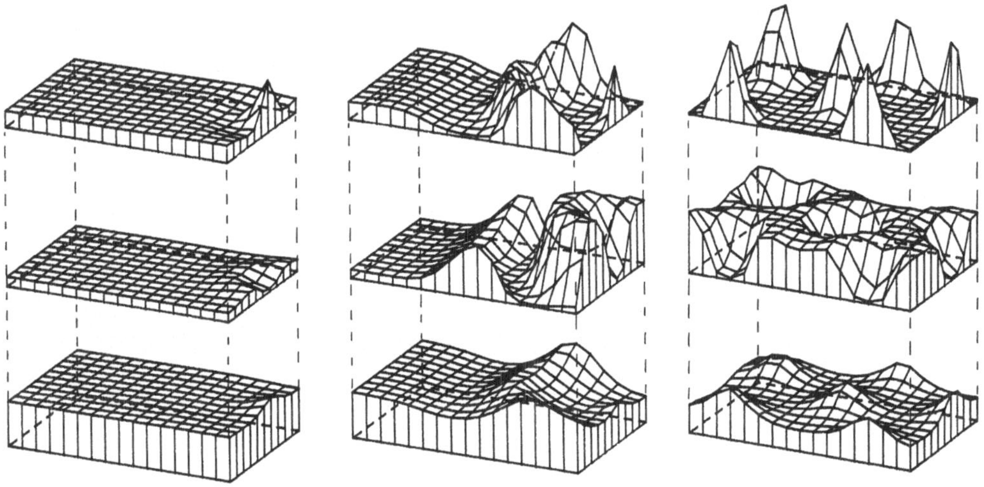

Fig.8 Minor changes in the equations (Eq. 6 versus Eq.5) lead to pattern of the lateral inhibition type; isolated maxima appear. Eq.6 with the same constants and initial conditions as in Fig.7 have been used (for comparison, see Fig.7).

THE FORMATION OF A REGULAR STRIPE-LIKE PATTERN UNDER THE INFLUENCE OF THE MORPHOGEN GRADIENT

The lateral activation mechanism, if initiated just by random fluctuations lead to a somewhat irregular stripe-like pattern (Fig. 6). The obvious means to get a regular pattern would be the organizing morphogen gradient mentioned above. Before explaining how this may work it is important to realize that in a lateral activation mechanism a homogeneous cell population has the tendency to oscillate between the two states (A and P) since if, for instance, all cells are in the A

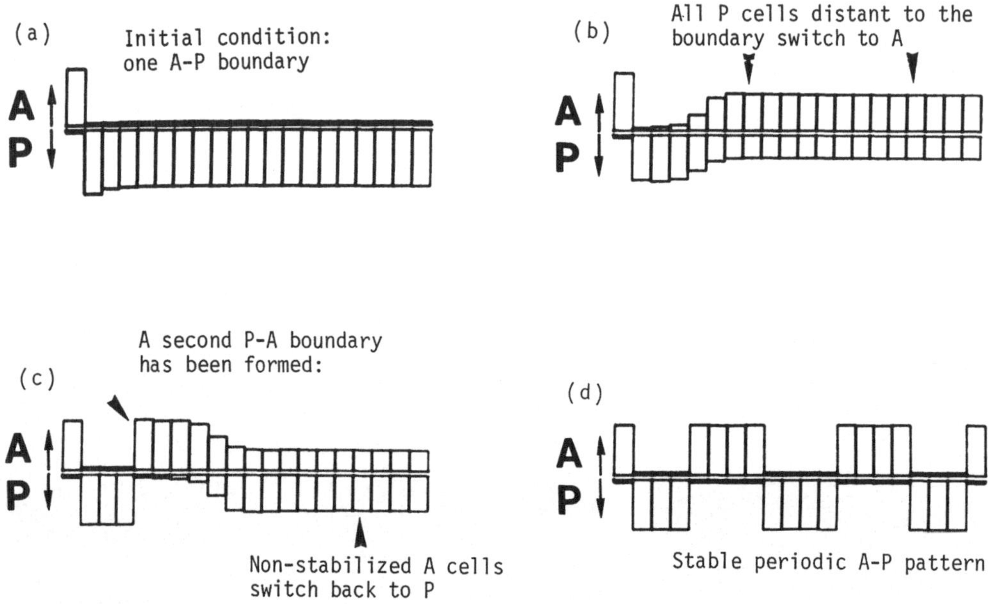
(a)
Initial condition:
one A-P boundary
A
P
(b)
All P cells distant to the
boundary switch to A
A
P
(c)
A second P-A boundary
has been formed:
A
P
Non-stabilized A cells
switch back to P
(d)
A
P
Stable periodic A-P pattern

Fig.9 Generation of a regular A-P pattern. (a) Cells around a primary A-P boundary stabilize each other. Cells distant to this boundary switch to the other state and create in this way a new boundary (b,c). The oscillation comes to a rest (d) if the total field is subdivided into A and P regions.

state, the P loop would be supported while the A loop remains without support. The cells would switch from A to P. After a further time interval and for the same reason, the P cells can switch back into the A-state. However, if one A-P boundary would exist in the system, the A and P cells around this boundary would stabilize each other. Thus, oscillation would continue only at distance from this boundary and each complete oscillation would create a new pair of a A and P cells. The oscillations come to a rest if the field is subdivided into A and P stripes which stabilize each other (Fig. 9).

How can the first boundary arise? Let us assume that initially all cells are in the P-state and that a certain threshold level of the morphogen gradient is required to switch from P to A. All cells above this threshold will switch from P to A, creating the first boundary. Cells distant from this boundary will switch back to A and so on. The formation of new stable A-P pairs spreads out from the initial boundary in a wave-like manner. The region of stable A-P stripes enlarges in expense of the oscillating cells until the total area consists of stable A-P stripes (Fig. 10).

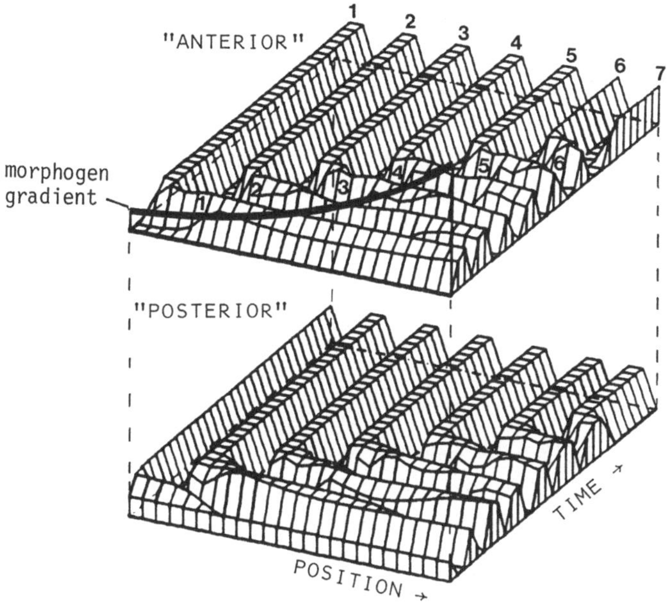

Fig.10 A-P-A oscillations under a control of a morphogen: A certain threshold is required for the first P-A transition. In this way, the first P-A border is formed. Further A-P oscillations lead to a regular AP pattern. The number of oscillations a cell has made in its history corresponds to its position in the field (after Meinhardt, 1982a).

THE ACTIVATION OF PARTICULAR CONTROL GENES UNDER THE INFLUENCE OF THE A-P OSCILLATION

How can these stripes become different from each other? From the mutants of the Bithorax complex (see Fig. 4) we know that particular genes become activated in this process. It is a property of the model as outlined so far that the number of A-P oscillations a cell has made in its history agrees with the number of AP stripe to which it belongs, counted from the first A-P boundary formed (Fig. 10). Thus, it would be appropriate if a cell switches under the influence of the A-P oscillations from one gene to the other. The mechanism may be compared with a pendulum-escapement mechanism of a grandfathers clock. Lifting up the weights initiate the periodic movement of the pendulum. With each complete left-right-left movement of the pendulum, the hand of the clock advances one unit. The periodic movement of the pendulum is the primary event. The sequential advancement of the hand is under its control.

To see how a sequential activation of control genes may work, let me use another analogy: Imagine a ship in a channel system with locks. A lock can be in two states. For instance, if the lower gate is open, the ship can enter the lock but it can pass only after a switch to the other state (lower gate closed and upper gate open). One state is characterized by the preparation of the transition but this transition is blocked. In the other state, the transition is no longer blocked, but no preparation of the next transition is possible. After a switch from one state to the other, the transition can take place. Therefore, I assume that in the P-state a substance is produced which tries to switch from gene g_i to g_{i+1} (crosses in Fig. 11) but in P, this transition is blocked. This block is released in the A state and the transition occurs. Fig. 11 shows a simulation of this pendulum escapement mechanism. A computer program for this and some other pattern forming reaction can be found elsewhere (Meinhardt, 1982a).

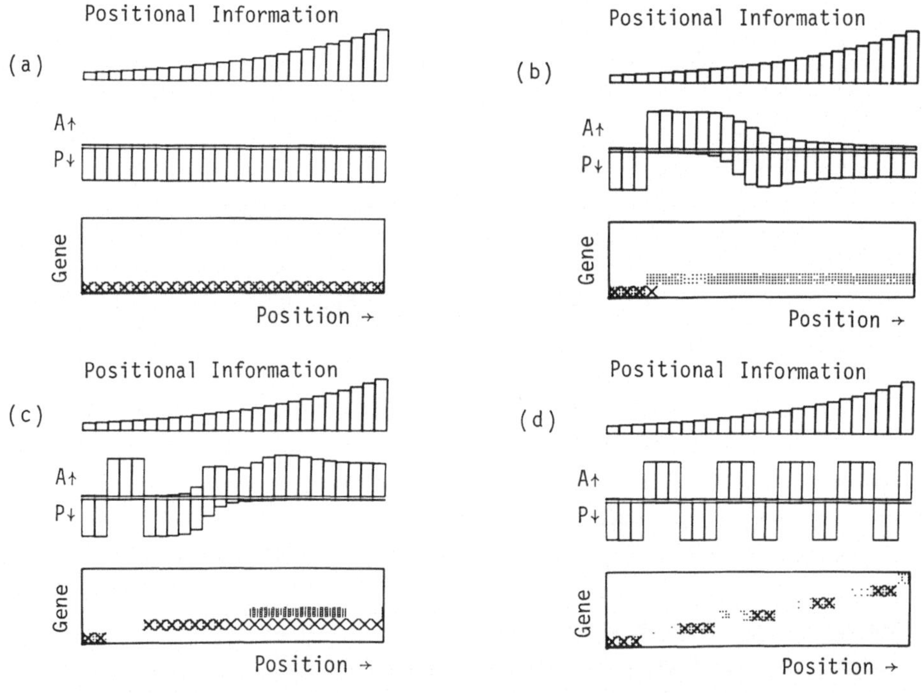

Fig.11 The oscillations between A and P (see Fig.9 and 10) can be used for a sequential activation of control genes. The activation of a subsequent control gene occurs simultaneously with a P-A transition. (a) Initial conditions. (b,c) Intermediate stages and (d) the finally stable steady state. The result is a periodic P-A-P pattern and in each AP pair, a particular control gene is active (for details of this simulation, see Meinhardt, 1982a). The A-P borders can be used to initiate the formation of legs and wings (see Fig.3).

This mechanism allows an assignment of functions to the known elements of the Bithorax gene complex. Not only the phenotypes of the mutations but also their cis/trans behavior as well as the phenotypes of double mutants are correctly described (Meinhardt, 1982a-c).

According to the model proposed, the sequential/periodic structure appears not simultaneously but in a time sequence. The somites of the vertebrates are formed in such a sequence (Fig. 12).

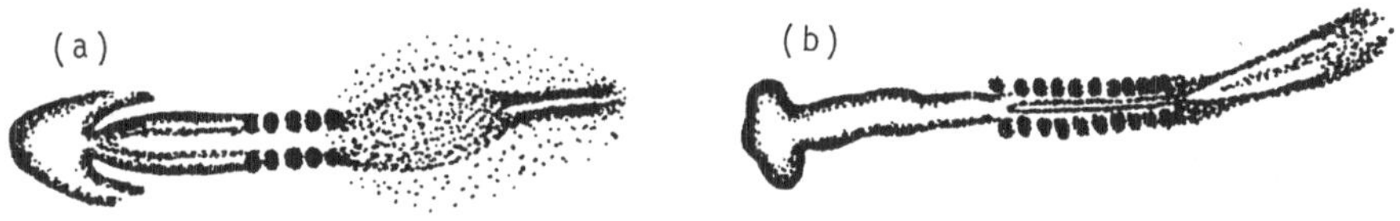

Fig.12 The formation of the basic sequential/periodic structure in vertebrates, the somites. Shown is a chicken embryo at 25 and 37 hours of incubation. New somites are separated from unstructured tissue in an anteroposterior sequence. This sequence is in agreement with the proposed mode of the formation of the periodic structure (Fig.9-11).

In several insect species, new segments are formed at the posterior end by a sprouting-like process. This can be accounted for by the model without any additional assumption: Whenever, let us say, the P region would become too large due to growth, the terminal P cells would be insufficiently supported and would switch to A and, after further growth, back to P. A similar sequential/periodic structure as described above would emerge.

SEGMENTATION REQUIRES THE INTERACTION OF THREE STATES

How is a segment border formed? It cannot be caused by the juxtaposition of A and P cells since the same juxtaposition would take place within each segment (Fig. 1,4). It can also not arise at juxtaposition of cells in which different segmental control genes are active since, as shown in Fig. 4d, even if meso- and metathoracic cells are confronted at an A-P boundary within a segment, this leads not to a segment boundary. Also, in an APAPAP pattern, the grouping of the stripes would not be defined; it can be AP/AP/AP or A/PA/PA/P. All these problems would disappear if one assumes that the primary oscillation takes place not between two but between three states. This would lead to the iteration of three stripes, let us say /SAP/SAP/. The segment border as well as the polarity within a segment would be unambiguously defined if, for instance, the confrontation between P and S cells leads to the induction of a segment border. This would explain further why limbs can be formed only within a segment. As mentioned above, an insect leg is formed over an A-P border (Fig. 3). If a three-fold subdivision takes place, no such border is present at the segment (P/S) border. However, such border can be experimentally provoked by surgical removal of an S-area with subsequent wound healing. This experiment has been done by Bohn (1974) who found that such an injury induces the formation of a seventh leg. This leg has had a reversed orientation and handedness, in agreement with the model proposed.

THE SITUATION IS STILL MORE COMPLICATED IN DROSOPHILA

The model, even if it appears already quite complicated, is certainly not sufficient to account for all related phenomena in Drosophila. For instance, after complete deletion of the Bithorax gene complex all posterior segments obtain mesothoracic specificity (in agreement with the model). However, the number of the segments remain unchanged. Thus, counting the segments and giving them particular names are two different processes. Further, recently obtained mutations (Nüßlein-Volhard and Wieschaus, 1980) indicate an intermediate formation of double segments. All these features have to be incorporated into a forthcoming theory of segmentation.

CONCLUSION

The precise mathematical formulation of molecular interactions has enabled us to develop models for complex developmental phenomena such as segmentation or limb initiation. Many models initially taken into consideration have turned out to be unable to account for the experimental observations. However, after finding a model consistent with the initially chosen set of experiments, such a model has in most cases also accounted for phenomena for which it was not originally designed, indicating that the models describe important features of what happens in reality. Discrepancies between the models and forthcoming experimental observations will certainly allow for improvements of these models and thus improve our picture of how development is controlled.

I wish to express my sincere thanks to Prof. A. Gierer for many years of fruitful collaboration.

REFERENCES

Bohn, H. (1974). Extent and properties of the regeneration field in the larval legs of cockroaches (Leucophaea maderae). I. Extirpation experiments. J. Embryol. exp. Morph. Vol. 31, 3, 557-572.

Garcia-Bellido, A., Ripoll, P., Morata, G. (1973). Developmental compartmentalization of the wing disk of Drosophila. Nature New Biol. 245, 251-253.

Gierer, A. (1981). Generation of biological patterns and form: Some physical, mathematical, and logical aspects. Prog. Biophys. molec. Biol. 37, 1-47.

Gierer, A., Meinhardt, H. (1972). A theory of biological pattern formation. Kybernetik 12, 30-39.

Lewis, E. B. (1978). A gene complex controlling segmentation in Drosophila. Nature 276, 565-570.

Meinhardt, H. (1977). A model for pattern formation in insect embryogenesis. J. Cell Sci. 23, 117-139.

Meinhardt, H. (1978). Space-dependent cell determination under the control of a morphogen gradient. J. theor. Biol. 74, 307-321.

Meinhardt, H. (1982a). Models of Biological Pattern Formation. Academic Press, London.

Meinhardt, H. (1982b). The role of compartmentalization in the activation of particular control genes and in the generation of proximo-distal positional information. Am. Zool. 22, 209-220.

Meinhardt, H. (1982c). Theory of regulatory functions of the genes in the bithorax complex. In: Embryonic Development, Part A: Genetic Aspects (M.M. Burger and R. Weber, Eds.). pp 337-348. Alan R. Liss, New York.

Meinhardt, H. (1983a). Cell determination boundaries as organizing regions for secondary embryonic fields. Dev. Biol. 96, 375-385.

Meinhardt, H. (1983b). A boundary model for pattern formation in vertebrate limbs. J. Embryol. exp. Morph. 76 (in press).

Meinhardt, H., Gierer, A. (1980). Generation and regeneration of sequences of structures during morphogenesis. J. theor. Biol. 85, 429-450.

Nüsslein-Volhard, C., Wieschaus, E. (1980). Mutants affecting segment number and polarity in Drosophila. Nature 287, 795-801.

Slack, J. M. W. (1976). Determination of polarity in the amphibian limb. Nature 261, 44-46.

Turing, A. (1952). The chemical basis of morphogenesis. Phil. Trans. B. 237, 37-72.

SPATIAL PATTERN FORMATION IN THIN LAYERS OF NADH-SOLUTIONS

S.C. Müller and Th. Plesser
Max-Planck-Institut für Ernährungsphysiologie,
Rheinlanddamm 201, D-4600 Dortmund, FRG

Introduction

Spatial pattern formation in a thin layer of initially homogeneous cell extract from yeast has been observed by Boiteux and Hess /1/. The patterns in the periodically glycolyzing extract are detected by changes in absorbance of reduced nicotinamide adenine dinucleotide (NADH). This organic compound is an important intermediate in the process of enzymatic degradation of glucose into alcohol and CO_2.

Today, the availability of modern digital video techniques for the recording of images considerably improves the quantitative investigation of pattern formation processes. In this paper such techniques are applied to the measurements and analysis of spatial structures in thin layers of simple aqueous NADH solutions. These layers can be considered as an experimental model system revealing the formation of structure due to hydrodynamic phenomena which may also occur in thin layers of reactive solutions, for instance yeast extract.

Equipment and Methods

Our equipment for experiments and image analysis is shown in Fig. 1. The apparatus is a two dimensional spectrophotometer provided that a specific wavelength is selected by appropriate optical filtering. The video camera measures the intensity of the incoming light as a function of time t and of the target coordinates x and y. The analog intensity pattern is converted into a raster image of 512 x 512 discrete picture elements (pixels). The light intensity measured at each pixel is digitized and converted into one out of $2^8 = 256$ grey levels. The sequence of digital images is stored on magnetic disk and thus accessible for further analysis of the temporal and spatial evolution of the intensity distribution in the sample dish. In order to account for the absorption properties of NADH with an absorption maximum at 340 nm, an UV sensitive VIDICON tube (Hamamatsu N983) is used. It is assumed, that the transmitted UV light intensity is determined only by the absorption properties of the sample and that Lambert-Beer's law holds for each pixel:

$$I(x,y,t) = I_o(x,y) \cdot 10^{-A(x,y,t)} \tag{1}$$

$I_o(x,y)$ is the reference intensity for each pixel;

$$A(x,y,t) = \varepsilon(x,y,t) \cdot c(x,y,t) \cdot d(x,y,t) \tag{2}$$

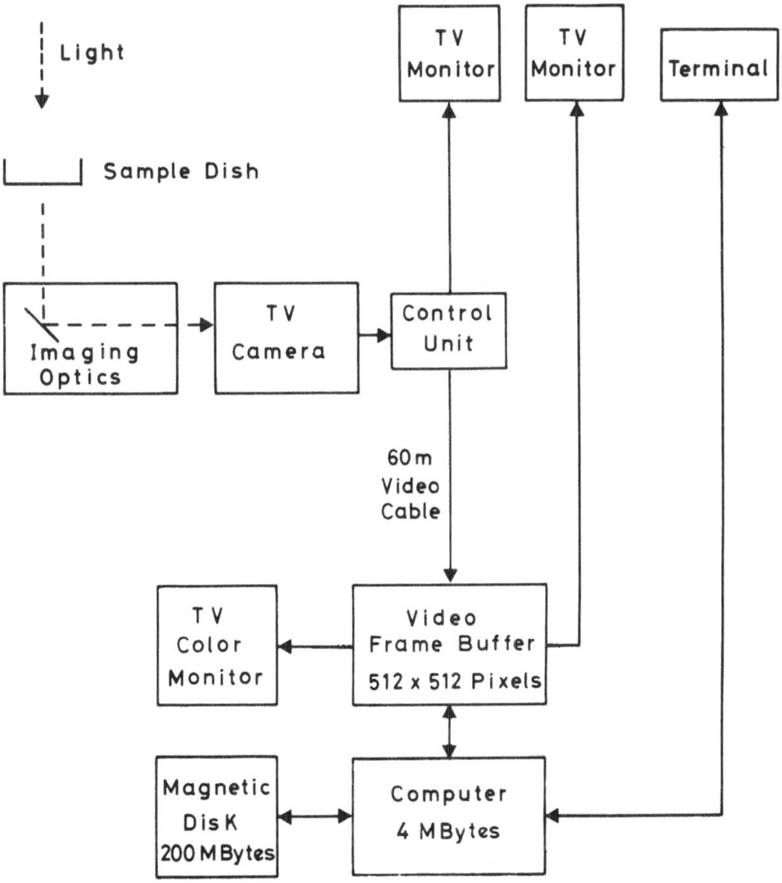

Fig. 1: Experimental setup with digital video system

is the absorbance or optical density of the sample where ε, c, and d denote the local value of the molar extinction coefficient, the molar concentration of absorbent, and the depth of the layer in cm, respectively.

Experiments and Results

The experiments on pattern formation were started by pouring a predetermined volume of NADH solution into a flat container (a petri dish of 1.6 cm radius). For concentrations ranging between 10^{-3} and 10^{-2}M rod like structures become detectable after about four minutes. An example is shown in Fig. 2. The diameter of the rods increases almost linearly with the depth of the layer, as measured in a 5.10^{-3}M NADH solution with a layer thickness ranging from 0.12 to 0.28 cm. The structure disappears after covering the dish with a glass plate (compare Fig. 3 for quantitative data). It arises again after removal of the plate. In the open dish the temperature close to the boundary is found to be approximately 0.3^{o}C higher than at the center. The temperature distribution in the covered dish is homogeneous within the experimental error of 0.05^{o}C. The occurrence of patterns is independent of the incident monitoring light.

With the recording technique shown in Fig. 1 structure is only detected for wavelengths, which are specific for NADH absorption (<410 nm). Therefore, the observed spatial inhomogeneities are predominantly caused by the absorption properties of the sample, whereas other effects, for example changes in refractive index, can be neglected. Inspection of the surface of the liquid layer by the naked eye shows no irregularities. Chemical interactions at the liquid gas interface were excluded by experiments under an argon atmosphere.

The experimental data presented so far suggest, that the observed pattern formation is coupled with convection and that the process is driven by temperature gradients produced by evaporation. From the hydrodynamic equations it is known that convection in thin liquid layers is driven by buoyancy and by surface tension. For reviews see /2,3/.

The critical dimensionless parameters in the hydrodynamic equations for the onset of convective flow are the Rayleigh number Ra and the thermal Marangoni number Ma. They are given by the expressions

$$Ra = \frac{g\alpha d^3\ \Delta T}{\kappa\ \nu} = 14\cdot 10^3\ d^3\ \Delta T, \tag{3}$$

$$Ma = -\frac{d\sigma}{dT}\frac{d\ \Delta T}{\rho\kappa\nu} = 11\cdot 10^3\ d\ \Delta T. \tag{4}$$

The symbols and their numerical values for water at 20°C are compiled in the following table:

thermometric conductivity	κ	=	$1.4\ 10^{-3}\ cm^2/sec$
kinematic viscosity	ν	=	$1.0\ 10^{-2}\ cm^2/sec$
volume expansion coefficient	α	=	$2.0\ 10^{-4}\ {}^{o}C^{-1}$
surface tension variation with temperature	$\frac{d\sigma}{dT}$	=	$-\ 0.16\ dyn/cm^{o}C$
acceleration of gravity	g	=	$981\ cm/sec^2$
density	ρ	=	$0.998\ gram/cm^3$

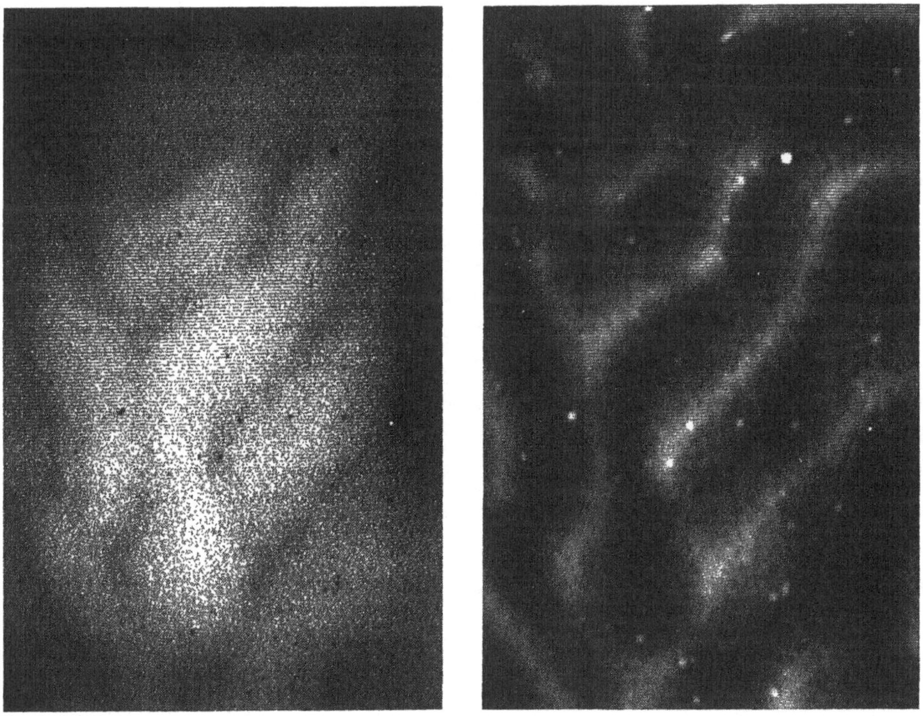

A B

Fig. 2: Pattern in a 0.18 cm layer of $5\cdot 10^{-3}$M NADH solution, observed at 380 nm. (A) Intensity of transmitted light I(x,y,t) seven and a half minutes after the start of the experiment. (B) Absorbance A(x,y,t) calculated from I(x,y,t). The maximum difference in absorbance is 1.4% of the average absorbance. This difference is represented by a grey level interval of thirty.

For a liquid layer with an insulating free surface and a rigid ideal heat conducting bottom plate the critical Rayleigh number is 669 and the critical Marangoni number is 79.6 /4/. From the observed maximum difference in temperature of $0.3^{O}C$ and the layer thickness of 0.18 cm follows Ra = 24.5 and Ma = 594. Since only Ma is well above the critical limit it turns out that the patterns are formed by surface tension forces.

It should be added, that surface tension is not only a function of temperature, but also of chemical composition /5/. The corresponding critical parameter in the hydrodynamic equations is the solutal Marangoni or elasticity number

$$Ma_s = -\frac{d\sigma}{dc}\frac{d\ \Delta c}{\rho D \nu} \qquad (5)$$

D is the mass diffusion constant of the solute. The parameter Ma_s is essential for the hydrodynamic stability of systems in which chemical or biochemical reactions are taking place. Note that the Marangoni numbers Ma and Ma_s do not depend on the absolute value of the surface tension σ.

The importance of surface tension is also supported by an experiment with the surfactant Sodium Dodecylsulfate (SDS), which does not undergo chemical reaction with NADH. A mixture of $5 \cdot 10^{-3}$M NADH and 0.17 M SDS was prepared and a volume of 10 μl of this mixture was added to a $5 \cdot 10^{-3}$M NADH solution in which a pattern had developed, following a procedure given in /6/. The addition was realized by carefully pipetting the surfactant mixture on the surface at the boundary of the dish. Immediately after the drop has touched the surface the pattern disappears. This experiment demonstrates clearly, that surface tension is the dominant factor in the formation of the observed pattern.

After getting an idea on the mechanism of pattern formation in these experiments, the question remains which property of the solution makes the modulation of absorbance in space and time detectable. According to equation (2) three factors have to be considered. Inhomogeneities in NADH concentration are excluded by the experiment with SDS: The rapid disappearance of the structures cannot be accomplished by diffusion. It would take several minutes to overcome the distance of 0.1 cm between the bright and dark areas in Fig. 2B. The molar extinction coefficient depends on the electron distribution within the NADH molecule. Its temperature dependence can be neglected /7/. Therefore the difference ΔA in absorbance at any pair of pixels with coordinates $(x_1 y_1)$ and (x_2, y_2), $\Delta A = A(x_1, y_1) - A(x_2, y_2)$, has to be attributed to the difference $\Delta d = d(x_1, y_1) - d(x_2, y_2)$ of the thickness of the liquid layer. From (2) follows:

$$\Delta d = \frac{\Delta A}{c_o \, \varepsilon_o} \tag{6}$$

The NADH concentration c_o and the molar extinction coefficient of NADH ε_o are constant quantities ($\varepsilon_o \simeq 10^3 M^{-1} cm^{-1}$ at 380 nm). The maximum difference in absorbance between the brighter and the darker areas of Fig. 2B leads to a Δd value of the order of 25 μm. The brighter areas are the elevated parts of the surface. The elevation is significantly larger than that of about 0.5 μ observed by Bénard on spermaceti (see /3/ page 587).

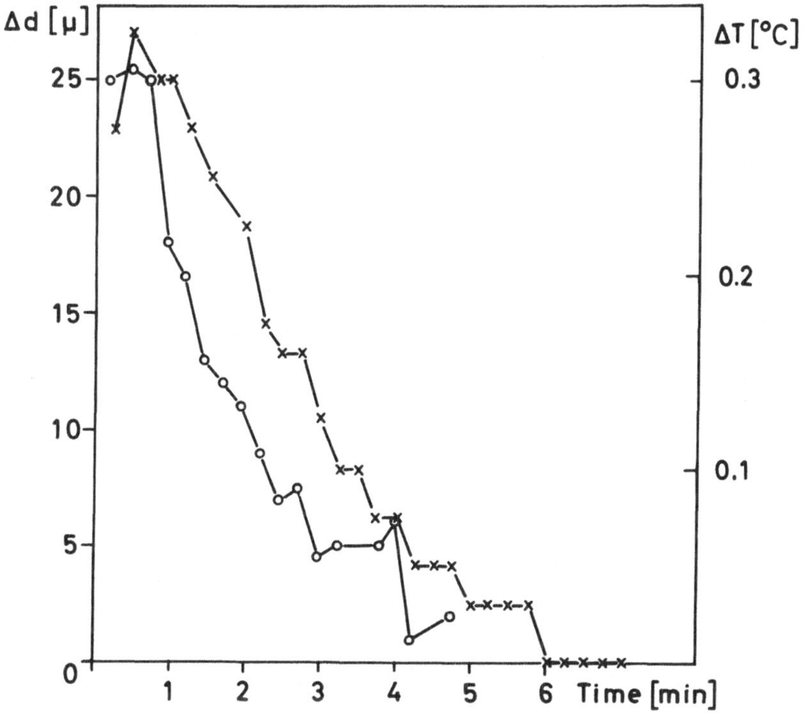

Fig. 3: Time dependence of Δd and ΔT after covering a dish, which contains a 5.10^{-3}M NADH solution of 0.18 cm layer thickness.

Fig. 3 shows the temporal evolution of the maximum value of Δd and the temperature difference ΔT in the bulk of the solution after suppression of evaporation by covering the petri dish with a glass plate. It is interesting, that there is a time lag of about one minute after which Δd drops abruptly and that this change precedes the temperature equilibration in the bulk.

The experiment with the surface active reagent SDS mentioned above leads to an unexpected transient pattern shown in Fig. 4. The organization of this pattern starts to be detectable about one and a half minutes after addition of SDS and lasts for about five minutes. Then this surfactant induced pattern disappears and a temperature driven rod like structure similar to that shown in Fig. 2 arises again. The pattern of Fig. 4 is a consequence of an unbalanced stress distribution at the surface during the transition of the mean surface tension for an aqueous NADH solution to the mean surface tension for the same solution

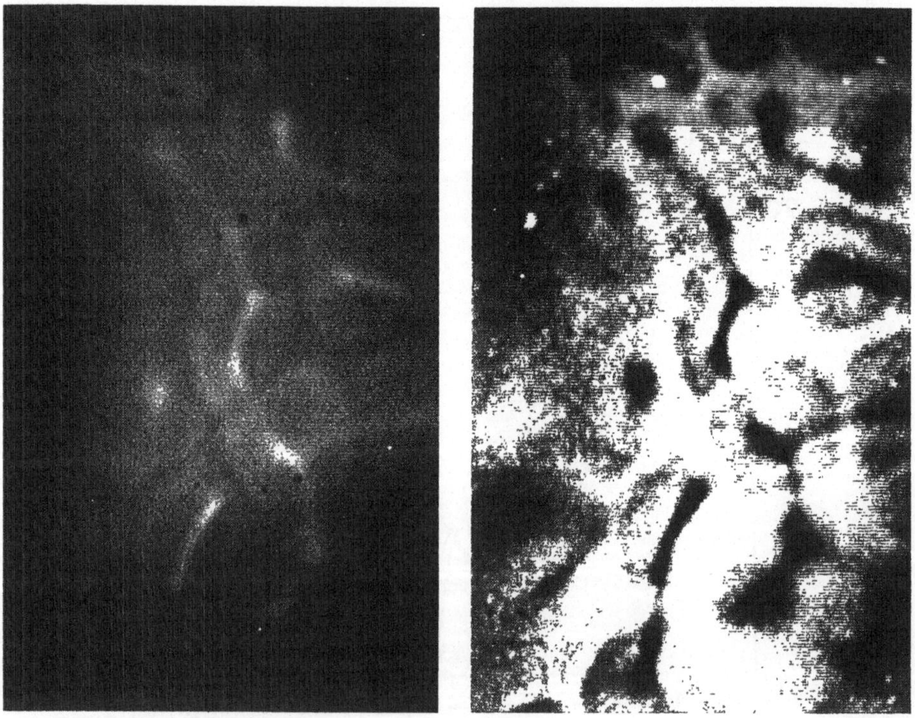

A B

Fig. 4: Pattern in a 5.10^{-3}M NADH solution three minutes after addition of surfactant. (A) Intensity of transmitted light I(x,y,t). (B) Absorbance A(x,y,t) calculated from I(x,y,t) as in Fig. 2.

with SDS. The reappearance of the temperature driven pattern indicates, that also with SDS the Marangoni number remains above the critical limit. It is noteworthy, that very similar patterns to those shown in Fig.4 are observed in systems with light induced photochemical reactions (see Kagan et al. these proceedings and Micheau et al. /8/).

Conclusions

A series of experiments with thin layers of NADH solutions shows, that modern digital video and computer techniques allow for quantitative analysis of pattern formation processes in space and time. The results obtained for NADH solutions reveal, that in order to understand spatial structures observed in yeast extract /1/, the hydrodynamic equations for surface tension driven convection must be taken into account. Instabilities at the surface may be driven by temperature gradients due to evaporation and by variation of the chemical composition in the bulk due to periodic changes of the activity of the enzymes in the yeast extract.

References

/1/ Boiteux, A. and Hess, B. (1980)
Ber. Bunsenges. Phys. Chem. 84, 392-398

/2/ Convective Transport and Instability Phenomena
(J. Zierep and H. Oertel eds.) G. Braun, Karlsruhe (1982)

/3/ Normand, Ch., Pomeau, Y. and Velarde, M.G. (1977) Rev. Modern Physics 49, 581-624

/4/ Nield, D.A. (1964)
J. Fluid Mech. 19, 341-352

/5/ Reference 3, page 595

/6/ Block, M.J. (1956)
Nature 178, No. 4534, 650-651

/7/ Ziegenhorn, J., Senn, M. and Bücher, Th. (1976)
Clin. Chem. 22, 151-160

/8/ Micheau, J.C., Gimenez, M., Brockmans, P. and Dewel, G. (1983)
Nature 305, No. 5929, 43-45

PATTERN FORMATION IN PRECIPITATION PROCESSES

S.C. Müller
Max-Planck-Institut für Ernährungsphysiologie, Rheinlanddamm 201, D-4600 Dortmund, FRG

G. Venzl
Institut für Theoretische Physik,
Physik-Department der Technischen Universität München,
D-8046 Garching, FRG

I. Introduction

Precipitation is a process of phase separation in multicomponent systems (mixtures or solutions). Common examples are liquid droplets in gaseous media (rain) and solid particles in liquid solutions. In this article the precipitation of weakly soluble inorganic salts in an aqueous solution is considered exclusively. Usually a gel-forming material is added to the solutions in order to prevent sedimentation and hydrodynamic convection which destroy the patterns we are interested in. For precipitation to occur the original solution has to become metastable with respect to phase separation so that nuclei of the solid phase can form either due to a thermally activated process (homogeneous nucleation) or epitaxial growth on impurities (heterogeneous nucleation). Generally, the nucleation rate is a rapidly increasing function of the amount of supersaturation or supercooling which are quantitative measures of metastability.

A metastable supersaturated solution can be produced if a soluble electrolyte diffuses into a solution containing another electrolyte and both together form a weakly soluble salt by chemical reaction. Under these conditions a variety of electrolytes of suitable concentrations do not precipitate continuously in space but produce bands parallel to the isoconcentration surfaces of the diffusing ions (Liesegang bands) /1/. Typical results are shown in Fig. 1 for an effectively one-dimensional pattern and a two-dimensional structure with roughly circular

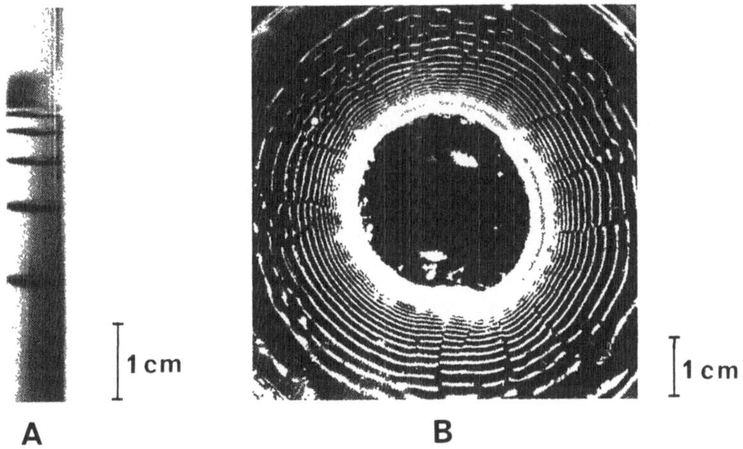

Fig. 1: Two examples of precipitation patterns in the presence of concentration gradients: (A) Parallel bands of $Mg(OH)_2$ after 2 days with initially 11.3 M NH_4OH in the upper portion and 0.4 M $MgSO_4$ in 9% gelatin in the lower portion of a tube. (B) Concentric rings of PbI_2 after 5 days with initially 0.12 M KI in the inner section and 0.012 M $Pb(NO_3)_2$ in the outer section of a petri dish, both containing 1% agar gel.

symmetry. During a period of time of the order of a day the bands form sequentially starting at the initial interface between the two electrolytes. If the concentration difference as well as the concentrations themselves are sufficiently high a large number of sharp bands at reproducible distances x_n from the interface are formed. Often an empirical spacing law x_{n+1}/x_n = constant is fulfilled but deviations have been reported /2,3/ (see also section II.1 of this article).

Alternatively, a metastable system can be prepared if a homogeneous solution is supercooled. In this way precipitation patterns have been obtained for lead iodide (PbI_2) solutions in an agar gel /4-7/. The structures are more random but often contain a dominating wavelength of 0.5 mm to 1 cm (see Fig. 6 below). Whereas the formation of Liesegang bands takes place in the presence of externally imposed gradients of electrolyte concentration no macroscopic concentration gradients are present in this experiment.

The experimental part of this article (section II) summarizes results of recent measurements reported in /3,6-9/. Some of the investigations are concerned with the details of the spatio-temporal evolution in a Liesegang system in particular prior to visible band formation /3,8/. The relation between Liesegang bands and "gradient-free" structures is studied by systematically varying the initial electrolyte concentrations in Liesegang systems towards small gradients /3,7/. In addition results are reported on patterns in initially uniform solutions /6/ as well as on structures more complex than those usually obtained/9/.

Various physico-chemical mechanisms have been proposed to explain the formation of precipitation patterns. They may be grouped into pre- and postnucleation theories. The first class of approaches considers the diffusion of ions in solution under the influence of external boundary conditions and precipitation bands which have already formed previously. A new precipitation arises if the critical condition for nucleation is exceeded. Nucleation is assumed to occur discontinuously in space which is rationalized by inhibition of further nucleation in the vicinity of bands. The following mechanisms leading to a depletion of supersaturation have been proposed: growth of the nucleated particles ("supersaturation theory" of Wi. Ostwald /10/), adsorption of electrolyte at the precipitate ("adsorption theory" of Bradford /11/), and the action of a third electrolyte produced in the reaction ("diffusion wave theory" of Wo. Ostwald /12/).

Among these prenucleation mechanisms only the supersaturation theory has been modelled mathematically /13,14/. If spatially discontinuous nucleation is assumed and introduced into the diffusion equations via repetitive boundary conditions the spacing law can be derived /13/. More recent work on precipitation /14/ and on microbial growth in chemical gradients /15,16/ uses threshold functions with hysteresis for the rate of precipitation and proliferation of bacteria, respectively. There is no doubt that these mathematical models can produce recurrent banding but their significance for colloidal nucleation and growth phenomena has to be questioned. The analysis of a model which avoids the assumption of additional boundary conditions or of phenomenological hysteresis functions shows a continuously advancing nucleation front /17/. Moreover, recent experimental studies (/3,8/ and section II) lead to the conclusion that precipitation patterns are a postnucleation phenomenon.

Assuming, alternatively, that the interdiffusion of electrolytes produces a continuous distribution of colloidal material one has to consider coarsening processes. In this context "colloidal" means: consisting of small particles of linear dimensions in the range 10^{-7} to 10^{-4} cm. Two basic coarsening mechanisms are coagulation due to encounters of particles and competitive particle growth (Ostwald ripening). The "coagulation theory" /18/ proposes that bands arise by means of coagulation of colloid, if certain critical electrolyte concentrations are locally exceeded. However, it has not been shown convincingly how the coagulation theory can lead to clear regions between bands. The "competitive particle growth theory" /5,19-22/ emphasizes the fact that due to surface free energy contributions the solubility of colloidal particles is a decreasing function of size. Mediated by diffusion of ions in solution large particles grow at the expense of smaller ones. This process is usually considered to occur on the length scale of the interparticle distance leaving the system homogeneous macroscopically. If the kinetics of particle growth is very slow competition occurs on a macroscopic scale /20/.

The analysis of mathematical models based on the competitive growth mechanism has suggested that a spatially homogeneous colloid is unstable against perturbations /5,19,20/ and that this instability leads to precipitation patterns by nonlinear interaction of modes (/21/ and section III). The theoretical section III is devoted to further analysis of the competitive particle growth theory. Inconsistencies inherent in previous treatments are removed and the issue of instability is clarified. So far the theory is exclusively concerned with the spatio-temporal evolution of a homogeneous colloid in response to perturbations. Nevertheless a qualitative comparison is possible with the experimental findings on Liesegang bands with small concentration gradients.

The purpose of this article is not only to present recent experimental and theoretical results on precipitation patterns but to discuss their mutual significance as thorough as possible. Our understanding is still at a preliminary stage. Therefore such an effort promises to be valuable.

II. Experimental Results

1. Liesegang precipitation patterns

For the experiments on the Liesegang phenomenon to be described here magnesium hydroxide and lead iodide were chosen as the precipitating salts. In order to obtain banded precipitation of $Mg(OH)_2$ a solution of $MgSO_4$ of relatively low concentration is placed in a cylindrical container to which gelatin is added as the gel-forming material. After pouring a solution of NH_4OH at high concentration on top of the gel the OH^- ions start to diffuse into the gel medium where they react with the Mg^{2+} ions to form $Mg(OH)_2$. The precipitation pattern evolving during the following hours is characterized by large and clear spaces between well defined and narrow white bands of $Mg(OH)_2$ (see Fig. 1A). Due to these properties and the high optical transparency of the gelatin gel this system is particularly suitable for investigations with optical techniques. The interdiffusing electrolytes for patterned precipitation of PbI_2 are solutions of $Pb(NO_3)_2$ and KI. Gels are formed in both electrolytes by adding a small amount of agar. Bands of PbI_2 can appear on both sides of the initial electrolyte junction depending on the particular choice of initial electrolyte concentrations. Since patterns are produced for a wide range of concentration combinations this system is appropriate for investigations of the dependence of the precipitation patterns on the initial concentrations of reactants, especially when these concentrations are systematically lowered. For details of our preparation procedure, including two-dimensional patterns, see refs. /8,9/.

The experimental study of the $Mg(OH)_2$ system was primarily concerned with the temporal and spatial evolution of the process of pattern formation. The measurements were designed to determine a sequence of events prior to band formation starting from the time when the two soluble salts are placed in contact with each other. By visual observation and by measurements of transmitted light, scattered light and of deflection of the transmitted light beam several phenomena were detected which lend support to the theoretical approach given in the following section. The visually recognized steps in the sequence of events are illustrated in Fig. 2. After addition of a suitable pH indicator to the $MgSO_4$ solution we could observe, from the start of the experiment, a distinct color change which continuously travels along the tube. It indicates the location where the diffusing OH^- ions raise the pH to a value of 9. For the example of Fig. 2 this color or pH front corresponds to an ion product $[Mg^{2+}][OH^-]^2$ three times larger than the solubility product of $Mg(OH)_2$. This is at least a sufficient condition for nucleation. After a given time following the pH front the onset of a weak turbidity was detected at any given location in the tube. Thus a broad turbidity region is established the front of which lags behind the pH front by

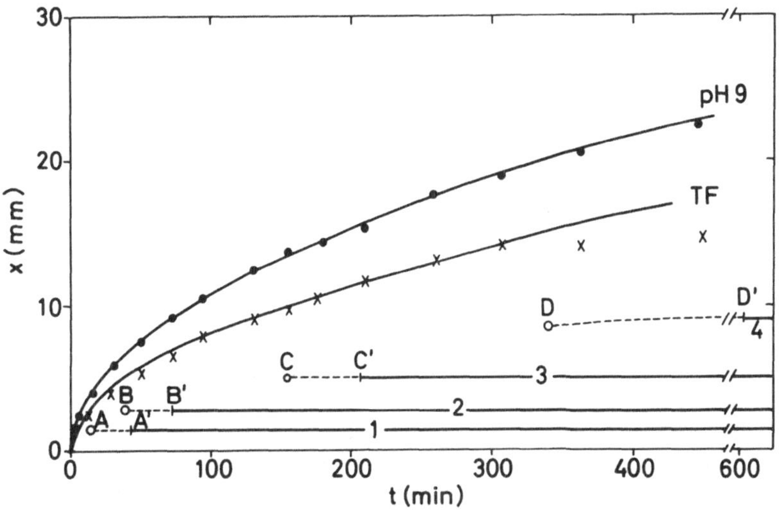

Fig. 2: Relationship in time and space between pH 9 front, turbity front (TF), onset of a localized refractive index gradient (A to D), and appearance of visible bands (A' to D') for $Mg(OH)_2$ band formation in 9% gelatin with 5.5 M NH_4OH and 0.4 M $MgSO_4$ as reagents.

a certain distance and time. Both fronts obey a simple diffusion law. The turbidity, which was further investigated by light scattering techniques (see below), is due to particles of considerable size, that is with a radius of at least several hundred angstroms. It is taken as an indication of the presence of colloid of $Mg(OH)_2$. Consequently, nucleation must have already occurred at a given point prior to the arrival of the turbidity front which is consistent with the estimated supersaturation at the pH front and is substantiated by the optical investigations presented below.

The turbidity region appears to be homogeneous without visually distinguishable spatial variations. This remains true until after a substantial time a strong change of index of refraction occurs localized at the positions where visible bands form later. The relationship between the different steps of the temporal sequence can be illustrated, for instance, for the first band in Fig. 2: When the formation of this band is completed, both the pH and turbidity front have already reached the location where later the third band will appear. Hence, a substantial supersaturation and, subsequently, the colloid exist a long time interval prior to the visible onset of structure.

The visual observations were reinforced by investigating the formation of $Mg(OH)_2$ bands with optical techniques described in detail in ref./8/. In Fig. 3 we show measurements of the scattered light intensity I_s along the spatial coordinate x of a square cuvette containing a $Mg(OH)_2$ system after 320 minutes. The dashed line in front of the completed third band represents the visual average of the measured points. It first passes through a minimum, then increases with increasing distance from the band and, for $x > 10$ mm, slowly decreases again until it reaches a constant level at $x \approx 15$ mm. This location roughly coincides with the visually observed turbidity front. (The constant level for $x > 15$ mm is due to a contribution from the gel). The smooth variation of the averaged scattering intensity (dashed line in Fig. 3) represents the essential characteristics of the spatial turbidity profile, while the local peaks in I_s, for instance at $x \approx 8$ mm, are not reproducible. They are probably due to isolated large particles immobilized by the gel network, as discussed elsewhere /8/. A series of measurements of scattered light at different instants of time reveals the same features: With decreasing x, starting at the turbidity front, the scattering due to larger colloidal particles increases smoothly to a certain distance away from the Liesegang band last formed. Within that distance we observe a depletion of colloidal material. The next band to be formed appears with-

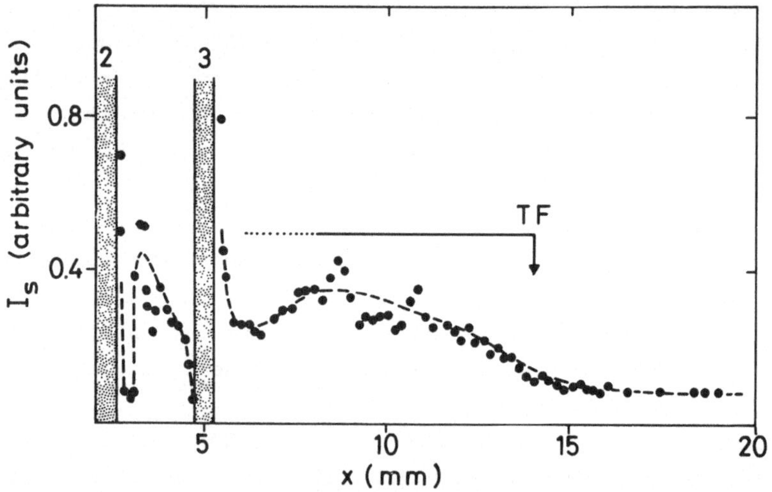

Fig. 3: Plot of the scattered light intensity I_s vs the coordinate x along the axis of a 1 cm^2 square cuvette containing a $Mg(OH)_2$ Liesegang system 320 min after the start of an experiment with reagents as in Fig. 2. The arrows mark the visually observed turbidity front, the horizontal line indicates the turbidity region (full line) and the depleted zone (dotted line). The bands are represented by the shaded areas.

in the relatively uniform turbidity region a substantial interval of time after the formation of colloidal particles.

The localized change in index of refraction observed visually (compare Fig. 2) was further investigated by measuring the amount of deflection D_x of a narrow laser beam traversing a square cuvette as a function of the coordinate x. A temporal sequence of such measurements prior to the completion of the third band of $Mg(OH)_2$ is shown in Figs. 4(a) to 4(f). The beam deflection is proportional to the gradient of refractive index (RIG) with respect to x and is determined by the local concentrations of the different solutes in the sample. As pointed out in ref./8/ some features of the measured curves can be interpreted in terms of the volume fraction of colloidal $Mg(OH)_2$ and these features will be discussed here qualitively. The graphs of Fig. 4 show that there exists a broad zone of continuously changing refractive index which extends far beyond the locations of the visible bands. It roughly coincides with the turbidity region the front of which is indicated by the arrows labeled TF.

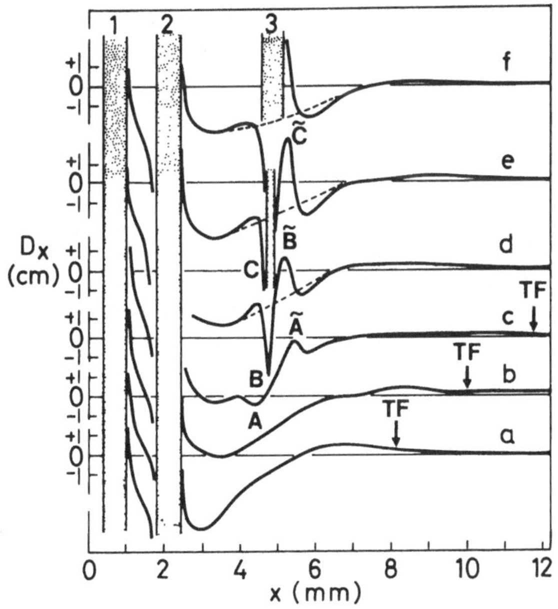

Fig. 4: Plots of the beam deflection D_x vs the coordinate x in a square cuvette containing a $Mg(OH)_2$ Liesegang system with reagents as in Fig. 2. The curves were measured during the following time intervals after the start of the experiment: (a) 107-112 min, (b) 165-170 min, (c) 180-186 min; (d) 195-205 min; (e) 215-225 min, (f) 260-270 min. The shaded areas indicate the regions of visible bands with number 1,2, and 3. Further details are explained in the text.

During a long interval of time the D_x curves are slowly varying functions of x without any detectable small-scale variations at the future location of the third band [graphs (a) and (b)]. The first indication of structure formation is a wavelike variation in the RIG around $x \approx 5$mm [extremes A and $\tilde{A}$ in graph (c)] which emerges from a previously smooth region within only a few minutes. This modulation represents a localized peak in the volume fraction of $Mg(OH)_2$ colloid superimposed on the measured background. Fifteen minutes later [graph (d)] this pattern has become more pronounced with sharper peaks at B and $\tilde{B}$ and a smaller distance between both extremes. Due to an enhanced growth rate at the band location a thin region of colloid soon turns into visible precipitate [graph (e)] and in Fig. 4(f) the band has grown to its final visible width. In graphs (d) to (f) dashed lines have been drawn to indicate the monotonically increasing background curve from which the structure evolves. A comparison of the measured D_x values with this background shows that, in addition to the sharp extremes labeled A,$\tilde{A}$ to C,$\tilde{C}$ there are small deviations between the two curves in the immediate vicinity of the extremes. These deviations are consistent with the observation that the regions on both sides of a band show a reduction in the scattering intensity I_s (Fig. 3), which corresponds to a reduction in colloid concentration.

As an additional result, the beam deflection measurements lead to the detection of a very small variation in the RIG the location of which is closely correlated with the location of the pH front and continuously moves through the system. This variation corresponds to a small step-like change in the refractive index. It is interpreted as an indication of the onset of homogeneous nucleation which can be expected to be close to the pH front as shown earlier in this section. The colloid concentration and the particle number density close to the nucleation site are estimated to be 10^{-2}mol/l and 10^{15} to 10^{16}cm^{-3}, respectively /8/.

The sequence of events prior to completion of the $Mg(OH)_2$ Liesegang system is summarized as follows. After diffusion of one electrolyte (NH_4OH) into the gel medium containing the second electrolyte ($MgSO_4$) results in an ion product larger than 3 times the solubility product, we observe the onset of homogeneous nucleation of colloidal particles. The nuclei gradually grow to a size of several hundred angstroms which gives rise to a sufficiently strong light scattering signal to surpass the scattering contribution from the gel (turbidity). Both nucleation and colloid formation take place continuously in space. While the fronts of these phenomena move through the system a broad and smooth distribution of colloidal material is established without variations on the spatial scale of the final precipitation bands. A substantial time interval after the passage of the front a localized RIG at the prospective band positions signals the onset of structure formation. While the RIG becomes more pronounced and narrower in space a depletion of colloidal material is observed in the immediate surroundings of the evolving structure.

Hence, the experimental evidence presented leads to the conclusion that the process of structure formation is a postnucleation phenomenon and comes about by a focusing mechanism by which a fraction of colloid is dissolved to form ions which then are removed from their original location to be incorporated into the growing band. Eventually this mechanism gives rise to a sharp band of visible precipitate, which is clearly separated from the preceding band.

These conclusions are also supported by experiments on the influence of gravity on the final ring locations. The motion of colloidal particles of sufficient mass is subject to gravitational forces provided that the size is small enough to allow motion in the network of the gel, whereas the motion of ions in a gel is essentially unaffected by gravity. From the measured extent of gravitational influence on a PbI_2 system it is estimated that colloidal particles of several hundred angstroms in size exist for a substantial fraction of the time required for the formation of a visible structure /8/.

In traditional Liesegang experiments structure formation takes place in the presence of strong initial gradients of concentrations as is the case for the investigated $Mg(OH)_2$ system. These patterns consist of a large number of sharp bands at reproducible locations. The issue of the relation of patterns with to those without imposed concentration gradients has been the motivation for a series of investigations on PbI_2 precipitation patterns which were concerned with the behaviour of the structures when the initial concentrations of reactants (KI and $Pb(NO_3)_2$) are systematically decreased. The determination of critical concentrations below which no patterns appear lead to the conclusion that the initial difference between the concentrations of the two electrolytes $\Delta = \frac{1}{2}[I^-]-[Pb^{2+}]$ and the initial concentration product $S = [Pb^{2+}][I^-]^2$ are useful parameters to characterize the onset of pattern formation. In the following we report results of experiments in which these two quantities were varied independently /3,7/.

A simple spacing law, as often cited in the literature /1/, is obeyed only in systems with high values of S and Δ, in which the number of bands is comparatively large. When the concentration gradients are lowered (small Δ or S) there are systematic deviations from this law and the structure is increasingly random in spatial location. The stochastic features of such PbI_2 patterns can be characterized by measuring the statistical distribution of band locations in a large number of tubes prepared under identical conditions. Especially for Δ close to zero and low S such distributions may be very broad /3/. This observation is considered to be important for elucidating the connection to inhomogeneous pattern formation in gradient-free precipitating systems where a stochastic element prevails (see section II.2).

A variety of structural details are found in patterns with low values of Δ and S. They are related to the formation of broad zones of colloid of PbI_2 and, subsequently, highly structured precipitate and thus emphasize that the essential step of structure formation occurs in the post-nucleation phase, in good agreement with the results for the $Mg(OH)_2$ system /3/. Here we focus on the description of an interesting phenomenon referred to as a "spatial bifurcation" of a single Liesegang band into two clearly separated bands of precipitate, both located within an extended region of colloid of PbI_2. The bifurcation occurs when the concentration difference Δ is increased from almost zero to a positive but comparatively small value while the ion product S is low and maintained constant /7/.

This is illustrated in Fig. 5 for a series of seven Liesegang systems with constant S arranged such that the initial value of Δ increases from left to right. The spatial distribution of PbI_2 is represented by plotting the intensity of light transmitted through the tubes, I_t, versus the space coordinate x. The seven graphs of Fig. 5 show all the qualitative features of the patterns in terms of the transparency of the colloid or precipitate of PbI_2 formed along each of the tubes. The triangles point to the centers of bands of high density precipitate. Most bands are well developed and sharply focused ($I_t = 0$, full triangles) but a few are not so well recognizable ($I_t \approx 0$, open triangles). Only one band appears in tubes 1 to 4 (7.1 mM < Δ < 16 mM). Its width slightly increases from left to right and in tube 4 it is essentially a broad zone of colloid of relatively large density. In all of the sys-

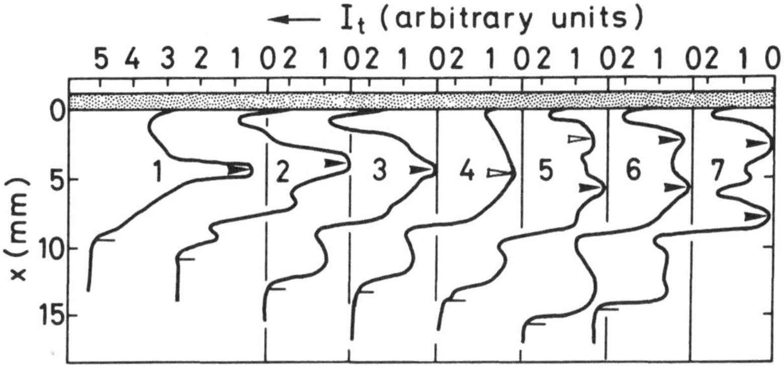

Fig. 5: Spatial bifurcation of precipitation bands of PbI_2. The intensity of transmitted light I_t is plotted vs the space coordinate x for a series of tubes containing PbI_2 Liesegang structures with constant initial ion product $S = 1.8 \times 10^{-3} M^3$ and initial concentration differences Δ = 7.1, 10.8, 14.0, 16.1, 18.2, 20.0, 22.6 mM for graphs 1 to 7, respectively. The shaded area represents the precipitation area at the initial junction between the two electrolytes. The triangles point to the centers of Liesegang bands. The bifurcation occurs between tube 4 and 5.

tems 5 to 7 ($\Delta > 18$ mM) two bands have formed.The higher Δ the more distinctively these bands are separated from each other. The bifurcation point Δ_{bf} lies between tubes 4 and 5 (16 mM < Δ_{bf} < 18 mM). In each of the tubes a zone of low density colloid of PbI_2 is distinguishable which extends from the initial electrolyte junction well beyond the band locations. It corresponds to the region between x = 0 and the point in space marked by horizontal bars in each of the plots of Fig. 5. The width of these zones tends to increase with increasing Δ. The spatial profiles of colloid density, as represented by the x dependence of the transmitted light intensity I_t, show some distinct structural features, for instance step-like variations in transparency, which have been discussed in detail elsewhere /7/. There is evidence for a second bifurcation to occur for one of the two bands in tubes with $\Delta > \Delta_{bf}$ if the concentration difference is further increased /7/.

The bifurcation point Δ_{bf} was determined for several series of PbI_2 systems with different low values of S (constant for each series). The quotient Δ_{bf}/S was found to be approximately constant. It turns out that, independently of the choice of S, the concentration of lead ions is roughly the same when Δ reaches the critical value Δ_{bf}; hence, this ion concentration is likely to be an essential parameter for the phenomenon. The issue of focusing of colloidal material into two instead of one narrow spatial region within one continuous zone of colloid is further discussed in the theoretical section of this article.

2. Pattern formation without imposed gradients of concentration

Spatially inhomogeneous distributions of precipitate may also arise from initially uniform colloids /4-7/. In order to produce such macroscopic structures PbI_2 is dissolved in water containing 1% agar-gel at $95^{\circ}C$ with salt concentrations of the order of 5 mM. The hot solutions are poured into a preheated petri dish to obtain a layer of 1 to 3 mm thickness of the solution. While the covered container is allowed to cool slowly the gel solidifies. Subsequently a uniform faint yellow color signals spatially homogeneous nucleation of colloidal PbI_2 induced by sufficient supercooling. Within the range of initial salt concentrations from 4.8 to 6.6 mM visible precipitate of PbI_2 is deposited, after a few hours, in rather stochastic than deterministic macroscopic patterns with a length scale of the order of 1 mm to 1 cm. For concentrations outside this range, both lower and higher, the distribution of colloidal particles of PbI_2 remains uniform macroscopically /6/.

Different varieties of structure have been reported /4-6/. Two examples are presented in Fig. 6 which can be roughly characterized as a wave-like pattern (Fig. 6A) and a two-dimensional network (Fig. 6B). The pictures clearly show that randomness is an inherent property of such gradient-free structures. A microscope photograph of a selected region of system B in the figure gives evidence that the visibly dense regions of PbI_2 consist of a large number of predominantly independent colloidal particles (Fig. 6C). The spaces separating dense regions of PbI_2 are essentially devoid of colloid except for a few relatively large particles which are randomly distributed. Representative sections of Fig. 6C were investigated with a microscope of high magnification. The particle size distribution was determined and was found to be peaked at 4.3 μm with a variance of 2 μm. Its shape is consistent with the main characteristics of reaction-limited first order kinetics /23/ (see also section III).

Further analysis of the structural properties of network type patterns with different initial salt concentrations shows that an increase in the initial degree of supersaturation results in a pronounced decrease in the macroscopic length scale as well as in the average particle size and the average interparticle distance in the dense regions of

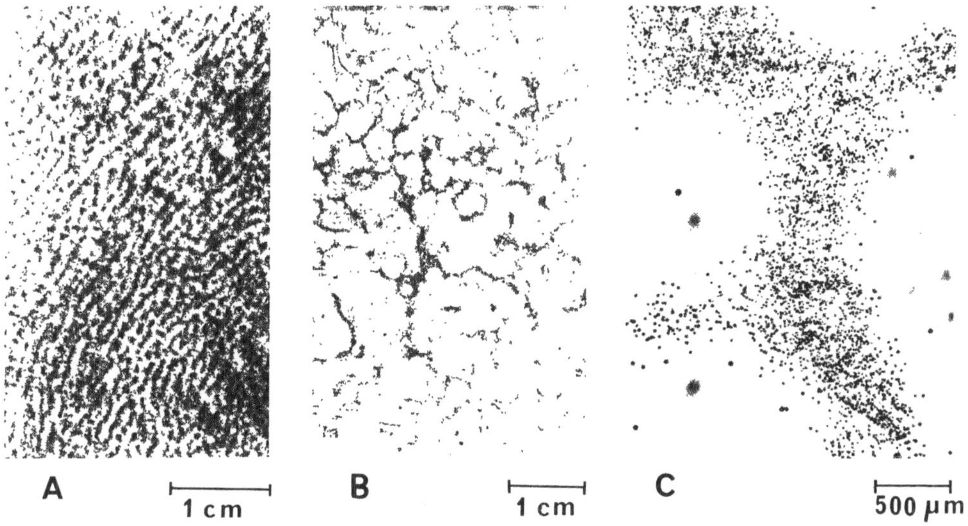

Fig. 6: Precipitation patterns of PbI_2 in agar gel without imposed gradients of concentration. The concentrations of PbI_2 and agar gel were, respectively: (A) 6.1 mM, 0.2%; (B,C) 5.2 mM, 1%. (C) is a microscope photograph of a part of (B) with magnification 28 x.

the structure. The similarity of that change with supersaturation in all of these three investigated length scales points to a strong correlation between the macroscopic and mesoscopic features of gradient-free precipitation patterns /6/.

3. Liesegang systems with complex structural features

We briefly describe observations of several types of Liesegang patterns that are more complex than the concentric rings or parallel bands most frequently found. A closer examination of the rings in Fig. 1B reveals that there are convex segments of PbI_2 which are separated from each other by small gaps without visually detectable precipitate. Note the roughly radial alignment of the gaps in subsequent rings. Another type of complex structure is a set of rings with superposed "dislocations" (Fig. 7A). They also extend out across the ring system in radial direction. Additional interesting structural properties within rings of two-dimensional Liesegang patterns have been reported /9/. Finally, in Fig. 7B, we show a "curiosity" in a one-dimensional experiment: the formation of a spiral in a $Mg(OH)_2$ system. Here a helicoidal structure

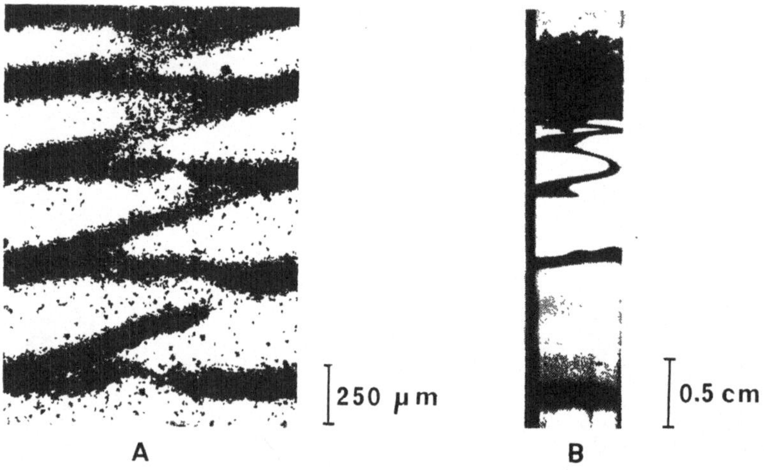

Fig. 7: (A) Dislocations within a set of concentric rings of PbI_2 after 6 days. Initial electrolyte solutions: 0.12 M KI (inner section of dish) and 0.015 M $Pb(NO_3)_2$ (outer section of dish), both containing 1% agar gel.

(B) Spiral band of $Mg(OH)_2$ after 4 days. Initial electrolyte solutions: 5.5 M NH_4OH (upper portion of tube) and 0.37 M $MgSO_4$ in 8% gelatin (lower portion of tube).

occupies the region in which the first three precipitation bands are usually located. We do not know the experimental conditions necessary to produce such complex structures which must be considered as events of low probability. These findings provide further evidence for the complexity of patterned precipitation.

4. Summary

The experimental findings presented in this section provide evidence that, for the examples investigated, spatial structure formation in precipitation processes is a postnucleation phenomenon. The sequence of events prior to the formation of visible Liesegang structures leads to the conclusion that, as diffusion occurs, nucleation and deposition of colloid take place without spatial patterning. The formation of precipitation bands comes about by a focusing mechanism which is associated with partial depletion in colloidal material in the neighbouring zones. In systems in which strong initial gradients of concentration are imposed, this mechanism gives rise to the formation of sharply defined bands of visible precipitate at reproducible locations which obey a simple spacing law. When these gradients are lowered the structures become increasingly random in spatial location and continuous zones of colloid may develop into two instead of one precipitation band (spatial bifurcation). It is suggested that the randomness of Liesegang patterns with low gradient is connected to the rather irregular than deterministic features of precipitation patterns which evolve from initially uniform colloids. Investigations of these gradient-free structures by microscopy point to a colloidal composition of the precipitated salt which indicates that colloidal growth leads to the final pattern. Furthermore, it is shown that standard Liesegang experiments may lead to pattern formations with unusual complex properties.

III. Theoretical Investigations

The experimental evidence that Liesegang band formation is a postnucleation phenomenon and the challenge to develop a theory valid also for the patterning in initially uniform colloids led us to an investigation of the competitive particle growth theory. We formulate the basic kinetic equations of the theory and point out the adequacy of a "hydrodynamic" description of the coarsening process. We present results based on the traditional approximation which assumes that the particle size distribution function is monodisperse /5,19-22/. This approximation is criticized and an improved model derived recently /24/ is analyzed.

The competitive particle growth theory considers the coarsening in an ensemble of colloidal particles distributed in space. We assume that due to the presence of a gel the particles do not move but change in size due to condensation/evaporation of monomers (or ions). For simplicity we work within the droplet model which characterizes individual particles by their radius R only. Introduce the particle distribution function $f(R,\vec{x},t)$ such that

$$\int_{\Delta V} d^3x \int_{R-\Delta R}^{R+\Delta R} dR' \; f(R',\vec{x},t)$$

is the number of particles within the (spatial) volume ΔV and with a radius between $R-\Delta R$ and $R+\Delta R$. On the scale of the average interparticle distance the distribution is a sum of delta-functions

$$f(R,\vec{x},t) = \sum_i \delta[R-R_i(t)]\,\delta^3(\vec{x}-\vec{x}_i)$$

where i numbers all individual particles in the ensemble. Instead we introduce a hydrodynamic description and use a continuous distribution function based on the following argument. Experimentally (see section II.2) the typical distance between neighbouring particles is much smaller than the macroscopic patterning length L. We introduce an intermediate length l such that l is large compared to the interparticle distance but smaller than the macroscopic length L. One obtains a coarse-grained distribution function after averaging over volume elements l^3. The relation between the length scales is expressed by the inequality $n^{-1/3} \ll l \ll L$ where $n(\vec{x},t) = \int_{-\infty}^{+\infty} dR\, f(R,\vec{x},t)$ is the coarse-grained particle density and $n^{-1/3}$ represents the average interparticle distance. Note that, for a given $\vec{x}$, $f(R,\vec{x},t)$ now describes an ensemble of particles which is macroscopic in that it contains many individuals ($nl^3 \gg 1$). This is important for the criticism of the traditional model to be discussed.

Introducing further a locally averaged monomer density $c(\vec{x},t)$ we have the following equations which are conservation laws in differential form

$$\frac{\partial}{\partial t} f(R,\vec{x},t) + \frac{\partial}{\partial R} V[R,c(\vec{x},t)]f(R,\vec{x},t) = 0 \tag{1}$$

$$\left(\frac{\partial}{\partial t} - D\nabla^2\right) c(\vec{x},t) + \frac{4\pi}{3v} \int_0^\infty dR\, R^2 V[R,c(\vec{x},t)]f(R,\vec{x},t) = 0. \tag{2}$$

One obtains a closed set of equations by specification of the particle growth law $V(R,c)$. We use simple first order kinetics,

$$V(R,c) = K[c-c_{eq}(R)]/c_{eq}(\infty) \tag{3}$$

with the size dependent solubility

$$c_{eq}(R) = c_{eq}(\infty)(1+\alpha/R). \tag{4}$$

The constants $D, v, K, c_{eq}(\infty)$, α are material parameters (see /20/ for details). The basis of the competitive growth mechanism is the following. Consider the supersaturation $\sigma = [c-c_{eq}(\infty)]/c_{eq}(\infty)$ and the related critical radius $R_c = \alpha/\sigma$. Then the growth law (3) is such that particles with $R < R_c$ dissolve whereas those with $R > R_c$ grow. If R_c depends on the coordinate $\vec{x}$ competition does not only occur locally (within volume elements l^3) but may lead to macroscopic coarsening.

Eqs. (1) to (4) are not convenient for a general analysis. The issue is to find physically reasonable approximations which lead to tractable mathematical models. We discuss two models in terms of ordinary reaction-diffusion equations. In each case the analysis will proceed in two steps: investigation of the linear stability of a spatially homogeneous state and numerical simulations of the nonlinear equations.

The model considered so far in the literature /5,19-22/ assumes a monodisperse distribution function

$$f(R,\vec{x},t) = n(\vec{x},t)\,\delta[R-\bar{R}(\vec{x},t)]. \qquad (5)$$

In this approximation spatially homogeneous steady states can be shown to exist. Introducing dimensionless variables one finds equations of motion

$$\frac{\partial \rho}{\partial \tau} = s-1/\rho$$
$$\varepsilon \frac{\partial s}{\partial \tau} = \nabla_\xi^2 s - 3\rho^2(s-1/\rho) \qquad (6)$$

where $\rho, s, \vec{\xi}, \tau$ are the suitably scaled average particle radius, supersaturation, length, and time. The steady states form a one-dimensional manifold $s\rho = 1$. If we choose $s = \rho = 1$ the dependence on the particular steady state is contained in the constant ε. The linear stability analysis yields two eigenvalues $z_\pm(k)$ depending on the wave vector $\vec{k}$ of the perturbations,

$$z_\pm(k) = -\frac{1}{2}(3-\varepsilon+k^2) \pm \frac{1}{2}\,[(3-\varepsilon+k^2)^2 - 4\varepsilon k^2]^{1/2} \qquad (7)$$

which for $\varepsilon \ll 1$ reduce to

$$z_\pm(k) \approx \begin{cases} k^2/(3+k^2) \\ -(3+k^2)/\varepsilon. \end{cases} \qquad (8)$$

One recognizes that z_+ is positive for $k > 0$ which indicates an instability of the homogeneous solution. Note that there is no control parameter which can be manipulated in order to drive the system from a stable to an unstable situation and allows a perturbative treatment in the vicinity of marginal stability. Furthermore, the linear stability analysis does not provide a length scale because the dispersion relation $z_+(k)$ has no maximum at finite k.

In order to gain more insight into the nonlinear evolution the equations (6) have been solved numerically (see also /21/). Fig. 8 shows the result of a simulation on a one-dimensional domain $0 \leq \xi \leq 20$ subject to zero-flux boundary conditions and a small localized perturbation centered at $\xi = 0$. A series of roughly equidistant bands form sequentially in time. The sharpening of the bands and the depletion of their neighbourhood during formation are reminiscent of the experimental observations discussed in section II.1. Fig. 9 contains the ε-dependence of the patterns at two different times. Two features are remarkable. One is the fact that, in scaled units, the patterns are essentially independent of ε for $\varepsilon < 1$. The differences at the later time (Fig. 9B) are due to competition of bands and boundary effects and may be discarded. The second is the splitting of bands for $\varepsilon = 10$. The doublets evolve out of local maxima. **The size (dotted lines** in Fig. 9A) within the earlier formed broader zones. This process has a striking similarity to the "spatial bifurcation" found experimentally (compare section II.1).

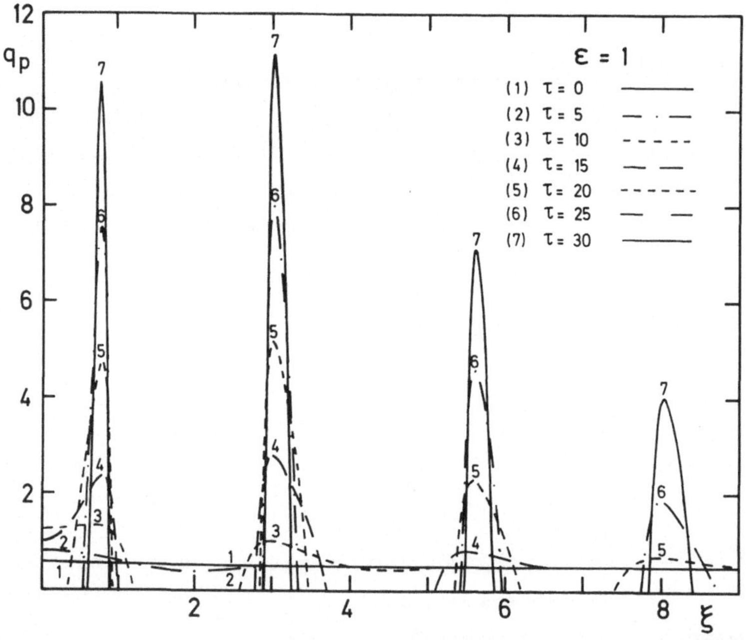

Fig. 8: Evolution of recurrent banding in model (6). Amount of matter in the colloid $q_p = \rho^3/(1+\varepsilon)$ as a function of the space variable ξ. Initial perturbation: Gaussian with amplitude $\Delta\rho = 0.05$ and width $\Delta\xi = 1$ centered at $\xi = 0$.

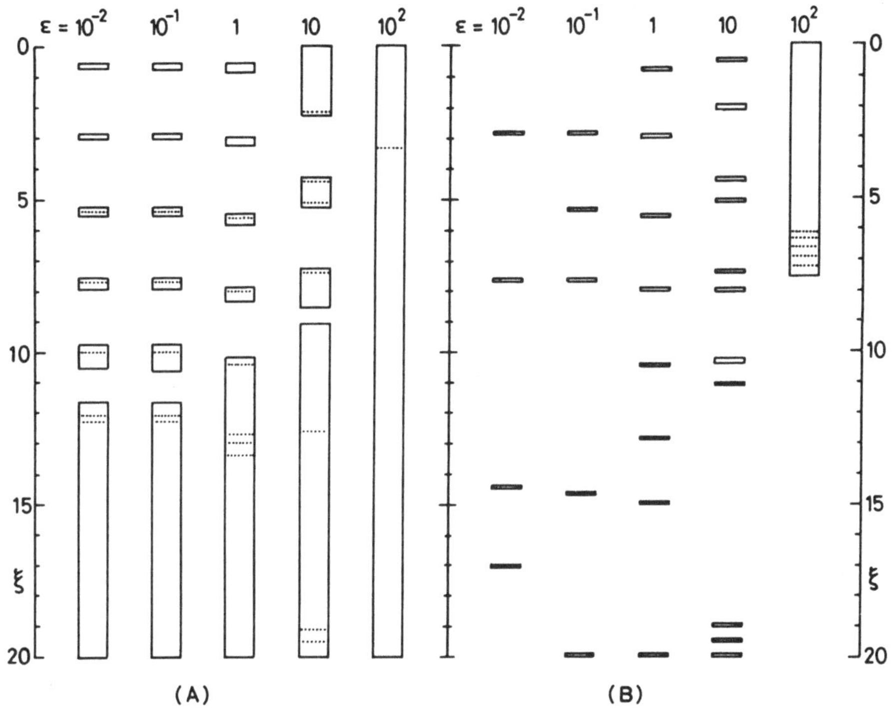

Fig. 9: Patterns in model (6) at times $\tau = 30$ (A) and $\tau = 100$ (B) for various ε indicated on top of the columns. Rectangles denote presence of colloid, dotted lines indicate maxima of particle size ρ. Initial conditions as in Fig. 1.

The mathematical model considered so far obviously has properties resembling those observed in experiments on precipitation patterns. It turns out, however, that the assumption of a locally monodisperse distribution, Eq. (5), is hardly reasonable physically. Remember that at fixed $\vec{x}$ the distribution function $f(R,\vec{x},t)$ represents an essentially homogeneous macroscopic system. It has been pointed out on theoretical ground that for such an ensemble of particles the distribution is neither narrow nor time-independent /23/. Instead it approaches an asymptotic form

$$f(R,\vec{x},t) \xrightarrow[t\to\infty]{} n(t) \ \Phi[R/R_c(t)] \tag{9}$$

independent of initial conditions. This is true at least for common growth laws like (3), (4), and has recently been confirmed by further theoretical analysis /25,26/. The width of Φ is of order unity and $R_c(t)$ and $n(t)$ are asymptotically known functions of time, usually power laws. The major point in the argument are the time scales involved. The rate at which an initially narrow distribution broadens is roughly the same as the (time dependent) rate of change of the homogeneous state. The latter may be characterized by $|\dot{\sigma}/\sigma|$. For the growth law (3),(4) it is given by $|\dot{\sigma}/\sigma| \approx \frac{1}{4} K\sigma^2/\alpha$. This rate has to be compared with the rate of unstable evolution in the crude model. The supremum of the positive eigenvalue is $z_+(\infty) \approx K\sigma^2/\alpha$ in real time units. Since these rates are of the same order the distribution cannot be considered to be narrow nor can the time-dependence of the homgeneous background be neglected during pattern formation. This theoretical conclusion is supported by the experimental observation of a broad size distribution in the dense regions of gradient-free structures (compare section II.2).

In order to achieve a more realistic evaluation of the competitive particle growth theory a model has been derived which takes into account the first three moments of the size distribution function /24/. Using dimensionless variables $\nu,\rho,\delta,s,\vec{\xi},\tau$ denoting particle density, mean particle radius, relative width of the distribution function, supersaturation, space, and time, respectively, the model equations read

$$\frac{\partial \nu}{\partial \tau} = -\frac{3}{4}\,\nu\phi(\delta)/\rho^2$$

$$\frac{\partial \rho}{\partial \tau} = (1-\delta/9)(\sigma-1/\rho) + \frac{1}{4}\,\phi(\delta)/\rho \tag{10}$$

$$\frac{\partial \delta}{\partial \tau} = [4\delta\,\{1-\delta/6-\tfrac{1}{2}\sigma\rho(1-\delta/3)\} - \tfrac{3}{4}(1+\delta)\phi(\delta)]\;/\rho^2$$

$$\varepsilon\,\frac{\partial s}{\partial \tau} = \nabla_\xi^2\, s - 3\nu\rho^2(1-\delta/9)(\sigma-1/\rho)\,.$$

They have to be supplemented by an equation for $\phi(\delta)$ which determines the local coarsening rate as a function of the relative width δ. A choice which reproduces the correct asymptotics (9) is

$$\phi(\delta) = (5\delta/3)^\gamma. \tag{11}$$

As long as $\gamma > 2$ holds the results of the model are insensitive to the particular choice of γ. We used $\gamma = 3$ in the numerical simulations. Note that for $\delta \equiv 0$ we recover the crude model, Eqs. (6). The improved model (10), (11) turns out to be an excellent approximation if compared with numerical solutions of the original equations (1)-(4) for special cases /24/. Applying this model to the problem of macroscopic coarsening one has to keep in mind that the spatially homogeneous solution is time-dependent. Therefore, there is some arbitrariness in choosing initial conditions. All the results to be presented were obtained for initial conditions close to the asymptotic solution (9). This is an "unbiased" choice but variations have to be considered in further work.

The result of the linear stability analysis is at first sight qualitatively similar to the crude model. An inspection of the time scales, however, reveals that the rate of change of the homogeneous background is by a factor of three larger than the maximal rate of unstable evolution. We conclude that the spatially homogeneous state is effectively stable against infinitesmal perturbations. This is confirmed by numerical simulations analogous to those leading to Figs. 8,9. For larger perturbations a precipitation band develops which is surrounded by an empty region (Fig. 10 shows an example). In accord with the experimental findings (section II.1) the band shows some amount of sharpening and a strong tendency to deplete its neighbourhood. However, we have not found recurrent banding. In Fig. 10 calculation up to $\tau = 10^3$ do not indicate any induction of satellites. We attribute this finding to the fact that in the improved model as well as in the crude approximation the central peak induces only a small growth-advantage some distance away. This small perturbation is too weak to develop into a new band whereas in the crude model infinitesmal perturbations can grow.

Let us summarize the results of our theoretical investigations on the basis of the competitive growth mechanism. Despite the fact that no analysis with external concentration gradients has been performed there is good evidence that the theory accounts for some of the experimental observations, notably focusing of colloidal material into bands and spatial bifurcation. We expect that the latter is not restricted to the crude model because the bifurcation phenomenon appears to depend mainly on the dynamics of maturation within distinct zones of colloid (see Fig. 9) and less on the interaction between zones which leads to the induction of satellites. On the other hand the significance of the competitive particle growth mechanism for gradient-free structures has to become questioned. The instability property lent itself as an explanation for this phenomenon since small fluctuations are always present in macroscopic systems. Our results show that perturbations of finite size are necessary to induce focusing of material into bands and that

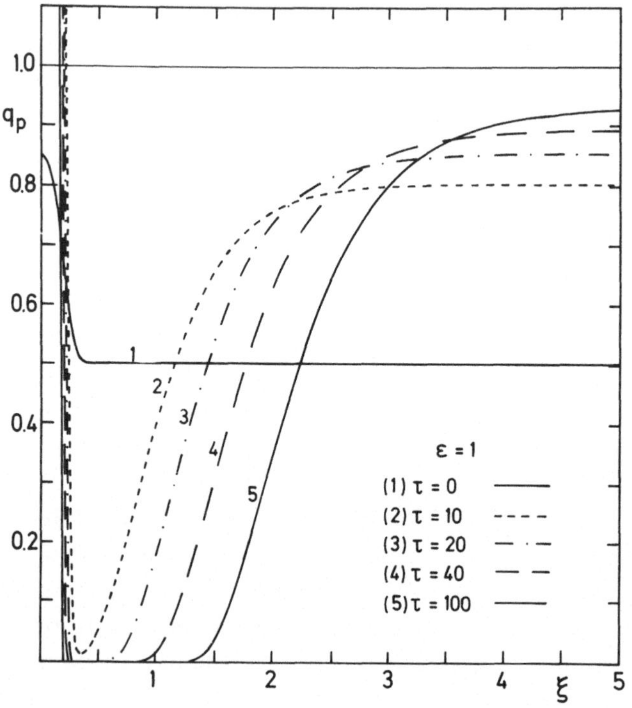

Fig. 10: Evolution of a band in model (10). Amount of matter in the colloid $q_p = \nu\rho^3/(1+\varepsilon)$ as a function of the space variable ξ. Initial condition for the homogeneous state: $\nu = \rho = \sigma = 1$, $\delta = 0.6$. Initial Gaussian perturbation at $\xi = 0$ with half width $\Delta\xi = 0.1$ and amplitudes $\Delta\rho = 2.3$, $\Delta\sigma$ and $\Delta\nu$ such that $\sigma\rho = (\varepsilon\sigma + \nu\rho^3)/(1 + \varepsilon) = 1$.

the response to these perturbations remains local without induction of recurrent banding. There is experimental evidence that some of the gradient-free structure is due to exceptionally large particles probably due to impurities ("greedy giant catastrophy" /22/). As a comprehensive explanation this does not seem to be sufficient.

IV. Conclusions

Experimental results on the formation of precipitation structures for two different salts, i.e. magnesium hydroxide and lead iodide, have been presented. We believe that the conclusions which can be drawn from these studies are valid for a more comprehensive class of precipitating substances. The experiments are novel in their focus on processes preceding the visual formation of structure. There is strong evidence now that the material to be precipitated is first deposited as a continuous distribution of colloid from which discrete precipitation bands evolve after a substantial time lag. The nucleation of the colloid does not predetermine the location of the bands. Given this result, all the pre-nucleation theories mentioned in the introduction are not adequate for the class of system investigated. The physical effects which these theories emphasize may certainly be important for the details of the process but on their own they cannot be responsible for structure formation.

Using optical probing techniques a focusing of colloidal material into bands of precipitate was identified as the decisive step for pattern formation. Its prominent features are the sharpening of bands during their evolution accompanied by depletion of their neighbourhood. The competitive particle growth mechanism which was investigated theoretically in this article can at least qualitatively account for these experimental findings. Numerical simulations also provide evidence that the "spatial bifurcation" of precipitation bands observed for the lead iodide system with small concentration differences is within the scope of this theory.

More complex precipitation patterns than the well-known regular Liesegang structures have been described: Formation of helicoidal structure in one-dimensional Liesegang patterns, dislocations superimposed on ring systems, and random structures in initially homogeneous supercooled lead iodide solutions in agar gel. These phenomena also point to a more subtle mechanism than that envisaged by the traditional supersaturation hypothesis.

The original idea of an instability in macroscopically uniform colloidal solutions which was thought to explain the gradient-free structures has to be modified in the light of the new theoretical results. If the coarsening is properly taken into account finite perturbations are necessary to break the homogeneous colloid up into discrete precipitation zones. The origin of these perturbations remains to be explained. On the other hand, the threshold behaviour which was found to be an inherent property of the competitive particle growth theory may, on a new physical basis, provide a connection to the mathematical models based on diffusion and thresholds /14-16/.

References

/1/ Hedges, E.S. (1954) Liesegang Rings and Other Periodic Structures, 1932, Chapman and Hall, London; Stern, K.H. (1954), Chem. Rev. 54, 79, and references therein.

/2/ Hedges, E.S. & Henley, R.V. (1955) J. Chem. Soc. 1928, 2714; Packter, A. (1955) Nature (London) 175, 556.

/3/ Müller, S.C., Kai, S. & Ross, J. (1982) J. Phys. Chem. 86, 4078.

/4/ Flicker, M. & Ross, J. (1974) J. Chem. Phys. 60, 3458.

/5/ Feinn, D., Ortoleva, P., Scalf, W., Schmidt, S. & Wolff, M. (1978) J. Chem. Phys. 69, 947.

/6/ Müller, S.C., Kai, S. & Ross, J. (1982) J. Phys. Chem. 86, 4294.

/7/ Kai, S., Müller, S.C. & Ross, J. (1983) J. Phys. Chem. 87, 806.

/8/ Kai, S., Müller, S.C. & Ross, J. (1982) J. Chem. Phys. 76, 1392.

/9/ Müller, S.C., Kai, S. & Ross, J. (1982) Science 216, 635.

/10/ Ostwald, W. (1897) Lehrbuch der Allgemeinen Chemie, Engelmann, Leipzig.

/11/ Bradford, S.C. (1922) Kolloid-Z. 30, 364.

/12/ Ostwald, W. (1925) Kolloid-Z. 36, 380.

/13/ Wagner, C. (1950) J. Colloid Sci. 5, 85; Prager, S. (1956) J. Chem. Phys. 25, 279; Kahlweit, M. (1962) Z. Physik. Chem. Neue Folge 32, 1; Zel'dovich, Y.A., Bärenblatt, G.I. & Zalganik, R.L. (1962) Sov. Phys. Doklady 6, 869.

/14/ Keller, J.B. & Rubinow, S.I. (1981) J. Chem. Phys. 74, 5000.

/15/ Jäger, W. & Pöppe, C., these proceedings.

/16/ Wimpenny, J.W.T. (1982) Phil. Trans. R. Soc. Lond. B. 297, 497; Wimpenny, J.W.T. & Jaffe, S., these proceedings.

/17/ Venzl, G. & Ross, J. (1982) J. Chem. Phys. 77, 1302.

/18/ Dhar, N.R. & Chatterji, A.C. (1922) Kolloid-Z. 31, 15; ibid. (1927) 37, 2; ibid. (1927) 37, 89; Shinohara, S. (1970) J. Phys. Soc. Japan 29, 1073.

/19/ Lovett, R., Ortoleva, P. & Ross, J. (1978) J. Chem. Phys. 69, 947.

/20/ Venzl, G. & Ross, J. (1982) J. Chem. Phys. 77, 1308.

/21/ Freeney, R., Schmidt, S.L., Strickholm, P., Chadam, J. & Ortoleva, P. (1983) J. Chem. Phys. 78, 1293.

/22/ Ortoleva, P. (1982) Z. Phys. B49, 149.

/23/ Lifshitz, I.M. & Slyozov, V.V. (1961) J. Phys. Chem. Solids 19,35; Wagner, C. (1961) Z. Elektrochem. 65, 581.

/24/ Venzl, G., in preparation.

/25/ Venzl, G. (1983) Ber. Bunsenges. Phys. Chem. 87, 318.

/26/ Marqusee & Ross, J., "Kinetics of Phase Transitions: Theory of Ostwald Ripening", preprint.

ON A MECHANICAL MODEL FOR MORPHOGENESIS: MESENCHYMAL PATTERNS

J.D. Murray
Centre for Mathematical Biology
Mathematical Institute
University of Oxford
Oxford OX1 3LB
England

Abstract

Motile embryonic mesenchymal cells can generate large traction forces which can deform the cells' environment which in turn affects the cells' motion. We derive a model mechanism, based on experimentally observed facts, for generating spatial patterns in cell populations, caused by the mechanical interaction between the cells and their extracellular matrix (ECM) sustrata. We derive the field equations for the model and demonstrate some of the pattern formation potentialities. Finally we apply the model to the patterns observed in (i) feather and scale primordia and (ii) cartilage condensations in limb development.

1. Introduction

In early embryogenesis there are two main classes of cells, mesenchymal and epithelial. This paper is concerned with patterns generated by the former while those involving the latter are discussed in the following paper by Oster (1983). Mesenchymal cells can migrate and spread within a substratum made up of other cells and the extracellular matrix (ECM) - a fibrous-like medium - and patterns in cell density are obtained. Here we propose a model mechanism, based on experimental facts, for generating such mesenchymal patterns and we apply it to two specific developmental problems of longstanding interest.

Various factors affect cell motion such as chemotaxis, random dispersal, contact inhibition, extracellular matrix guidance and so on. Recently, however, possibly the most important factor of all has been added by the work of Harris et al. (1981): see also the preceding paper in this volume by Harris et al. (1983). They have shown that motile mesenchymal cells can generate remarkably strong traction forces which act on the extracellular matrix substrata. These forces can affect the substrata over lengths of the order of hundreds of cell diameters. The consequent deformations can guide the movement of distant cells.

The mechanism we describe and analyse here models the traction-producing cells migrating within an elastic matrix, which is also secreted by the cells. The mechanism models the various factors, all based on experimental evidence, and shows how they interact in such a way as to produce a variety of non-random spatial patterns of cell and matrix densities. An earlier version of this model has been given by Murray et al. (1983) and a more biologically oriented version with specific embryological applications by Oster et al. (1983).

Our mechanism is an alternative to the widely studied reaction-diffusion (Turing) systems for biological pattern formation based on Turing's (1952) concept of diffusion-driven instabilities. The spatial patterns of chemical (morphogen) concentrations obtained by such a process are then reflected by the cells response in, for example, a positional information (Wolpert, 1971) way. Such reaction-diffusion models have been applied to a variety of real biological situations, for example Murray (1979, 1981a,b) and Bard (1981) applied them to animal coat patterns and Meinhardt (1982) to a variety of early patterning problems. (A pedagocial discussion is given in the book by Murray (1977).) A serious drawback of such an approach has been the elusive nature of the morphogens involved and hence the inability to get estimates for the various parameters of the model.

Important advantages of the mechanical model here are (i) the actual spatial patterns appear in the cell densities directly and (ii) the parameters are experimentally measurable.

In Section 2 we describe the various types of cell motion and how they interact with the ECM to effect pattern generation. We then derive the model equations for the cell motion, the cell-matrix interaction and the conservation of ECM. In Section 3 we demonstrate some of the pattern formation potentialities of the field equations and indicate the mathematical problems. In Section 4 we apply the model to two practical biological situations, namely, (i) feather germ patterns and (ii) cartilage condensations in limb morphogenesis. The model predictions compare well with the observed patterns.

2. Cell motions, cell-matrix interactions and the model field equations

Let $n(\underset{\sim}{x},t)$ and $\rho(\underset{\sim}{x},t)$ represent the cell and ECM densities respectively and $\underset{\sim}{u}(\underset{\sim}{x},t)$ the displacement vector of the matrix: a point in the matrix which was initially at a position $\underset{\sim}{x}$ is displaced to $\underset{\sim}{x} + \underset{\sim}{u}$. The three field equations are: (i) a conservation equation for the cell density; (ii) the mechanical balance of cellular forces and those generated in the matrix; and (iii) a conservation law for the ECM. We consider each of these in turn.

Cell equation

Generally $n(\underset{\sim}{x},t)$ satisfies a conservation equation of the form

$$\frac{\partial n}{\partial t} = -\nabla . \underset{\sim}{J} + M, \tag{1}$$

where M is the source of cells through mitosis and $\underset{\sim}{J}$ is the vector flux of cells per unit area. A variety of phenomena contribute to the flux term, the principal ones of which are random dispersal, mechano- or haptotaxis and convection by the ECM; of these convection is probably the most important. In this paper we do not include any contribution from chemotaxis and contact guidance. Contact guidance is the phenomenon whereby cells can follow geometric cues in their substrata such as tending to move in directions aligned with fibres in the matrix.

Mesenchymal cells have relatively long filopodia which extend beyond their nearest neighbours. In vivo such cells are fairly densely packed. So cells can sense the density beyond their immediate neighbourhood. A diffusional flux appropriate to such a situation has been discussed by Cohen and Murray (1981), using a Landau-Ginzburg approach, and is given by

$$\underset{\sim}{J}_D = \underset{\sim}{J}_{\substack{\text{random}\\\text{dispersal}}} = -D_1 \nabla n + D_2 \nabla(\nabla^2 n), \tag{2}$$

where D_1 and D_2 are positive constants with D_1 the usual Fickian diffusion coefficient. D_2 is a measure of the non-local effect of neighbouring cells. Since random dispersal results from variations in local and non-local averages a more appropriate derivation of (2) based on these has been given by Othmer (1983). In (2) D_1 and D_2 are both taken to be positive and hence both terms have a stabilizing effect on spatial heterogeneities.

The cells, by exerting traction, deform the ECM and generate gradients in the matrix density ρ. If we think of the matrix as providing attachment sites for the cells to hold on to we suppose that the greater the matrix density the greater the number of sites. Thus there is a tendency for cells to move up a gradient, $\nabla\rho$, in the ECM. As a first approximation we take this haptotaxis contribution to the flux of cells to be

$$\underset{\sim}{J}_H = \underset{\sim}{J}_{\text{haptotaxis}} = \alpha n \nabla \rho \tag{3}$$

where α is a positive constant. Although we do not do so here we could reasonably include a long range haptotactic effect similar to that in (2): it results in an extra term proportional to $n\nabla(\nabla^2\rho)$ in (3).

If the ECM is compressed or stretched, cells sitting on the matrix are passively transported. This convection contribution to the flux is equal to the average substratum velocity $\partial \underset{\sim}{u}/\partial t$ times the cell density, namely

$$\underset{\sim}{J}_C = \underset{\sim}{J}_{\text{convection}} = n\frac{\partial \underset{\sim}{u}}{\partial t} \tag{4}$$

Here and below we assume the linear strain approximation is a reasonable approximation to the matrix deformation.

With (2) - (4) it is now clear how further effects such as chemotaxis can easily be included in $\underset{\sim}{J}$ in (2).

We shall take the cell division mitotic source term M to be approximated by a simple logistic law for simplicity. We finally have as the cell conservation equation, from (1) - (4),

$$n_t = D_1\nabla^2 n - D_2\nabla^4 n - \alpha\nabla.n\nabla\rho - \nabla.n\underset{\sim}{u}_t + rn(N-n), \tag{5}$$

where r, the linear growth rate, and N are positive constants.

Cell-matrix interaction equation

In embryonic development the size scale of the patterns, with which we shall be concerned, is very small and motion takes place over the order of hours. We thus can neglect inertial effects and assume that the various forces are in equilibrium. We shall assume the simplest practical mechanical coupling which implies that the elastic restoring forces produced by the matrix plus the external forces on the ECM balance the traction forces generated by the cells. The equation is then

$$\nabla.\underset{\approx}{\sigma} + \rho\underset{\sim}{F} = 0, \tag{6}$$

where the stress $\underset{\approx}{\sigma}$ is the sum of

$$\left.\begin{aligned} \underset{\approx}{\sigma}_{ECM} &= \mu_1\underset{\approx}{e}_t + \mu_2\theta_t\underset{\approx}{I} + \frac{E}{(1+\nu)}(\underset{\approx}{e} + \hat{\nu}\theta\underset{\approx}{I}) \\ \underset{\approx}{\sigma}_T &= \underset{\approx}{\sigma}_{\text{cell traction}} = \tau(n)(\rho + \beta\nabla^2\rho)\underset{\approx}{I} \end{aligned}\right\} \tag{7}$$

where in $\underset{\approx}{\sigma}_{ECM}, \mu_1, \mu_2$ are the shear and bulk viscosities, $\underset{\approx}{e}(=\frac{1}{2}(\nabla\underset{\sim}{u} + \nabla\underset{\sim}{u}^T))$ the linear strain, $\theta(=\underset{\sim}{\nabla}.\underset{\sim}{u})$ the dilation, E and ν the Youngs modulus and Poisson ratio with $\hat{\nu} = 2\nu/(1-\nu)$: this is simply the usual linear visco-elastic stress tensor. In $\underset{\approx}{\sigma}_T$, $\tau(n)$ is the traction force generated by cells with density n, for example τn, where τ is a constant, or, more in keeping with the observation that there is some cell-cell contact inhibition, $\tau n/(1+\lambda n)$ where τ,λ are positive parameters. The β-term is a measure of the, by now usual, long range contribution.

Since in this paper we shall consider dermal mesenchymal patterns there is an external force which arises from the elastic attachment of the dermal ECM to a substratum. This body force we take as being proportional to the matrix strain $\underset{\sim}{u}$, that is $\underset{\sim}{F} = s\underset{\sim}{u}$ where s is a positive constant.

Finally the force equation (6) is then

$$\nabla\cdot\left\{\mu_1\underset{\approx}{e}_t + \mu_2\partial_t\underset{\approx}{I} + \frac{E}{(1+\nu)}\left(\underset{\approx}{e} + \hat{\nu}\theta\underset{\approx}{I}\right) + \tau(n)(\rho + \beta\nabla^2\rho)\underset{\approx}{I}\right\} = s\rho\underset{\sim}{u}. \qquad (8)$$

For simplicity at this stage we shall take $\tau(n) = \tau n$ in the rest of the paper.

Matrix conservation equation

Although mesenchymal cells secrete matrix, with the time scale we are considering (of the order of hours) we shall neglect this. Thus the conservation equation for ρ is simply

$$\rho_t + \nabla\cdot(\rho\underset{\sim}{u}_t) = 0. \qquad (9)$$

Equations (5), (8) and (9) are the model field equations for the dependent variables $n,\underset{\sim}{u},\rho$, which govern the cell-ECM mechanical interaction.

As always, it is convenient to non-dimensionalize the equations before analysis and this is done by introducing

$$\left.\begin{array}{l} \underset{\sim}{x}^* = \underset{\sim}{x}/L,\ t^* = t^*/T,\ \underset{\sim}{u}^* = \underset{\sim}{u}/L,\ \theta^* = \theta,\ \underset{\approx}{e}^* = \underset{\approx}{e} \\ D_2^* = D_2T/L^4,\ D_1^* = D_1T/L^2,\ \alpha^* = \alpha\rho_0T/L^2 \\ r^* = rNT,\ \mu_i^* = \mu_i(1+\nu)/ET,\ i = 1,2 \\ \tau^* = \tau\rho_0N(1+\nu)/E,\ \rho^* = \rho/\rho_0,\ n^* = n/N, \\ s^* = s\rho_0L^2(1+\nu)/E,\ \beta^* = \beta/L^2, \end{array}\right\} \qquad (10)$$

where T,L are a typical time and length. We could for example choose T to be the mitotic time $1/rN$ in which case $r^* = 1$. Similarly we could have $\alpha^* = 1$ or $\mu_i^* = 1$ and so on.

The equation system we consider in the following Section 3 is, on substituting (10) into (5), (8) and (9),

$$\left.\begin{aligned} n_t &= D_1\nabla^2 n - D_2\nabla^4 n - \alpha\nabla.n\nabla\rho - \nabla.(n\underset{\sim}{u}_t) + rn(1-n), \\ \nabla.\{\mu_1\underset{\approx}{e}_t &+ \mu_2\theta_t\underset{\approx}{I} + \underset{\approx}{e} + \hat{\nu}\theta\underset{\approx}{I} + \tau n(\rho+\beta\nabla^2\rho)\underset{\approx}{I}\} = s\rho\underset{\sim}{u}, \\ \rho_t &+ \nabla.(\rho\underset{\sim}{u}_t) = 0, \end{aligned}\right\} \qquad (11)$$

where for convenience we have omitted the asterisks. Henceforth all variables will be dimensionless.

Note in (11) that the model parameters, ten in all, are divided into those associated with the cell properties, namely $\tau, D_1, D_2, \alpha, r$ and β, and those relating to matrix properties, $\mu_1, \mu_2, \hat{\nu}$, and s.

3. Linear analysis and pattern formation potentialities

The non-trivial uniform steady state of (11) is

$$n = \rho = 1, \ \underset{\sim}{u} = 0. \qquad (12)$$

Linearizing about this solution we look for solutions of the linearized system in the form

$$\begin{pmatrix} n-1 \\ \rho-1 \\ \underset{\sim}{u} \end{pmatrix} \propto e^{\sigma t + i\underset{\sim}{k}.\underset{\sim}{x}} \qquad (13)$$

which gives the dispersion relation $\sigma = \sigma(k^2)$, where $\underset{\approx}{k}$ is the wave vector, as solutions of

$$\left.\begin{aligned} &\sigma[\mu k^2\sigma^2 + b(k^2)\sigma + c(k^2)] = 0 \\ &b(k^2) = \mu D_2 k^6 + (\mu D_1 + \beta\tau)k^4 + (1+\mu r - 2\tau)k^2 + s \\ &c(k^2) = \tau\beta D_2 k^8 + [\tau(\beta D_1 - D_2) + D_2]k^6 + \\ &\qquad + [\tau(r\beta - D_1 - \alpha) + D_1 + sD_2]k^4 + (r + sD_1 - r\tau)k^2 + rs. \end{aligned}\right\} \qquad (14)$$

Here $\mu = \mu_1 + \mu_2$ and τ, μ and s replace $\tau/(1+\hat{\nu})$, $\mu/(1+\hat{\nu})$ and $s/(1+\hat{\nu})$.

The uniform steady state is unstable if for any wave number k there is a corresponding $\sigma(k^2)$ such that $R\ell\sigma > 0$. From (14) if $k^2 = 0$, $b > 0$, $c > 0$ and so $\sigma = -c/b < 0$ and hence stability obtains. Thus for $R\ell\sigma(k^2) > 0$ to exist we must have $k^2 \neq 0$ and so solutions (13) with such wave numbers are linearly unstable and grow exponentially with time. The assumption, which is borne out by numerical simulation is that such unstable heterogeneous linear solutions evolve into finite amplitude spatially structured solutions. The predictive ability of the linearized solutions will probably be comparable to that in reaction diffusion mechanisms and discussed briefly by Murray (1981b).

We assume that all of the parameters in (14) are non-negative so, for a solution with $R\ell\sigma > 0$ to exist, we must have $\tau > 0$ at least. That is the cell traction is an essential ingredient if the mechanism is to generate spatial patterns. From (14) we require $b(k^2) < 0$ or $c(k^2) < 0$ for some $k^2 \neq 0$. The variation in the behaviour of σ, as a function of k^2, for various values or ranges of the parameters, is large. This is, in marked contrast to the limited variation possible in reaction diffusion systems, at least up to three species models.

As an illustrative example, suppose $\beta = \mu = r = 0$, then (14) gives $\sigma = 0$ and

$$\sigma = \frac{-c(k^2)}{b(k^2)} = \frac{-k^2[(1-\tau)D_2k^4 + \{(D_1+sD_2) - (D_1+\alpha)\tau\}k^2 + sD_1]}{(1-2\tau)k^2 + s}$$

If $\tau < \frac{1}{2}$, $b(k^2) > 0$ so $\sigma > 0$ only if $c(k^2) < 0$, that is when the bracketed quadratic in k^2 is negative. This gives as conditions on the parameters

$$(D_1+\alpha)\tau - (D_1+sD_2) \geq 0, \quad [(D_1+\alpha)\tau-(D_1+sD_2)]^2 \geq 4sD_1D_2(1-\tau). \qquad (15)$$

If we fix all the parameters but τ and let it vary as the bifurcation parameter, Fig. 1 illustrates the $\sigma(k^2)$ behaviour as τ passes through a critical valve τ_c. As the traction increases beyond τ_c certain wave

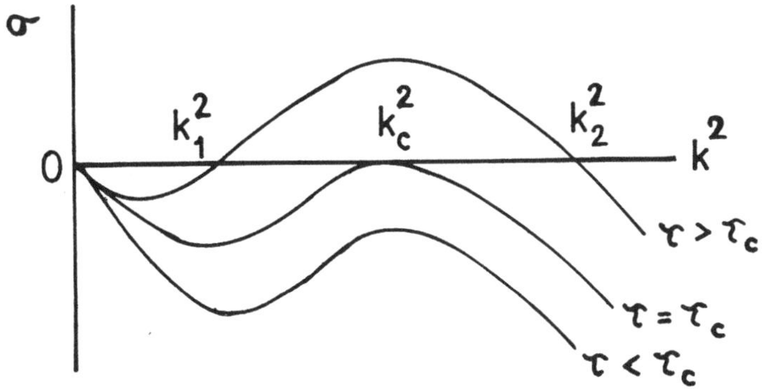

Figure 1
Growth factor σ as a function of k^2 as τ passes through the bifurcation traction value. For $\tau > \tau_c$, $\sigma > 0$ for all wave numbers $k_1^2 < k^2 < k_2^2$.

lengths become linearly unstable. There is experimental justification for this since the cell traction increases with age in certain circumstances.

Simulation of the full nonlinear system, for parameters in the range where certain modes are unstable, indicates that the unstable linear modes evolve into a finite amplitude steady state inhomogeneous solution. Using τ as the bifurcation parameter Murray and Maini (1985) have carried out a nonlinear stability analysis in the vicinity of τ_c and found steady state heterogeneous solutions.

The field equations (11) are still quite general. Certain simpler models will generate spatial structure. A general classification and analysis is given in the paper by Murray and Oster (1983). What is required is a $\sigma(k^2)$ from (14) such that $R\ell\sigma > 0$ for some $k^2 \neq 0$. For example, a particularly simple but interesting model has no diffusion, $D_1 = D_2 = 0$, no haptotaxis, $\alpha = 0$, and negligible matrix viscosities, $\mu_1 = \mu_2 = 0$. The resulting field equations, given by (11), in, for example, one space dimension are

$$\left.\begin{aligned} & n_t + (nu_t)_x = rn(1-n) \\ & \frac{\partial}{\partial x}[u_x + \tau n(\rho + \beta\rho_{xx})] = s\rho u \\ & \rho_t + (\rho u_t)_x = 0 \end{aligned}\right\} \qquad (16)$$

with $\sigma(k^2)$ from (14). The $\sigma - k^2$ behaviour is illustrated in Fig. 2 where again we have taken τ as the varying parameter.

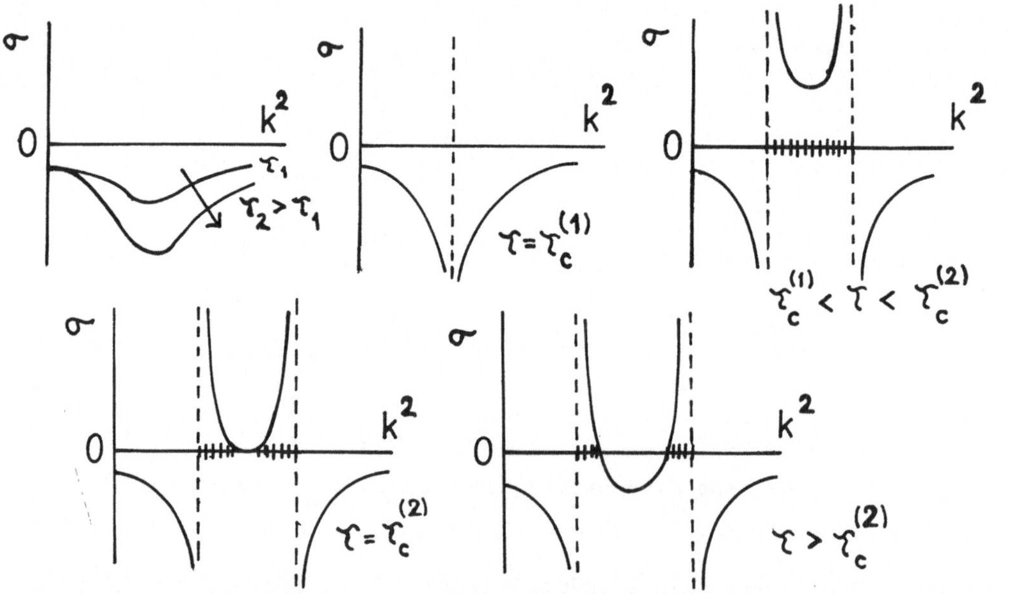

Figure 2

Dispersion relation $\sigma \sim k^2$: ++++++ denotes unstable wave numbers

The two-dimensional situation is particularly interesting biologically as described in Section 4. Here we look for periodic solutions of the field equations, that is solutions such that $\psi(\underset{\sim}{x} + m\underset{\sim}{\omega}_1 + \ell\underset{\sim}{\omega}_2) = \psi(\underset{\sim}{x})$ where ψ is $n, \underset{\sim}{u}$ and ρ, m and ℓ integers and $\underset{\sim}{\omega}_1$ and $\underset{\sim}{\omega}_2$ independent vectors. Such solutions give a tessellation in the plane. The linearized system from (11) is (taking the divergence of the second equation)

$$\left.\begin{aligned} & n_t + D_2\nabla^4 n - D_1\nabla^2 n + \alpha\nabla^2\rho + \theta_t + rn = 0 \\ & \nabla^2[\mu\theta_t + (1+\hat{\nu})\theta + \tau n + \tau\rho + \tau\beta\nabla^2\rho] = s\theta \\ & (\rho + \theta)_t = 0 \end{aligned}\right\} \qquad (17)$$

At the very least solutions include the periodic eigenfunctions of

$$\nabla^2\psi + k^2\psi = 0 \qquad (18)$$

with $(\hat{\underset{\sim}{n}}.\nabla)\psi = 0$ on the periodic boundaries where $\hat{\underset{\sim}{n}}$ is the unit normal vector on the boundary. Regular plane tessellation solutions of (18) are squares, hexagons (and hence triangles) or rhombi. The hexagonal pattern is particularly relevant, as we see below. Even the linear problem for the eigenfunctions of (17) is far from trivial and the resulting set will include all of those from (18). The relevant solutions of (11) or even the simpler model (16) pose a challenging analytical problem. The class of solutions will be even greater than that for reaction-diffusion systems. The key mathematical problem is to determine the way the linearly unstable modes predict the finite amplitude spatial solutions.

4. Biological applications

The specific applications where the generation of regular patterns is crucial are cartilage condensations which evolve into the skeletal elements in the vertebrate limb and mesenchymal cell condensations in the dermis. These latter presage the epidermal placodes which gives rise to feathers and scales.

Limb cartilage patterns

In the developing limb cells primarily come from the progress zone at the distal tip of the limb. The cell properties, that is parameters, are set as they leave this zone. Refer now to Fig. 3 in which we approximate the cross-section of the proximal limb bud shape to be a circle. The first pattern to evolve from our model is a central condensation of cells which we suggest become the chondrocytes - cartilage forming cells. The traction of this aggregation of cells deforms the

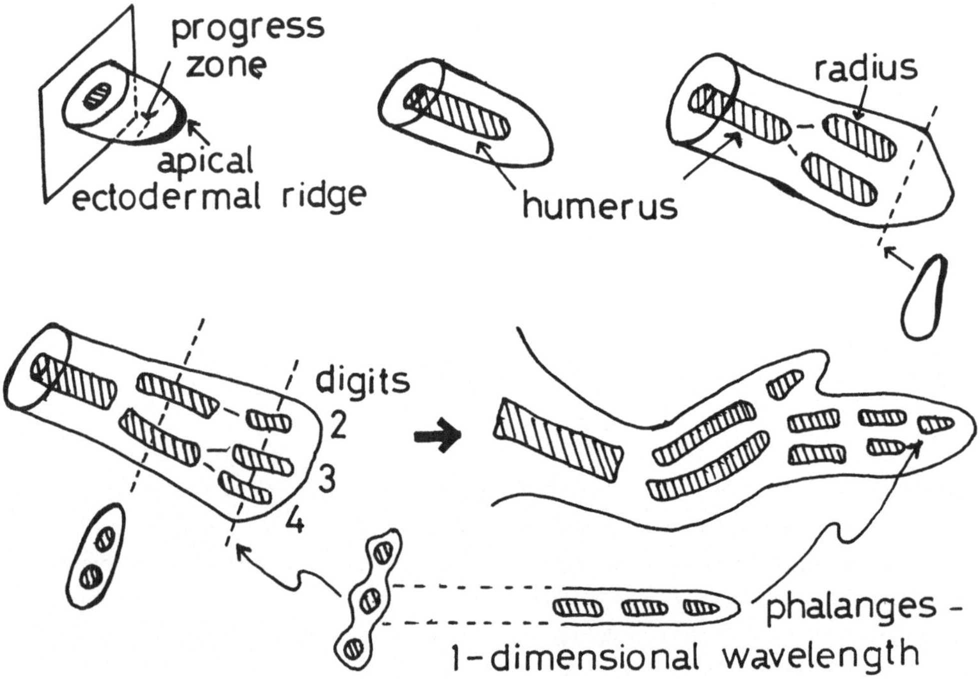

Figure 3. Chick limb bifurcation scenario.

boundary so that it is more elliptical and the model predicts by virtue of a geometric parameter change the excitation of the eigensolution consistent with two condensations. The analysis is similar to that described by Murray (1981a). As each pattern is set up it has within the resulting arrangement the potential for geometry deformation to encourage the sequential pattern of cartilage observed: see Fig. 3. The bifurcation from one bone (humerus) to two (radius and ulna) occurs as the parameters in the system (11) cross a bifurcation surface. The bifurcation space is that of all the model parameters plus those related to the domain geometry and scale.

Periodic patterns of feathers and scales

Feather and scale primordia begin as local thickenings of the epithelial cells which are cued by local periodic condensations in the dermis: see, for example, the recent work by Davidson (1983). It will be helpful to refer to Fig. 4.

The model predicts a periodic array of cell aggregations in the dermis as in stage 1 in Fig. 4: for example, the critical k_c at τ_c in Fig. 1. This occurs along the dorsal mid-line. These generate

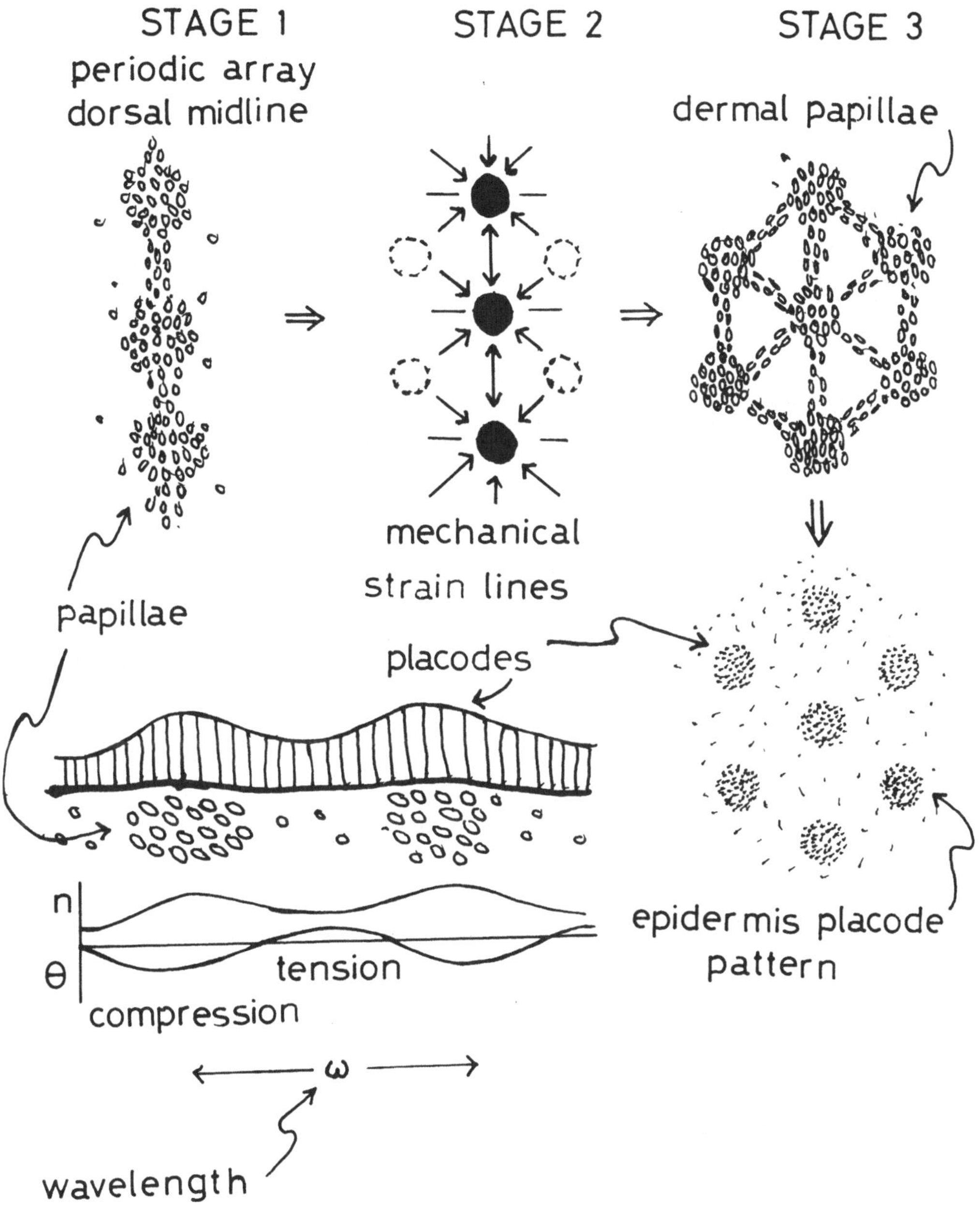

Figure 4

Morphogenesis of feather/scale primordia. The papillae wave length ω is a function of the ECM parameters (μ, ρ_0, E, ν, s) and the cell parameters $(\tau, \alpha, \beta, D_1, D_2, r, N)$. Epidermis mitosis is enhanced and inhibited by the underlying tension and compression respectively.

mechanical strain lines as in stage 2 and these in turn encourage the next 'wave' of condensations. In this way a regular hexagonal array can be formed as in stage 3. The pattern wave lengths are functions of the parameters of the model: geometry is not relevant here.

Of course morphogenesis is very complex and a host of mechanisms probably operate at various stages. What we suggest in this paper is that mechanical models such as we have described do indicate how the cells can organise themselves into patterns which are required. Unlike reaction-diffusion models, our mechanism relies on experimentally observed facts with experimentally measurable parameters. The class of potential patterns and phenomena possible with this model, with field equations (11), is *very* large. Finally, such a mechanical model has built in to it the mechanism for generating its own morphological form, for example as in the cartilage condensation patterns. There the gradual flattening of the limb towards the distal end is typical of vertebrate limbs and the geometric flattening could plausibly be the bifurcation trigger for initiating more structure.

References

Bard, J.B.L. (1981) A model for generating aspects of zebra and other mammalian coat patterns. J.Theor.Biol. 93, 363-386.

Cohen, D.S. and Murray, J.D. (1981) A generalized diffusion model for growth and dispersal in a population. J.Math.Biol. 12, 237-249.

Davidson, D. (1983) The mechanism of feather pattern development in the chick. J.Embryol.Exp.Morph. 74, 245-273.

Hall, B.K. (Editor) (1983) Cartilage Vol.2 Development, Differentiation and Growth. Academic Press, New York.

Harris, A.K., Stopak, D. and Warner, P. (1983) Generation of spatially periodic patterns by a mechanical instability: a mechanical alternative to the Turing model (this volume).

Harris, A.K., Stopak, D. and Wild, P. (1981) Fibroblast traction as a mechanism for collagen morphogenesis. Nature 290, 249-251.

Meinhardt, H. (1982) Models of Biological Pattern Formation. Academic Press, London.

Murray, J.D. (1977) Nonlinear Differential Equation Models in Biology. Clarendon Press, Oxford.

Murray, J.D. (1979) A pattern formation mechanism and its application to mammalian coat markings. In 'Vito Volterra' Symposium, Accademia dei Lincei, Rome, Dec. 1979. (Proceedings: Springer-Verlag Lecture Notes in Biomathematics 1980, pp.360-399.)

Murray, J.D. (1981a) A pre-pattern formation mechanism for animal coat markings. J.Theor.Biol. 88, 161-199.

Murray, J.D. (1981b) On pattern formation mechanisms for lepidopteran wing patterns and mammalian coat markings. Phil.Trans.Roy.Soc. London B295, 473-496.

Murray, J.D. and Maini, P. (1985) A nonlinear analysis of a mechanical model for biological pattern formation. (To appear)

Murray, J.D., Oster, G.F. and Harris, A.K. (1983) A mechanical model for mesenchymal morphogenesis. J.Math.Biol. 17, 125-129.

Murray, J.D. and Oster G.F. (1984) Cell traction models for generating pattern and form in morphogenesis. J. Math. Biol. (in press).

Oster, G.F. (1983) A mechanical model for plasmodial oscillations in physarum (this volume).

Oster, G.F., Murray, J.D. and Harris, A.K. (1983) Mechanical aspects of mesenchymal morphogenesis. J.Embryol.Exp.Morph. 78, 83-125.

Othmer, H.G. (1983) Personal communication.

Turing, A.M. (1952) The chemical basis for morphogenesis. Phil.Trans. Roy.Soc. London B237, 37-72.

Wolpert, L. (1971) Positional information and pattern formation. Current Topics in Dev.Biol. 6, 183-224.

EVERY MULTI-MODE SINGULARLY PERTURBED SOLUTION RECOVERS ITS STABILITY
- FROM A GLOBAL BIFURCATION VIEW POINT -

Y. Nishiura
Kyoto Sangyo University
Kyoto 603, Japan

§1. Introduction.

It is a well-known phenomenon in nonlinear diffusion system that several static or dynamic patterns coexist for a specified value of parameters. If we change parameters, the number of patterns may increase or decrease through, for example, a bifuration or a limit point like Fig. 1.1. The question is "Is there a possible way to understand the whole structure of the set of these solutions?". It might be possible to get (almost) all solutions for a particular value of parameters and even possible to trace solutions numerically when parameters vary, and eventually we might be able to obtain the whole structure of the set of solutions. However, this method is much harder than we can imagine, and the structure may change drastically when we change the nonlinearity. Therefore, we take another point of view which gives us a good perspective of the total structure, i.e., find an organizing center which creates a global structure of the set of solutions. For the stationary problem, several works have been done along this line for the following reaction-diffusion system of activator-inhibitor type with nonlinearity (f,g) like Fig. 1.2 (see [5] for the details of the assumptions of the nonlinearity and for examples):

$$\begin{cases} u_t = d_1 u_{xx} + f(u,v) \\ v_t = d_2 v_{xx} + g(u,v) \end{cases} \quad x \in I \ (\equiv (0,L)) \qquad \text{(P)}$$

$$u_x = 0 = v_x \qquad x \in \partial I .$$

In [2], it was shown that the set of multiple bifurcation points is one of the organizing centers to produce a global bifurcation picture. Another kind of organizing center was presented in [3] called singular shadow edge which we are mainly concerned about here. However, there are no rigorous justifications for this sort of organizing center so far. This note may be the first attempt to justify at least partially the usage of the terminology "organizing center".

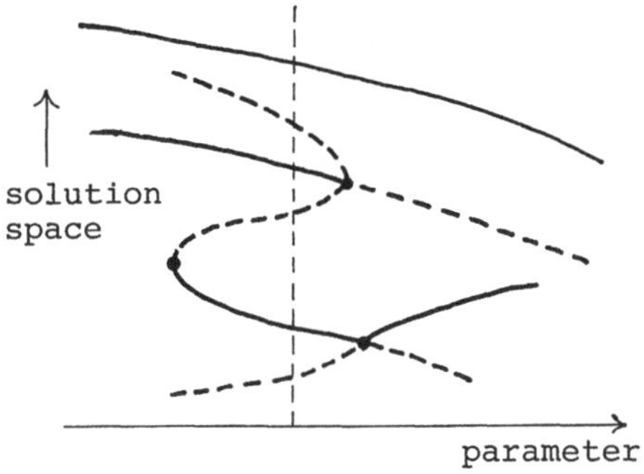

Fig. 1.1

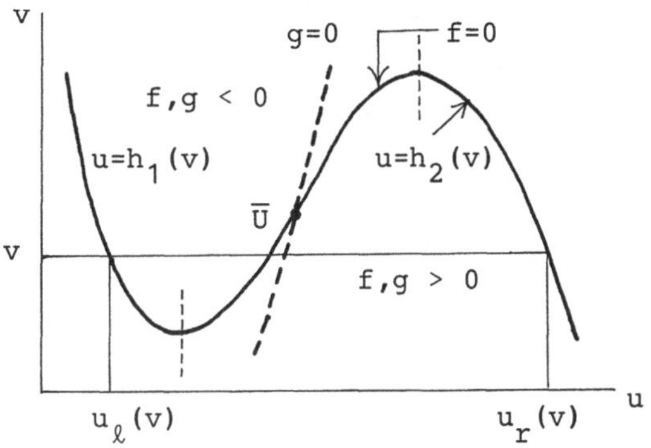

Fig. 1.2

On the other hand, there is a different problem which is important in pattern formation of reaction-diffusion system, i.e., stability of singularly perturbed solutions. The existence of such solutions was first shown by Fife [1] for Dirichlet boundary condition, and using his method, Mimura et al. [4] obtained the similar kind of solutions for the Neumann boundary conditions which has inner transition layers. However, there are several difficulties for the stability problem: Firstly, singularly perturbed solutions are of large amplitude and have transition layers each of which squeezes to one point as a parameter goes to zero. This makes the linearized eigenvalue problem difficult. Secondary, if the boundary conditions are of Neumann type, we can get infinitely many number of singularly perturbed solutions with different mode numbers (see §2). Therefore, it seems rather complicated to study the stabilities of all these solutions simultaneously. However, we can overcome these difficulties by looking at the relation between singularly perturbed solutions and a global bifurcation diagram.

The aim of this note is to present one of the organizing centers called singular shadow edge, and to show that it reveals the phenomenon of the recovery of stability of multi-mode singularly perturbed solutions. In fact we can prove that every multi-mode singularly perturbed solution gains its stability when the diffusion coefficient d_1 becomes sufficiently small for a large fixed d_2, which is one of the keystones to understand the coexistence phenomenon of multiple stable states.

We start from the summary of construction of singularly perturbed solution in the following section, and in §3 we consider the relation between singularly perturbed solutions and the global bifurcation diagram where singular shadow edge comes up. Finally in §4 we show the recovery of stability of singularly perturbed solutions of multi-mode type.

§2. Singularly Perturbed Solutions ([1], [4]).

If d_1 is sufficiently small and d_2 is sufficiently large, then we can construct a large amplitude solution with transition layers by singular perturbation method. We rewrite the diffusion coefficients as follows:

$$\begin{aligned} \varepsilon^2 u_{xx} + f(u,v) &= 0 \\ \frac{1}{\alpha} v_{xx} + g(u,v) &= 0, \end{aligned} \tag{SP}$$

where $\varepsilon = \sqrt{d_1}$ and $\alpha = d_2^{-1}$.

The first approximation (u_0, v_0) to the solution of (SP) is given by that of the following reduced problem:

$$\begin{aligned} f(u,v) &= 0 \\ \frac{1}{\alpha} v_{xx} + g(u,v) &= 0 \end{aligned} \tag{RSP}$$

The first equation is no longer a differential equation. The solution set of (RSP) is much bigger than that of (SP), however we are only interested in the solutions of (RSP) which are the limits of those of (SP) as $\varepsilon \downarrow 0$. One of the special solutions of $f = 0$ of such kind is given by

$$u = h^*(v) = \begin{cases} h_1(v) & \text{for } v < v^* \\ h_2(v) & \text{for } v > v^*, \end{cases} \tag{1}$$

where $h_1(v)(h_2(v))$ is the left(right) branch of $f = 0$ respectively, and v^* is the unique zero of the function $J(v) = \int_{u_\ell(v)}^{u_r(v)} f(s,v)\,ds$ Here $u_\ell(v)(u_r(v))$ is the left (right) zero of $f = 0$ respectively (see Fig. 1.2). Substituting this into the second equation, we obtain a reduced scalar equation for v:

$$\frac{1}{\alpha} v_{xx} + g(h^*(v), v) = 0.$$

Solving this equation, we can get C^1-solution $v_0(x)$ of one mode. Then using (1), we can obtain $u_0(x) = h^*(v_0(x))$ which is discontinuous at $x = x^*$ (see Fig. 2.1a). Finally, we can smooth out this discon-

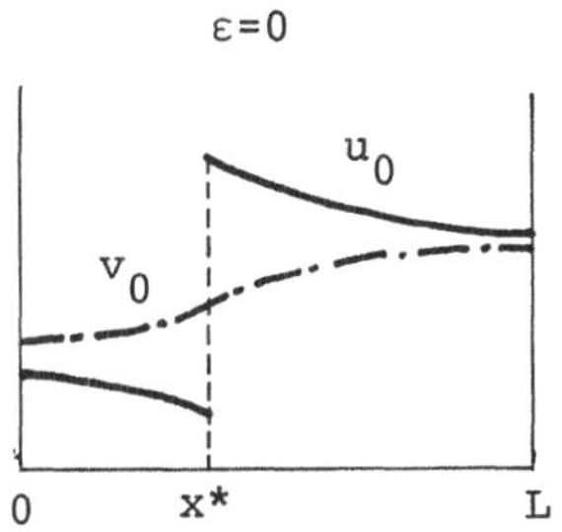

Fig. 2.1_a

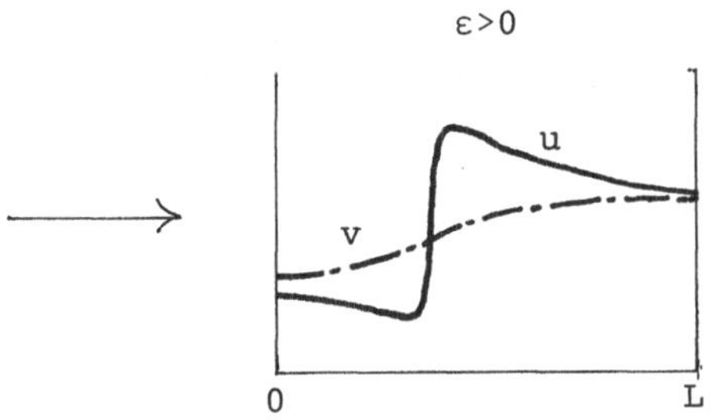

Fig. 2.1_b

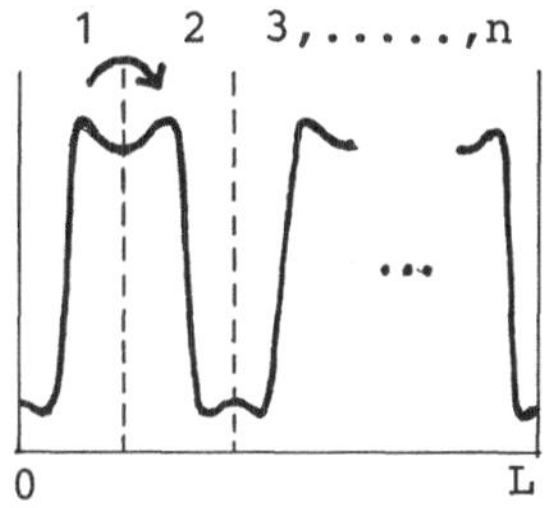

Fig. 2.2

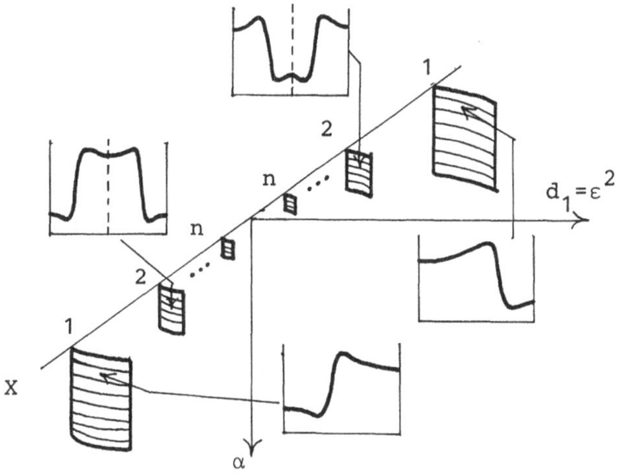

Fig. 2.3

X: a solution space for (SP)

tinuity for a positive ε due to the particular choice of solution (1) and obtain the smooth one-mode solutions $(u(x;\varepsilon,\alpha),\ v(x;\varepsilon,\alpha))$ for $0 < \varepsilon < \varepsilon_0$, $0 < \alpha < \alpha_0$ (see Fig. 2.1b).

Since the boundary conditions are zero flux, n-mode singularly perturbed solutions are easily constructed from one-mode solution by folding up with changing the diffusion coefficients from $(\varepsilon^2,\ 1/\alpha)$ to $(1/n)^2(\varepsilon^2,\ 1/\alpha)$ (see Fig. 2.2). Therefore, we get infinitely many families of solutions with different number of modes which are shown pictorially in Fig. 2.3. Here the natural question arises: "Which one is stable?" The most standard way to determine the stability is the principle of linearized stability. However, it might be quite difficult and complicated if we try to apply this principle to each solution without any perspective, because we have to consider infinitely many solutions simulteneously and the stability of multi-mode solution may depend on ε and α sensitively. In order to get a good perspective, we consider the relation between singularly perturbed solutions and a global structure of bifurcating solutions in the next section.

§3. Global Bifurcation Diagram.

Global bifurcation diagram in the limit $\alpha \downarrow 0$ ([5], [6]). Multiplying α on both sides of the second equation of (SP) and let α go to zero, we obtain $v_{xx} = 0$, which leads from our boundary conditions that v is a constant function. On the other hand, we can get the integral relation $\int_I g(u,v)dx = 0$ by integrating the second equation over I, which holds independently of α. Therefore, the stationary problem in the limit $\alpha \downarrow 0$ is given by

$$\begin{aligned} & d_1 u_{xx} + f(u,\xi) = 0 \\ & \int_I g(u,\xi) = 0, \end{aligned} \qquad \text{(SS)}$$

where $v \equiv \xi$ is a constant function. We call this the shadow system. The constant state $\bar{U}$ is a trivial branch for all d_1 and the Turing's instability occurs at each $(d_1^{c,n}, \bar{U})$ corresponding to the number n of Fourier mode for u (see Fig. 3.1). There are two important results concerning the global behavior and the stability of the bifurcating solutions emerging from $(d_1^{c,n}, \bar{U})$.

1) Each bifurcating branch from $(d_1^{c,n}, \bar{U})$ exists globally with respect to d_1 (we call this D^n-branch or, more precisely, D^n-shadow branch), and is connected to the singularly perturbed solutions of the same mode number as $d_1 \downarrow 0$ (Fig. 3.1). This connection result holds also for small positive α([5]).

2) Generically, there are no secondary static and Hopf bifurcations along each global branch D^n. Therefore, the number of the unstable eigenvalues of the linearized eigenvalue problem along D^n is essentially preserved and is equal to $n-1$ for small d_1 (Fig. 3.1)

One consequence of the above two results is that only one-mode singuarly perturbed solution is stable and all other solutions never recover their stabilities in the limit $\alpha \downarrow 0$. Therefore, we cannot observe the coexistence phenomenon in the limit $\alpha \downarrow 0$. The next question is what happens to the stability of each singularly perturbed solution when α becomes positive. It seems to be difficult to set up a plausible conjecture for this problem without any additional informations. Here we need a global bifurcation diagram for positive α to get an insight into this problem.

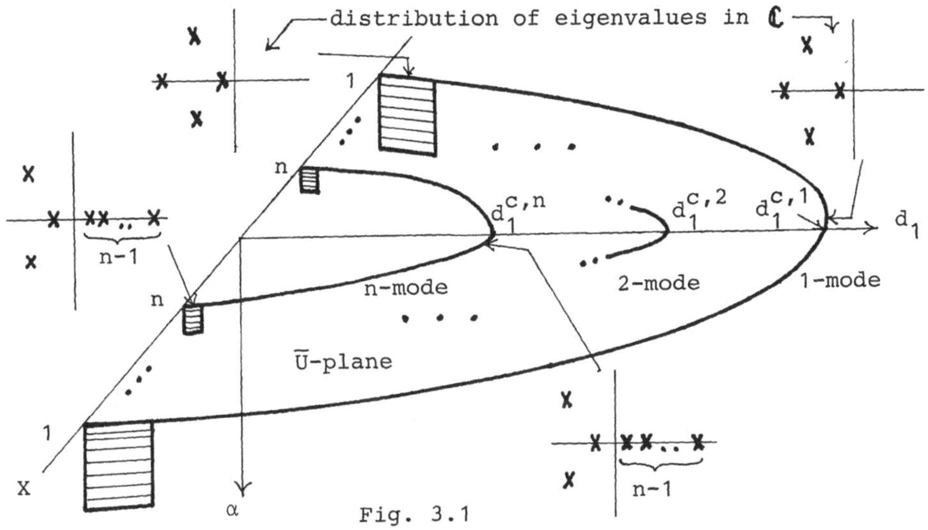

Fig. 3.1

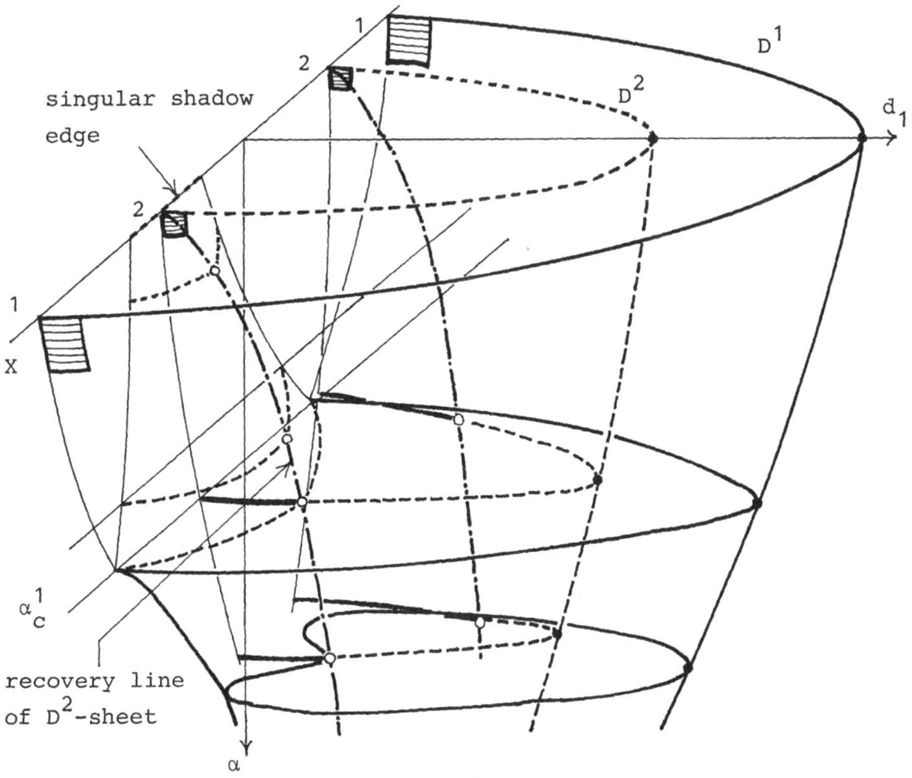

Fig. 3.2

● : primary bifurcation point

○ : secondary bifurcation point

Global bifurcation diagram for positive α ([2], [3]). Fig. 3.2 is a part of the global bifurcation diagram which shows us a inter-relation between D^1-and D^2-branches for all (d_1, d_2). We can see that D^1-branch falls into the D^2-branch for $\alpha > \alpha_c^1$, i.e., D^1-branch is just a secondary branch if we stand on the D^2-branch, and D^2-branch recovers its stability through this secondary bifurcation point. D^1-branch splits into two parts when α becomes smaller than α_c^1: inner one is a secondary branch of D^2-branch and outer one is a global branch which connects the local bifurcating solutions with singularly perturbed solutions in §2. It is suggested numerically that the secondary bifurcation point of D^2-branch seems to approach to the edge $\varepsilon = \alpha = 0$ when α goes to zero, i.e., the singularly perturbed solution of two mode type seems to recover its stability from the singular shadow edge (we call the edge $\varepsilon = \alpha = 0$ the singular shadow edge). Moreover, the secondary branch from D^2-primary one seems to converge to the one-parameter family of solutions in the singular shadow edge (which we call edge continuum, see [3]). Therefore it is worth trying to prove the following claim: The recovery line of D^2-sheet emerges from the singular shadow edge. Here the recovery line is the trace of secondary bifurcation points on D^2-sheet (—·—· in Fig. 3.2).

§4. Every Multi-Mode Singularly Perturbed Solution Recovers its Stability.

Last claim in §3 is true. In fact we can prove a more general one.

THEOREM ([7]) *The singularly perturbed solution of n-mode recovers its stability from the singular shadow edge* $\varepsilon = \alpha = 0$, *i.e., there exist n-1 number of recovery lines on* D^n*-sheet which emerge from the singular shadow edge (see Fig. 4.1).*

Outline of proof. As we mentioned in §3, there are $n - 1$ number of unstable eigenvalues of the linearized eigenvalue problem along D^n-shadow branch. It is necessary for D^n-branch to recover its stability from the singular shadow edge that all these unstable eigenvalues converge to zero as ε goes to zero. This actually occurs, in fact we can show that all unstable eigenvalues along D^n-shadow branch converge to zero with the order $O(K\exp(-\gamma/\varepsilon))$, where K and γ are positive constants which do not depend on ε (see Fig. 4.1).

REMARK 4.1. *There is another real eigenvalue along each shadow branch which converges to zero from below as $\varepsilon \downarrow 0$. However, this is not dangerous to the stability for $\alpha > 0$ because it can be shown that this eigenvalue moves to the left-half plane when α becomes positive.*

The next step is more subtle, i.e, how unstable eigenvalues accumulated at zero as $\varepsilon \downarrow 0$ behave when α becomes positive. Let us consider two linearized eigenvalue problems.

$$\begin{cases} \varepsilon^2 w_{xx} + f_u w + f_v \eta = \sigma^n w \\ \frac{1}{|I|}\int_I (g_u w + g_v \eta)\,dx = \sigma^n \eta \end{cases} \quad \text{(ESS)} \qquad \begin{cases} \varepsilon^2 w_{xx} + f_u w + f_v z = \lambda^n w \\ \frac{1}{\alpha} z_{xx} + g_u w + g_v z = \lambda^n z . \end{cases} \quad \text{(ESP)}$$

The first (second) one is a linearized eigenvalue problem at the D^n-shadow (D^n-) branch, respectively (all partial derivatives are evaluated at the corresponding solution). We already know that there are (n-1) number of positive real eigenvalues $\{\sigma_\ell^n(\varepsilon)\}_{\ell=1}^{n-1}$ of (ESS) which go to zero with the order $K\exp(-\gamma/\varepsilon)$ as $\varepsilon \downarrow 0$. Let us denote by $\lambda_\ell^n(\varepsilon,\alpha)$ the eigenvalue of (ESP) which approaches $\sigma_\ell^n(\varepsilon)$ when $\alpha \downarrow 0$ (each $\sigma_\ell^n(\varepsilon)$ is a real simple eigenvalue for positive ε). We are interested in the behavior of $\lambda_\ell^n(\varepsilon,\alpha)$ for positive α. Let us consider the difference $\tau_\ell^n(\varepsilon,\alpha) = \lambda_\ell^n(\varepsilon,\alpha) - \sigma_\ell^n(\varepsilon)$ for small ε and α. Note that $\tau_\ell^n(\varepsilon,0) = 0$ from its definition. After some asymptotic computation, it is possible to show that $\tau_\ell^n(\varepsilon,\alpha)$ is negative for small positive (ε,α) enough to beat the positive $\sigma_\ell^n(\varepsilon)$ in magnitude. This implies that $\lambda_\ell^n(\varepsilon,\alpha)\,(= \tau_\ell^n(\varepsilon,\alpha) + \sigma_\ell^n(\varepsilon))$ crosses the 0-level from positive to negative when α becomes positive. Therefore, we can define the ℓ-th recovery line $\alpha = R_\ell^n(\varepsilon)$ emerging from the singular shadow edge which is derived from the implicit relation $\lambda_\ell^n(\varepsilon,\alpha) = 0$. Thus, the n-mode singularly perturbed solution recovers its stability successively when the parameters cross the lines $\alpha = R_\ell^n(\varepsilon)$ $(\ell = 1,2,\ldots,n-1)$, and eventually recovers its total stability when the parameters are below all recovery lines, i.e., the shaded region on D^n-sheet in Fig. 4.1.

The above THEOREM is important to understand the coexistence phenomenon: If we fix α and take ε sufficiently small, we can see that there are several stable singularly perturbed solutions of different mode numbers (see Fig. 4.1), and the number of stable solutions goes to infinity when $\varepsilon \downarrow 0$. Moreover, it suggests to us that the stable region on D^n-sheet spreads to larger α-valued region and form an open band on D^n-sheet surrounded by recovery and losing lines, which causes the coexistence phenomenon for general values of parameters.

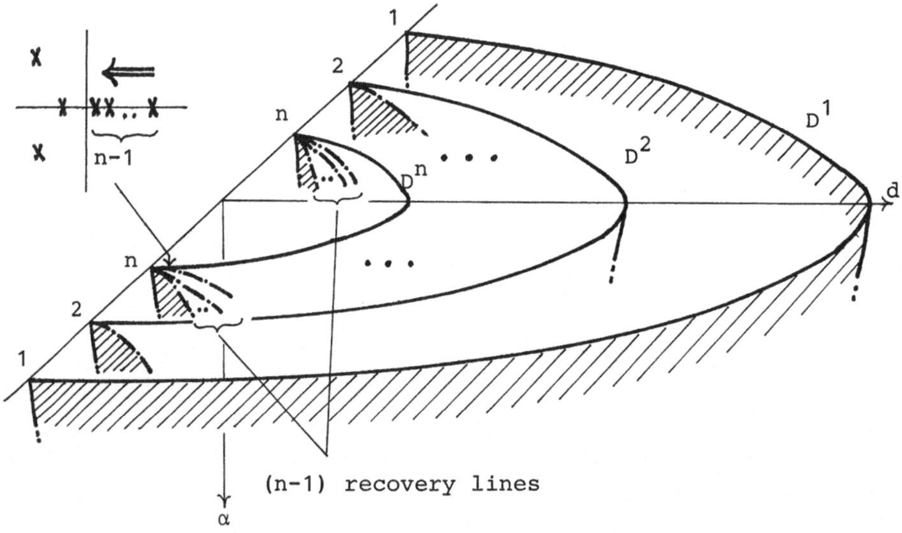

Fig. 4.1

///// : stable region

REFERENCES

[1] Fife, P. C., Boundary and interior transition layer phenomena for pairs of second-order differential equations, J. Math. Anal. Appl., 54(1976), pp. 497-521.

[2] Fujii, H., Mimura, M.,and Nishiura, Y., A picture of the global bifurcation diagram in ecological interacting and diffusing systems, Physica D-Nonlinear Phenomena, 5D(1982), pp. 1-42.

[3] Fujii, H. and Nishiura, Y., Global bifurcation diagram in Nonlinear diffusion systems, to appear in Math. Studies, LNNAA, 5 North Holland, 1983.

[4] Mimura, M., Tabata, M. and Hosono, Y., Multiple solutions of two-point boundary value problems of Neumann type with a small parameter, SIAM J. Math. Anal., 11(1980), pp. 613-631.

[5] Nishiura, Y., Global structure of bifurcating solutions of some reaction-diffusion systems, SIAM J. Math. Anal., 13(1982), pp. 555-593.

[6] Nishiura, Y., Global structure of bifurcating solutions of some reaction-diffusion systems and their stability problem, Computing Methods in Applied Sciences and Engineering, 5 Glowinski, R., Lions, J. L. (editors), North-Holland, 1982.

[7] Nishiura, Y., Stability of multi-mode singularly perturbed solutions of reaction-diffusion systems, in preparation.

A MECHANOCHEMICAL MODEL FOR PLASMODIAL OSCILLATIONS IN *PHYSARUM*

G.F. Oster
Department of Biophysics, University of California, Berkeley, CA 94720

G.M. Odell
Department of Mathematics, Rensellaer Polytechnic Institute, Troy NY 12181

ABSTRACT

Plasmodial strands of the slime mold *Physarum* exhibit rhythmic mechanical contractions. These contractions drive an alternating flow of cytoplasm, so-called "shuttle streaming", which underlies the movement of the plasmodium. Here we construct a mechanical model of the plasmodial cytogel which can mimic the periodic mechanical and chemical behavior of the plasmodial rhythms.

The authors were supported by NSF grants #MCS-8110557 (GFO) and #MCS-8301460 (GMO).

1. INTRODUCTION

The slime mold *Physarum* displays remarkable rhythmic activity: the plasmodial strands contract synchronously, creating an oscillating pressure head which drives an alternating flow of cytoplasm through the central core of the plasmodium (Sauer, 1982).

Many experiments have established that the contraction rhythms are based on the contractile properties of the actomyosin gel which forms the cortex of the plasmodium. This gel consists largely of actin and myosin, in both monomeric and polymer form, along with membranous vesicles which sequester and release calcium into the cytoplasm. Calcium, it turns out, is the principle chemical regulator of the contraction cycle, and its concentration in the cytoplasm oscillates in a fixed phase relationship with the mechanical contractions (Wohlfarth-Bottermann, 1979; Kamiya, 1981).

In this paper we will outline a mathematical model for the cortical cytogel of the plasmodium. The model is based on the mechanochemical properties of actomyosin gels, and may be generally applicable to other types of cell motion. The oscillatory behavior of the model results from the cross-coupling between the cortical deformations caused by the contractile protein machinery of the cytogel, and the biochemical reactions that control the contractile state of the gel.

2. THE MODEL

Mechanochemical Properties of Cytogel

Here we list the properties of cytogel that we must incorporate into the model; a more complete discussion of the biology can be found in Oster & Odell (1983).

1. Cytogel consists of a fibrous network of actin polymers cross-linked by actin binding proteins.

2. The activity of the actin binding proteins is regulated by the local concentration of Ca^{++} ions. Thus the sol-gel equilibrium of the network is regulated by Ca^{++} - generally in the micromolar concentration range.

3. Myosin assembly, phosphorylation and actin-binding affinity is also regulated by Ca^{++}.

4. The contractile force of an actomyosin gel fiber is generated by a sliding filament mechanism roughly similar to that in striated muscle.

Mathematical Formulation

We shall model the cytogel as a viscoelastic continuum. Since the contractile motions we study proceed very slowly we can neglect inertia forces. The equations of motion are then

$$\nabla \cdot \sigma + \underline{F} = 0 \qquad 1$$

where σ is the viscoelastic stress tensor of the cytogel, and $\underline{F}$ is the body force. The stress tensor will depend on the strain, the strain rate, and the local Ca^{++} concentration. To model σ we shall proceed as follows.

The Viscoelastic Stress Tensor for Cytogel

Consider a strip of cytogel of length L_o and cross-sectional area A_o. When it is stretched to length L, the cross-section decreases to A; the (Lagrangian) strain is defined as

$$\varepsilon \equiv (L - L_o)/L_o \qquad 2$$

When the cylinder is stretched to length $L = L_o(1+\varepsilon)$, its cross-section decreases to A. If the gel were incompressible, $A_oL_o=AL$, and so $A = A_o/(1+\varepsilon)$. In general, $A/A_o = \psi(\varepsilon)$; e.g. $\psi(\varepsilon) = 1+2\nu\varepsilon$, where ν is Poisson's ratio. However, we shall restrict our attention here to the simpler case where the gel is considered incompressible, and the strain is uniform along the cylinder.

Next, we define a fiber density flux, N, as

$$N = \text{number of fibers which pierce an area A } (\#/cm^2) \qquad 3$$

In 3 dimensions, N is a second order tensor; however, here we are restricting our discussion to the case of a 1-dimensional strip. The total fiber volume in the cytogel cylinder is

$$\begin{aligned}\text{Fiber volume} &= (A_oN_o)(aL_o)\\ &= \int aN(x)A(x)dx \qquad 4\end{aligned}$$

where a is the cross-sectional area of a fiber. In a small cylinder, the strain is nearly uniform: $N(x) \approx N$, $A(x) \approx A$; thus $NA = N_oA_o/(1+\varepsilon)$, and we can write the net axial force as

$$F = \phi NA = \phi N_oA_o/(1+\varepsilon) \qquad 5$$

and the axial stress as

$$\sigma = N_o \frac{\psi(\varepsilon)}{1+\varepsilon} \phi \equiv N_o\psi(\varepsilon) \qquad 6$$

where $\phi(.)$ is the axial force per fiber. As we shall see shortly, $\phi(.)$ depends not only on the strain, ε, but on the chemical state of the gel as well. Since we are considering only the case where the gel is incompressible, $\psi(\varepsilon)/(1+\varepsilon) = 1$, and the fiber count remains constant, $N = N_o$,

σ is proportional to ϕ, and so we will write the constitutive relation in terms of $\sigma(\varepsilon,c)$ directly.

We model the cytogel by a Kelvin body, so that we write σ as

$$\sigma = \sigma_V + \sigma_E \qquad 7$$

where σ_V and σ_E are the viscous and elastic contributions, respectively.

The viscous contribution to the stress tensor we assume to have the form

$$\sigma_V = \mu \, \partial\varepsilon/\partial t \qquad 8$$

where μ is the shear viscosity. However, the viscosity is not constant: because the cytogel can undergo a Ca^{++} stimulated sol-gel phase transition, μ depends on c in the sigmoidal fashion shown in Fig. 1.

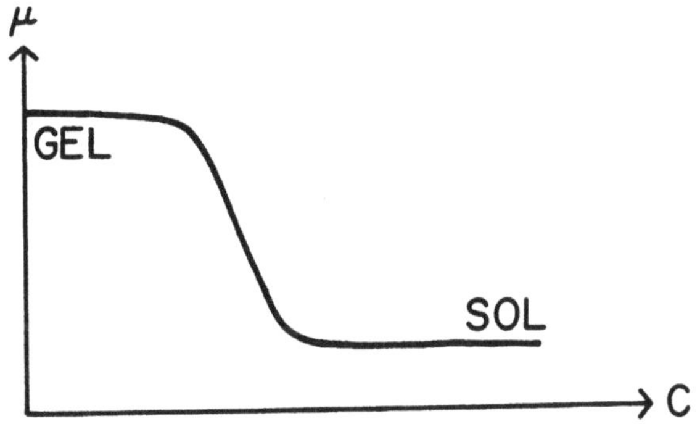

Fig. 1 The dependence of viscosity on calcium concentration. The sigmoidal curve is characteristic of the sol-gel phase transition as the local calcium concentration increases.

Since the gel consists of a fibrous network which can actively contract, the elastic contribution to the stress tensor is more complicated. We write the elastic stress as

$$\sigma_E = \sigma_P + \sigma_A \qquad 9$$

where σ_P and σ_A are the passive and active contributions to the elastic stress, respectively.

The qualitative form of σ_P is shown in Fig. 2a. In region I, the cytogel cylinder is in compression; the ensure that the cylinder cannot be compressed to zero volume, the stress-strain curve falls off to $-\infty$. In region II the cylinder is in tension, and the curve rises faster than linearly as the fibers commence to align. Finally, in region III, as the fibers commence to yield, the curve peaks and then falls off.

Figure 2

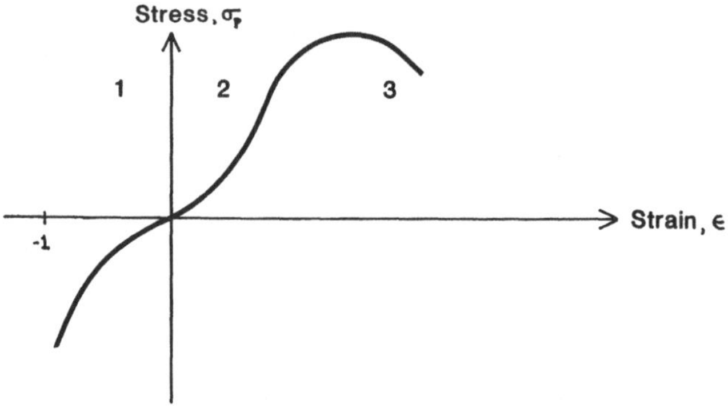

a

b

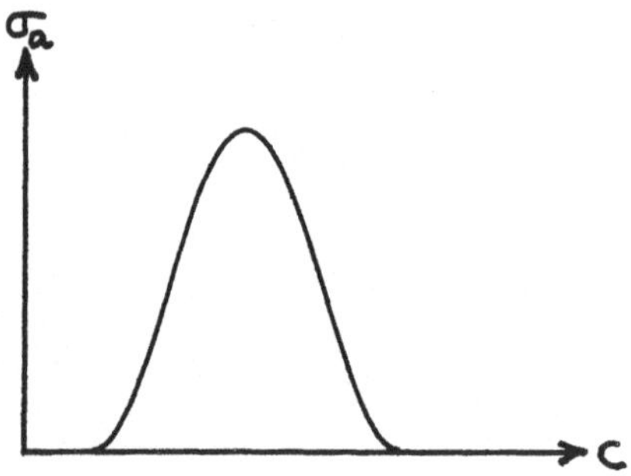

c

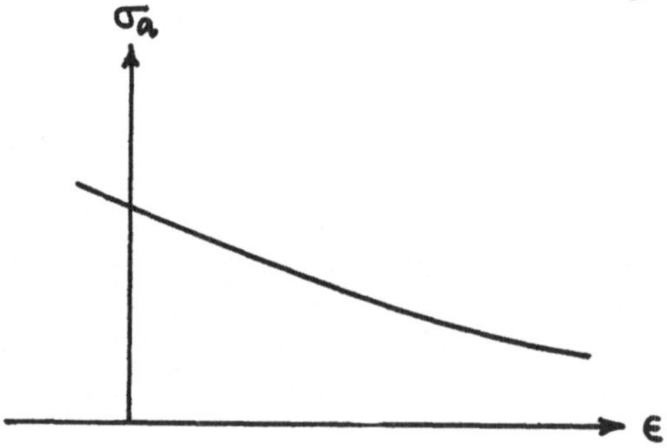

d

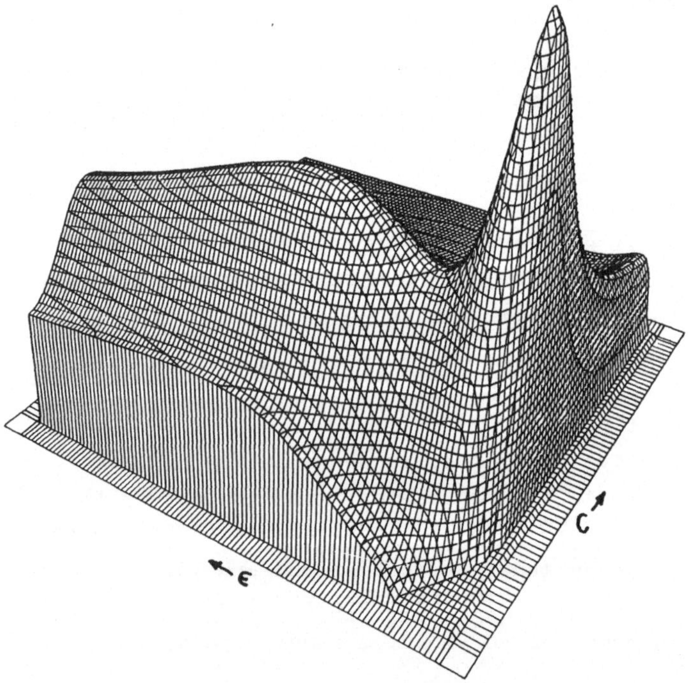

Fig. 2 (a) The passive elastic stress, σ_P, as a function of axial strain, ε. In region 1 the cytogel is in compression, and $\sigma(\varepsilon)$ falls off to $-\infty$ as $\varepsilon \to -1$, since the cytogel is ultimately incompressible. In region 2 the curve rises faster than linearly as the fibers align and the effective elastic modulus increases. In region 3 the curve levels off and ultimately falls as the fibers commence to break.

(b) The active stress, σ_A, falls with increasing strain since the traction developed by the cytogel arises from the myosin bridges acting in overlap regions of the actin and myosin fibers.

(c) The elastic stress as a function of calcium concentration. The curve first rises as the contractile machinery is activated, then falls as the cytogel solates due to the action of the solation enzymes.

(d) The mechanical response surface of the cytogel strip, $\sigma(\varepsilon,c)$. We regard the surface as an empirically determined constitutive relation which can be parameterized in various ways.

The active component of the stress, σ_A, is an increasing function of the fiber density, n, and a decreasing function of strain, ε. This latter dependence is shown in Fig. 2b; it arises from the sliding filament mechanism which generates the active stress: as fibers contract, the region of overlap between actin and myosin molecules, and thus the number of active cross-bridges, increases. The shorter a fiber gets, the stronger it becomes.

Both the active and passive stress components depend on the local Ca^{++} concentration, c. This is because of calcium's dual role of regulating contraction (by activating the kinase which phosphorylates the myosin light chain) and the sol-gel equilibrium (by activating the solation and gelation enzymes). Fig. 2c shows how the active stress varies with c: at first it increases as the myosins are activated; finally it decreases as the fiber network becomes so solated that it cannot sustain much tension. Similarly, the passive elastic response of the gel decreases at high c levels as the network solates.

If we assemble Figures 2a,b and c we can construct the response surface, $\sigma(\varepsilon,c)$, shown in Fig. 2d. This surface is the basic experimental data required to model cytogel dynamics. (Unfortunately, the necessary data has not yet been gathered in this form, and so we have substituted a plausible estimate of what measured data would look like.) Thus we regard $\sigma(\varepsilon,c)$ as an empirical quantity, and parametrize it in whatever fashion is most convenient. For example, we can modify the usual linear Hookian solid: $\sigma = E\varepsilon$, where E is the elastic modulus, by writing

$$\sigma_E = \{\underbrace{E(\varepsilon,c)\varepsilon}_{\text{Passive}} + \underbrace{\tau(\varepsilon,c)}_{\text{Active}}\}N_O \qquad 10$$

where τ is the traction per fiber.

Therefore, we can write the stress tensor for the contractile cytogel as

$$\sigma = \{\mu(c)\varepsilon_t + E(\varepsilon,c)\ \varepsilon + \tau(\varepsilon,x)\ \}N_o \qquad 11$$

where the function m(.), E(.) and τ(.) are chosen to fit the curves in Figures 1 and 2.

Thus the mechanical equations of motion for the cytogel are

$$\frac{\partial}{\partial x}\{\mu(c)\frac{\partial \varepsilon}{\partial t} + E(\varepsilon,c)\ \varepsilon + \tau(\varepsilon,x)n\ \} = 0 \qquad 12$$

It turns out that, in some circumstances, nonlocal effects can be important (Oster, Murray & Odell, 1983). In Appendix A we sketch how 12 must be modified to include longer range stress effects.

The Chemical Trigger for Cytogel Contraction

Embedded in the cortical cytogel are membrane vesicles which can store and release Ca^{++} - the chemical trigger for contraction. The details of this process are quite complex (c.f. Sauer, op. cit.; Oster & Odell, op. cit.); however, the principle event which we must model is the release and resequestering of Ca^{++}. There are two ways in which the release can be triggered:

(a) A calcium stimulated calcium cascade (Jaffe, 1980): when a small amount of Ca^{++} leaks into the cytogel - either from the sequestering vesicles, or from the external medium through the plasmalemma - this initiates an autocatalytic release of the stored Ca^{++}. That is, the permeability of the plasmalemma and/or internal vesicle membranes to Ca^{++} is itself an increasing function of the Ca^{++} concentration outside of the vesicles.

(b) Stretch activation: mechanically straining the cytogel can initiate the Ca^{++} cascade. This probably occurs due to a strain-induced depolarization of the membranes of the sequestering vesicles.

We model this in the simplest possible way as follows. Let c denote the Ca^{++} concentration external to the sequestering vesicles (i.e. free to interact with the contractile proteins), and C be the concentration inside the vesicles. Then

$$\partial c/\partial t = \underbrace{P(c,\varepsilon)(C - c)}_{\text{leakage rate from vesicles}} - \underbrace{rc}_{\text{resequestration rate}} \qquad 13$$

where r is the first-order resequestering rate and $P(c,\varepsilon)$ is the permeability of the membrane compartment to Ca^{++}. $P(c,\varepsilon)$, in turn, can be broken into three parts: (i) a constant leakage rate; (ii) an autocatalytic "ligand-gated" release which is triggered by external Ca^{++}; (iii) a strain-induced leak:

$$P(c,\varepsilon) = \underset{\text{Leakage}}{\lambda} + \underset{\text{Autocatalytic Release}}{A(c)} + \underset{\text{Strain-Induced Leak}}{\gamma\varepsilon} \qquad 14a$$

The gated release term, A(c), has the sigmoidal form shown in Fig. 3a. For the simulations we have used a standard expression for autocatalytic kinetics

$$A(c,\varepsilon) = \frac{\alpha c^2}{1 + \beta c^2} \qquad 14b$$

We shall assume that the Ca^{++} concentration internal to the vesicles is much greater than the external concentration: $C \gg c$; therefore, we can approximate $C - c \approx C \simeq$ constant, and write the kinetic equation governing the chemical trigger as

$$\frac{\partial c}{\partial t} = \frac{\alpha c^2}{1 + \beta c^2} - rc + \gamma\varepsilon + \lambda \qquad 15$$

where we have absorbed C into the definitions of α, γ and λ. Thus the kinetics has the S-shaped form shown in Fig. 3b, with one or two stable equilibria, depending on the values of the parameters.

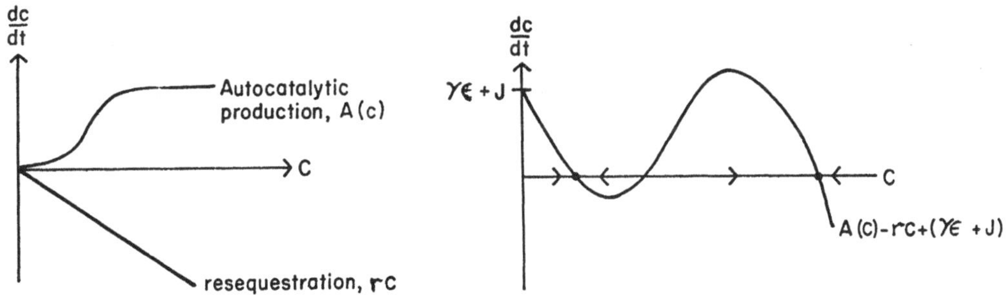

Fig. 3 (a) The components of the membrane compartment permeability as a function of calcium. The autocatalytic release is sigmoidal, while the strain leak and resequestration rate are assumed linear.
(b) The reaction kinetics $\partial c/\partial t = f(c,\varepsilon)$. There are one or two stable attractors, depending on the values of the parameters and the strain, ε.

Equations 12 and 15 constitute the simplest model for a 1-dimensional strip of active cytogel.

3. OSCILLATIONS IN A MODEL PLASMODIUM

The plasmodial strand of *Physarum* consists of a hollow cylinder of cytogel encased in a plasmalemma. Rhythmic contractions of the gel cortex drive the alternating "shuttle streaming" of cytosol back and forth through the lumen of the plasmodium. Reviews by Kessler (1982), Tyson (1982), Kamiya (1981) and Sauer (op. cit.) document the many experimental investigations on this system. In this section we shall construct a simplified model of the plasmodial rhythms; a more complete description can be found in Oster & Odell (1983).

We shall focus our attention on 3 phenomena associated with the plasmodial rhythms: (a) The contraction rhythms of the gel cortex; (b) the shuttle streaming of cytosol along the central core; and (c) the phase relationship between the contraction rhythms and the free Ca^{++} levels within the cortex.

Experiments have demonstrated that plasmodial rhythms will not persist unless cytogel is allowed to flow between regions. Moreover, the local contraction rhythms in the gel cortex will not become sufficiently synchronized over the plasmodial length to pump cytosol unless the mechanical coupling between segments remains intact. The simplest model which captures these aspects of the oscillating plasmodium is shown in Fig. 4. Here we have approximated two segments of the plasmodium by two cylindrical segments separated by a connecting tube. Each segment consists of a cortical shell of cytogel: a viscoelastic shell whose constitutive relations are given by equations 12 and 15 . Within the cavity flows a cytosol: a viscous fluid whose chemical properties are given by equation 15 . Thus the plasmodium model can be represented by a set of 5 ordinary differential equations

$$\frac{\partial x}{\partial t} = \underline{F}(\underline{x}) \qquad 16$$

where the variables contained in the vector $\underline{x}$ are

R_1 = radius of shell 1 17a

c_i = Ca^{++} concentration in the cytogel in shell i = 1,2 17b

e_i = Ca^{++} concentration in the endoplasm in shell i = 1,2 17c

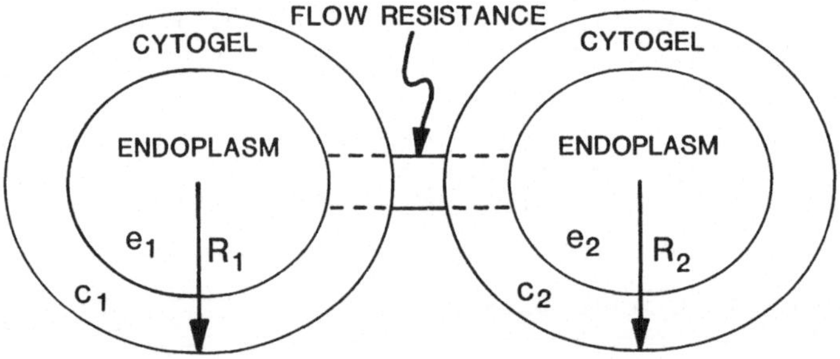

Fig. 4 The model plasmodium consists of two compartments separated by a flow resistance.

In Appendix B we give the exact form of equations 16 ; these constitute a finite dimensional approximation to a distributed parameter model, which we discuss in detail in Odell & Oster (1983).

In Fig. 5 we show some results of numerical simulations of the model equations. For a fixed set of parameter values, the system can sustain several limit cycle oscillations with the following properties (c.f. Appendix C):

(1) The periodic contractions of the two segments are out of phase, and pump the cytosol in an alternating shuttle stream.

(2) If the cytosol is replaced by an inert fluid into which Ca^{++} can diffuse, the oscillations damp out.

(3) The mechanical and chemical oscillations can be far out of phase, the amount depends on the parameters.

(4) Stretch activation is necessary to synchronize the oscillations and drive shuttle streaming: if the flow of cytosol is stopped the compartments will not synchronize.

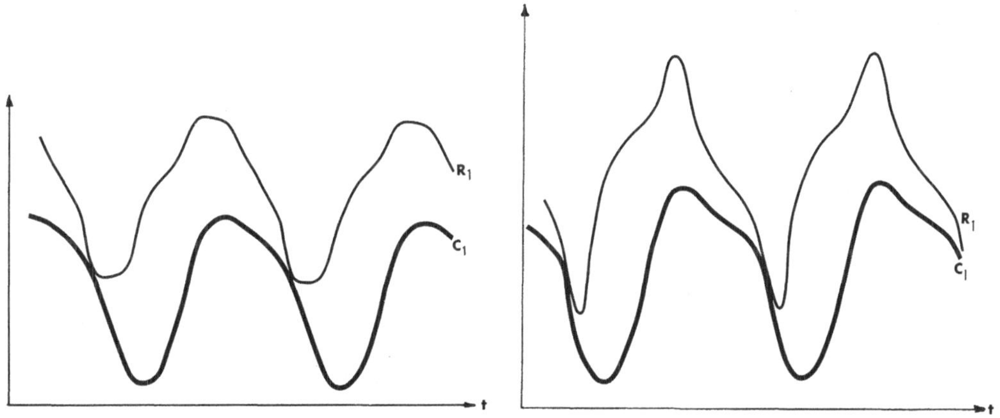

Fig. 5 (a) A numerical simulation of the model system for the case of spherical cortices (m = 3). The radial oscillations drive an alternating shuttle stream.
(b) The same simulation for cylindrical cortices (m = 2).

Each of these properties is characteristic of plasmodial rhythms in vivo. A full description and mathematical analysis will be published elsewhere.

4. DISCUSSION

The mechanical activity of cytoplasm stems from the contractile properties of actomyosin gels. We have commenced a study of the mechanochemistry of cytogels by constructing a simple model for the plasmodial rhythms in the slime mold *Physarum*. The model is finite dimensional: two hollow cytogel shells filled with endoplasm and connected by a flow resistance. The equations of motion consist of 5 nonlinear ODE's whose dynamical behavior exhibits many of the characteristics of the biological system: mechanical contractions synchronize to drive an alternating shuttle stream of endoplasm, with approximately the correct phase relationship between the chemical and mechanical oscillations.

APPENDIX A

Nonlocal Effects can be Important in Cytogel

Since a gel is a fibrous material the stress at a point, $\underline{x}$, depends not only on the strain at $\underline{x}$, but the strain at nearby nodal points of the network as well. Therefore, we must include nonlocal effects in the constitutive relation $\sigma(\varepsilon,c)$. The simplest way of doing this is to consider the stress at $\underline{x}$ to be proportional to a weighted average of the strains in the neighborhood of $\underline{x}$:

$$\sigma_{av} = E\langle\varepsilon(\underline{x})\rangle \equiv \int_V W(\underline{r})\,\varepsilon(\underline{x}+\underline{r})\,d\underline{r} \qquad \text{A1}$$

where $W(\underline{r})$ is the weighting factor and V is the volume over which the average is taken - roughly the nodal spacing of the network.

If the integrand of A1 is expanded about $\underline{x}$ and second order terms retained, we obtain an approximate nonlocal stress expression (c.f. Othmer, 1969)

$$\sigma = E(\varepsilon + \beta\varepsilon_{xx}) \qquad \text{A2}$$

APPENDIX B

The Model Equations

As illustrated in Fig. 4, the model consists of two thick cytogel shells connected by a flow resistance. In order to adjust for the effect of curvature we have included a parameter, m, which allows the geometry of the shell to vary between spherical (m = 3) and cylindrical (m = 2).

The model equations are derived by computing the strain rate in the cytogel shell along with the concentrations of Ca^{++} in the shells and in the endoplasm. Since the endoplasm is incompressible, we need only compute the radius of one shell; the other one is then obtained algebraically. In principle, one could simply insert the stress tensor 11 into the standard thick shell equations and perform the appropriate averages over the shell thickness. However, since the equations become rather cumbersome due to geometrical nonlinearities, we shall sketch the procedure explicitly for the case of spherical shells.

Table A1 summarizes the various parameters; from these we define the following dimensionless parameters (* denotes dimensionless quantities).

Table A1

R_1 = outer radius of cortical shell i, i=1,2
c_i = concentration of Ca^{++} in cortex i, i=1,2
e_i = concentration of Ca^{++} in the endoplasm within cortex i=1,2
R_o = original outer radius of the cortex
d_o = original thickness of the cortex
m = parameter controlling the cortex geometry: m = 3: sphere
= 2: cylinder
N_o = fibers/unit cross-sectional area of cortex
μ = viscosity parameter of the cortical cytogel
μ_o = shear viscosity of the endoplasm
D = diffusivity of Ca^{++} between cortical gel and endoplasm sol
r = resequestration rate of Ca^{++}
γ = release rate of Ca^{++} from the vesicles per unit strain
λ = leakage rate of Ca^{++} from the vesicles
a,β= parameters controlling the release kinetics of Ca^{++}
w,r= length and radius of the flow resistance tube connecting the shell

---

$$\Omega_m^* = \frac{3}{8} \quad \frac{m-1}{m(m+1)} \quad \frac{N_o r^4 \mu}{w R_o^3 \mu_o} \quad : \text{ flow resistance between shell 1 and 2.}$$

$$2 \leqq m \leqq 3$$

$$D_m^* = \frac{D}{R_o r} \quad \frac{6(m-1)}{m+1} \quad : \text{ cortex to cytosol diffusivity.}$$

$$d^* = d_o/R_o \qquad \alpha^* = \alpha/r\beta^{1/2} \qquad \gamma^* = \gamma\beta^{1/2}/r$$

$$\lambda^* = \lambda\beta^{1/2}/r \qquad \sigma^* = \sigma/r\mu \qquad c^* = c\beta^{1/2}$$

The empirical input to the model is the stress surface $\sigma(\varepsilon,c)$ shown in Fig. 2d. We shall denote the dimensionless elastic stress by $J = N_o\phi(\varepsilon,c^*/\beta^{1/2})/r\mu$, and henceforth drop the asterisks from the dimensionless variables. Thus, at each time step in the computation, J can be computed from the current values of ε and c.

The equations of motion may now be computed recursively from the following recipe.

(1) First, define the functions $\chi(u)$ and $\xi(u)$ by
$\chi(u) \equiv u/\rho(u)$,
where $\rho(u) \equiv u^m-\delta \ ^{1/m}$ = inner radius of cortex whose outer radius is u.

$\delta \equiv 1 - (1-d)^m$

$z(u) \equiv B(u) + \chi(u)m^{-1}b(u)/1-\delta)$

where $B(u) \equiv 1 - \ln \chi(u) / \chi(u)-1$

$b(u) \equiv \chi(u) \ln \chi(u) / \chi(u)-1 \quad -1$

(2) Let: $H(u) \equiv (2-u^m)^{1/m}$

Then: $R_2 = H(R_1)$, and

$\rho(R_1) = \rho_1$ = inner radius of cortex 1

$\rho(R_2) = \rho_2$ = inner radius of cortex 2

$J(R_1-1,c_1) = F_{R_1}$ = stress on outer surface of cortex 1

$J(R_2-1,c_2) = F_{R_2}$ = stress on outer surface of cortex 2

$J(\rho_1/(1-d)-1,c_1) = F_{\rho_1}$ = stress on inner surface of cortex 1

$J(\rho_2/(1-d)-1,c_2) = F_{\rho_2}$ = stress on inner surface of cortex 2

We compute dR/dt as follows:

$$dR/dt = F_1 = \frac{B(R_2)\cdot F_{R_2} + b(R_2)\cdot F_{\rho_2} - B(R_1)\cdot F_{R_1} - b(R_1)\cdot F_{\rho_2}}{R_1^{m-1}/\Omega_m + \xi(R_1) + (R_1/R_2)^{m-1} \cdot \xi(R_2)} \qquad \text{B1}$$

(3) Compute the chemical equations as follows:

Let: $\Theta(\varepsilon,c) \equiv \alpha c^2/(1+c^2) - c + \gamma(R-1) + \lambda$

$$\text{Then, } dc_1/dt = F_2 = \Theta(R_1-1,c_1) + D_m/\delta \; \rho_1^{m-1}(e_1 - c_1) \qquad \text{B2}$$

$$dc_2/dt = F_3 = \Theta(R_2-1,c_2) + D_m/\delta \; \rho_2^{m-1}(e_2 - c_2) \qquad \text{B3}$$

(4) Since convective flow of cytosol carries the upstream concentration of Ca^{++}, in order to define the Ca^{++} concentration in the endoplasm, (e_1, e_2), we must know which way the cytosol is flowing. We do this as follows.

If $\partial R_1/\partial d > 0$, then $E_{F_1} = mR_1^{m-1}F_1(e_2 - e_1)/\rho_1$

and $E_{F_2} = 0$

If $\partial R_1/\partial t < 0$, then $E_{F_1} = 0$

and $E_{F_2} = mR_1^{m-1}F_1(e_2 - e_1)/\rho_2$

The endoplasm equations are

$$\partial e_1/\partial t = F_4 = D_m/\rho_1 \; (c_1 - e_1) + E_{F_1} + \Theta(1,e_1) \qquad \text{B4}$$

$$\partial e_2/\partial t = F_5 = D_m/\rho_2 \; (c_2 - e_2) + E_{F_2} + \Theta(1,e_2) \qquad \text{B5}$$

Equations B1 - B5 define the vector field $\underline{F}(\underline{x})$ in equation 16 .

APPENDIX C

As Parameters are Varied, a Limit Cycle, Generated by a Hopf Bifurcation, Cleaves into two Limit Cycles

Time oscillatory behavior of our model corresponds to stable limit cycles of the vector field in our ODE system. From numerical experimentation, we believe these limit cycles arise through Hopf bifurcations at the critical point $r_1 = r_2 = c_1 = c_2 = 1.0$.

Figure C1 depicts a projection of the five-dimensional phase space into the three-dimensional (r_1, c_1, c_2)-subspace using perspective stereo pairs. The large dot locates the critical point just mentioned.

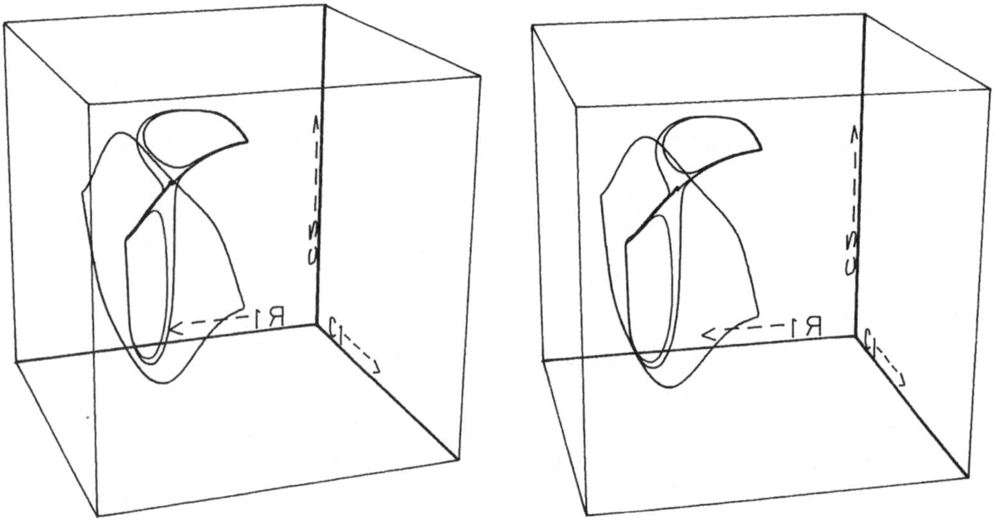

Fig. C1 Perspective stereo pair view of limit cycles of the fifth order system projected into the (r_1, c_1, c_2)-subspace. From the viewer's vantage, each limit cycle is traversed in the counterclockwise sense. In the figure, the left eye's view is drawn in the left frame, the right eye's view in the right.

We show limit cycles for three different parameter sets, all overlaid in the same figure. Only the cortex-endoplasm diffusivity parameter, D, is varied to generate the three sets of limit cycles.

As seen from the viewer's eye, the "lowest" limit cycle occurs when D = 0. In this isolated case the fifth order ODE system uncouples, so that all interesting behavior occurs in the 3-dimensional (r_1, c_1, c_2) phase space. For small, but positive, D, the limit cycle lies very close to the three-dimensional manifold of $(r_1, c_1, c_2, e_1, e_2)$ space, defined by $e_1 = e_2 = 1.0$, shown in the figure.

As D increases, the limit cycle lifts (towards the viewer's eye) and deforms into almost a "figure 8" shape, with the narrowest part drawn in toward the critical point, and with the two lobes of the "figure 8" lying in almost orthogonal planes. One such limit cycle is shown in the figure.

As D increases to a certain value, the single limit cycle becomes a true "figure 8", consisting of two homoclinic orbits attached to the critical point. As D is further increased, these two homoclinic orbits become separate stable limit cycles. A pair is shown in the figure. The domains of attraction of these two limit cycles touch at the critical point; trajectories flowing from near this critical point to these two limit cycles are shown.

REFERENCES

Jaffe, L. (1980). Calcium explosions as triggers for development. Ann. N.Y. Acad. Sci. 339, 86-101.

Kamiya, N. (1981). Physical and chemical basis of cytoplasmic streaming. Ann. Rev. Plant Physiol. 32, 205-36.

Kessler, D. (1982). Plasmodial structure and motility. In: Cell Biology of *Physarum* and *Didymium*, Vol. 1, pp. 145-208. H. Aldrich, J. Daniel (eds.) New York, Academic Press.

Oster, G., G. Odell, J. Murray (1983). In preparation.

Oster, G., G. Odell (1983). The mechanics of cytogels I: Plasmodial oscillations in *Physarum*. Cell Motility (submitted).

Othmer, H. (1969). Interactions of reaction and diffusion in open systems. Ph.D. Dissertation. Chem. Engr. Dept., Univ. of Minnesota.

Tyson, J. (1982). Periodic phenomena in *Physarum*. In: Cell Biology of *Physarum* and *Didymium*, Vol. 1, pp. 61-109. H. Aldrich, J. Daniel (eds.) New York, Academic Press.

Wohlfarth-Bottermann, K. (1979). Oscillatory contraction activity in *Physarum*. J. Exp. Biol. 81, 15-32.

GENOMIC CONTROL OF GLOBAL FEATURES IN MORPHOGENESIS

Robert Rosen
Department of Physiology and Biophysics
Dalhousie University
Halifax, N.S., Canada B3H 4H7

INTRODUCTION

Genetics and morphology provide an almost classical example of biological complementarity, in the sense of Bohr: neither is complete without the other, yet every attempt to relate the two has failed. Such attempts have themselves exhibited complementarity; they either work upwards from the genes towards form, or else they work downwards from form to the genes. The remarks which follow belong to the latter category, which seems to me infinitely easier and more natural; indeed, as we shall see, Nature seems to come to our aid by going quite a long way in this direction already.

The development to follow is necessarily hurried, informal, and proceeds by examples. It should be stressed, however, that our concern is not with technical devices to facilitate morphogenetic modelling; rather, it is with some rather radical conceptual innovations designed to incorporate the basic features of "genetic information" into morphogenesis from the beginning.

THE GENES

To the early geneticists, including Mendel, the physiological role of the genes was the determination of form in whole organisms. Not only relatively simple quantitative features (e.g. eye color) but gross qualitative ones: the shape of a leaf; five fingers; bilateral symmetry; all these were supposed to be determined by the genes. The output of the genes was anatomy.

Consequently, until relatively recently, it was widely supposed that insights into primary genetic mechanisms would have to come from developmental biology; i.e. from the manner in which genes modulate the dynamical processes which create form. Ironically, this has not proved to be the case at all. Instead, our present views about primary genetic activity come exclusively from the study of systems like bacteria and viruses; systems for which the concepts of form and development are vacuous. Indeed, as everyone knows, we regard the genes today exclusively as modulators of intracellular chemistry; the genes themselves are regarded as chemicals of a special type. There is thus an enormous gap between what the genes were originally supposed to do, and what we now believe they actually do; a gap which has made the problems of morphogenesis harder rather than easier. The basic difficulty is precisely that there is no obvious relation between intracellular chemistry and gross geometry.

PHENOCOPY

In the pursuit of links between primary genetic activities and development, a number of most interesting experimental systems were explored. Perhaps the most striking of these was the phenomenon called phenocopy, which was much studied in the 1920's and 1930's, but has today, as far as I know, been completely abandoned.

If genetically normal (wild-type) *Drosophila* larvae are exposed to somatic stresses (e.g. heat shocks, cold shocks, oxygen deprivation) of appropriate timing, intensity, and duration, and then allowed to complete their development, it is found that the resultant flies can be anatomically altered. These alterations are not random; it was found that such flies are anatomically indistinguishable from one or another mutant. Such a fly, which is genetically normal but phenotypically mutant, is called a phenocopy of the corresponding mutation. It was natural to conjecture that the effect of the somatic shock was to interfere with the primary activity of the gene whose mutation is mimicked by the phenocopy, thus providing a direct probe of primary genetic activity in development (cf. Goldschmidt, 1958).

GENETIC AND EPIGENETIC CONTROL

We will introduce one further bit of useful terminology. The biologist David Nanney (1958) has drawn a distinction between two kinds of intracellular control processes; those which he terms genetic and those which are epigenetic. Intuitively, genetic control processes are those which directly involve the genetic templates (DNA, RNA), while the latter involve interactions between proximate or remote gene products (proteins, metabolites). A cautionary note: the word epigenesis was perhaps unfortunately chosen by Nanney; it has a long history in the literature of developmental biology. Indeed, the preformation-epigenesis controversy of the 18th century is in some sense still actively with us today. To enlarge on this would take us too far afield; henceforth, when I use the term epigenetic, I will mean it in Nanney's sense.

EPIGENETIC PATTERN-GENERATION METAPHORS

Let me say first what I mean by the term metaphor. A metaphor is a class of formal (i.e. mathematical) systems defined by some commonality of structure (e.g. satisfying the same axioms, etc.) but not necessarily well-enough defined to constitute a category. A real system (e.g. an organism) or a real process (e.g. morphogenesis) is said to be represented by a metaphor if it has a model which belongs to the metaphor. Thus, we can utilize the mathematical properties of such a metaphor to provide information about any real system or process represented by the metaphor. We shall say that two real systems or processes are analogous if they are represented by a common metaphor; i.e. if they have models which are in some sense homomorphic. Note that we do not have to know the models explicitly to use these ideas.

It should be obvious that the bulk of theoretical work pertaining to biological morphogenetic mechanisms is metaphorical in this sense, and has been so for a long time. Indeed, for nearly half a century, the dominant metaphor has been the "open system metaphor", introduced by von Bertalanffy and others in the 1930's. von Bertalanffy observed that the remarkable stability properties of developing organisms, so baffling to proponents of naive mechanism, resembled those exhibited by trajectories of general dynamical systems moving towards stable (point) attractors. These <u>open systems</u> could readily be reconciled with mechanism, but in the process, gaping voids in contemporary physics, supposedly the science of mechanism, were disclosed. To the present day, theoretical research into morphogenetic mechanisms has consisted essentially of identifying and exploring submetaphors of the open-system (or dynamical system) metaphor of von Bertalanffy.

Let us for definiteness exhibit such a metaphor; one which subsumes many others, and thus reveals a number of unexpected relations between apparently diverse biological and physical phenomena. Intuitively, this metaphor deals with the generation of patterns in populations of spatially extended units. The units may have autonomous internal dynamics; they may interact with each other, and with their environment. Quite generally, a member of the metaphor is characterized by the following data:

1. A set $N = \{n_i\}$, $i \in I$, of <u>elements</u>.

(The index set I, which is arbitrary, should be thought of as a set of spatial coordinates, so that n_i represents "the element at position i".)

2. A set S of <u>internal states</u>.

A <u>pattern</u> is a mapping $f : N \rightarrow S$. Intuitively, we should think of every element $n_i \in N$ as receiving a copy of this common state set S, so that we have in effect a fiber space with N as base space and S as fiber. A pattern is geometrically a cross-section.

Now to generate patterns, we need some dynamics on the set H(N, S) of all patterns. The first item of business is to specify

3. An ordered group T of _time instants_.

In practice, we will always choose $T = \mathbb{Z}$ (discrete time) or $T = \mathbb{R}$ (continuous time). Formally, a dynamics on H(N, S) can now be represented as a one-parameter group of automorphisms on H(N, S), parameterized by T. However, we need to proceed more explicitly. Let us then specify

4. A map $U : N \rightarrow 2^N$, the _neighborhood map_.

Intuitively, $U(n_i)$ represents the set of elements with which n_i interacts; or more specifically, those elements whose states at an instant affect the change of state of n_i at that instant. Thus $U(n_i)$ consists of those elements afferent to n_i.

5. _The local state transition rule_.

We recognize two cases:

a. $T = \mathbb{Z}$ (discrete time):

$$n_i(t+1) = \Phi(n_i(t), U(n_i)(t), \vec{\alpha}(i, t))$$

or, more specifically,

$$= \varphi(n_i(t)) + \psi(n_i(t), U(n_i)(t)) + \theta(n_i(t), \vec{\alpha}(i, t)) .$$

Here the first term represents the autonomous dynamics of the element n_i; the second term represents the interaction of n_i with its neighbors, and the last term is the interaction of n_i and the environment $\vec{\alpha}(i, t)$.

b. $T = \mathbb{R}$, S a manifold (continuous time):

Same as (a), except replace $n_i(t+1)$ by $dn_i(t)/dt$.

A mathematical structure of this type will be called a _morphogenetic network_ (cf. Rosen, 1981). These morphogenetic networks, as we have noted, subsume an enormous number of other pattern generating metaphors in biology and physics: Ising models, reaction-diffusion networks, excitable nets of various kinds, and many others.

As noted above, many unsuspected relations between diverse kinds of systems are revealed through the fact that they are represented by a common metaphor. But this fact cuts two ways: the very fact that so many different kinds of systems, organic and inorganic, are represented by morphogenetic networks seems an insuperable obstacle to assessing the relevance of the metaphor to developmental biology. That is: if the metaphor can be so easily realized by frankly inorganic systems like Ising networks or, more concretely, by Belousov-Zhabotinskii reactions, what special significance can these networks have for biology?

In a nutshell, the key lies in the fact that the morphogenetic networks are entirely epigenetic, in the sense of Nanney. There is no reference to genes, genetic control, or genetic information anywhere in the above discussion. This will not do; somehow, a germ of genetic control must be located in these networks; a germ which can be used to characterize the difference between biological morphogenesis and simple pattern generation. We shall now turn to this question; following that, we shall utilize a simple example to probe the actual role of the genome, and of "genetic information" in general, in the generation of patterns.

GENOME, PHENOTYPE, AND ENVIRONMENT IN GENERAL SYSTEMS

We already know that the concept of the gene is not a simple one. Indeed, the gene has diverse roles to play in biology: physiological roles, in the modulation of chemistry and form, and in which the flow of information is from genotype to phenotype; and hereditary roles, in which the flow of information is from generation to generation. If we restrict ourselves to morphogenesis, however, the hereditary aspects become superfluous, and we can assert that the role of the genes is to characterize the species of system with which we are dealing. That is, in purely physiological terms, <u>the role of the genome is to determine identity</u>. This is in itself an enormous simplification. But in fact much more than this is true; we are now going to argue that in this sense not only organisms, but <u>any real system at all, can be thought of as possessing a genome, which determines corresponding phenotypes</u>. This basic fact will allow us not only to assess the biological significance of epigenetic metaphors like the morphogenetic networks, but also to throw entirely new light on the intuitive notion of "genetic information" and its ramifications. This is one of the "radical epistemological innovations" of which we spoke earlier.

It is best to proceed by means of an example. Let us choose one that I particularly like, and which is as far removed from biology as it is possible to go. This example is the familiar van der Waals equation, a thermodynamic equation of state characterizing the equilibria of a certain family of non-ideal gases. This equation may be written as

$$(p + a/v^2)(v - b) = rT$$

where p, v, T are the thermodynamic state variables (pressure, volume, and temperature respectively) and a, b, r are parameters. Mathematically, this equation is a single function of six arguments. We can rewrite this relation as a three-parameter family of functions of three arguments; i.e. in the form

$$\Phi_{abr}(p, v, T) = 0 \quad .$$

Further, if we arbitrarily pick environmental conditions such that two of the arguments of this three-parameter family are determined, then the value of the third is uniquely specified. Thus, we may rewrite the three-parameter family of <u>relations</u> as a three-parameter family of <u>mappings</u>; say as

$$\Phi_{abr} : (p, T) \longrightarrow (v) \quad .$$

We have thus partitioned the original set of six arguments of the van der Waals equation into three classes:

i. the parameters (abr), which now play the role of coordinates in a function space, which we shall call <u>genome</u>;

ii. the variables p, T, whose values are determined by external processes not obeying the system laws, which we shall call _environment_;

iii. the variable v, whose values are determined when the other arguments are specified, which we shall call _phenotype_.

What we claim is that the above example is in fact perfectly general; i.e. that _any mathematical relation describing a real system can be rewritten as a genome-parameterized family of mappings from environments to phenotypes_. This rewriting is not simply a technical trick; it has many deep theoretical consequences, which we cannot enter into here. Many of these are of course related to the _stability_ of the genome-parameterized family $\{\Phi_{abr}\}$. However, for present purposes, we shall simply call attention to one feature: suppose we keep the genome (abr) fixed, and change environments: $(p, T) \mapsto (p', T')$. Then if

$$\Phi_{abr}(p, T) = v; \qquad \Phi_{abr}(p', T') = v' \quad ,$$

we can ask if there is a _species_ (a'b'r') such that

$$\Phi_{a'b'r'}(p, T) = v' \quad ;$$

i.e. whose phenotype in the _original_ environment is the same as that of the original species in the new environment. If this is the case, it is reasonable to say that $\Phi_{abr}(p', T')$ is a _phenocopy_ of $\Phi_{a'b'r'}(p, T)$. From such considerations, we can see how to reformulate apparently intrinsic biological concepts so that they are meaningful quite generally; in fact, so that they are universal properties of general system descriptions.

Armed with these ideas, we can now return to our epigenetic morphogenetic metaphors; the morphogenetic networks. The above argument shows that there are indeed genomes implicit in these epigenetic metaphors; genomes for the individual elements, and for the network as a whole. Let us investigate what it is that these genomes actually do within the metaphor, and contrast this with what the genome should do.

EXAMPLE: THE SIMPLEST REACTION-DIFFUSION NET

Once again, we proceed with the simplest of examples, one which can simply generate a polarity or gradient. In the language of morphogenetic nets, let us choose:

1. $N = \{n_1, n_2\}$;

2. $S = \mathbb{R}^+$, positive reals ;

3. $T = \mathbb{R}$;

4. $U(n_1) = \{n_2\}$; $U(n_2) = \{n_1\}$;

5. $$\begin{cases} dx_1/dt = -ax_1 + D(x_1 - x_2) + S \,, \\ dx_2/dt = -ax_2 + D(x_2 - x_1) + S \,. \end{cases}$$

It can easily be verified that for this system, if $D < a/2$, there is a stable node at $x_1 = x_2 = S/a$, corresponding to a homogeneous situation. However, if $D > a/2$, the stable node bifurcates to a saddle point, and one or the other of the cells empties out. More specifically: any initial deviation from perfect homogeneity under these circumstances grows autocatalytically.

So far so good. But this epigenetically generated polarity suffers from a grave defect when compared to true biological morphogenesis. Namely, if at any time the constituent cells are separated, each of them will separately revert to its original steady-state phenotype S/a. In mathematical terms, the pattern generated epigenetically is not stable to perturbations of the neighborhood map U. But biologically, such pattern generation is in effect accompanied by a <u>change of species</u> of the cells concerned; a change which is then stable to any subsequent perturbation of the pattern. In our epigenetic metaphor, there is no change in the <u>species</u> of our cells; what happens, in effect, is the epigenetic production of <u>phenocopies</u> of differentiated cell types, while the genome of the elements in the network remains fixed.

To generate a true change of species in our elements, we must generate appropriate paths in the genome space (here the space of pairs (a, D)) as well as epigenetically in the state spaces; we must in a sense force phylogeny to recapitulate ontogeny. And to do this, we must introduce more genome. In a precise sense, this new genome, required to stabilize the pattern, belongs to the network as a whole, while the genomes (a, D) belong to the individual elements.

In the present case, we can stabilize the pattern by doing two things: (a) forcing $D \to 0$ as the pattern develops, so that each cell in the network becomes refractory to the other cells; (b) forcing the catabolic parameter a to move inversely to the way its state is moving; to become small in cells for which x becomes large, and conversely. Thus, cells with large x will maintain large x autonomous; cells with small x will maintain small x. This can be accomplished by adding two new equations to our original local state transition rules:

$$\begin{cases} da/dt = -g_1 D(x_1 - x_2) \\ dD/dt = -g_2 D(x_1 - x_2)^2 \end{cases} .$$

Here, the new parameters (g_1, g_2) correspond to a new layer of genome, which for want of a better term we can call deep genome to distinguish it from the original genomic parameters (epigenome). As noted before, the latter determine the species of cell within a morphogenetic network; the former pertain to the species of the network as a whole.

The above is, of course, not intended as a model for any specific biological situation. It does, however, illustrate several basic features which need to be incorporated into any metaphorical studies of biological morphogenesis, or alternately, in the biological realization of such metaphors. We wish to emphasize that our introduction of a new layer of genome ("deep genome") is again not simply a technical device, but reflects a fundamental difference between organic and inorganic systems. In the latter, there appears to be no real distinction between the genomes of populations and those of their constituents; in the former, there is.

GENETIC INFORMATION, AND INFORMATION IN GENERAL

Considerations of the above type force us to drastically reconsider what is meant by "genetic information", and in general the role of information in modulating dynamical processes like morphogenesis. The fact is that "information" is never manifested statically (e.g. as in a sequence of nucleotides) or represented in terms of configurational coordinates; it always appears in an active role, in the form of rate constants, or more generally in the determination of a vector field imposed on a manifold of underlying configurational variables. Thus, in general, "information" must be characterized in terms of dynamical interactive capability.

One way to do this is to regard "information" symbolically as an answer to a question of the form: if x is changed, what happens to y? All measurements answer questions of this form; e.g. if x is changed from nothing to something in the vicinity of a meter, what happens to the dial of the meter? If this is so, then "information" resides symbolically in quantities which may be denoted formally by $\partial y/\partial x$.

Recall now that we have partitioned the arguments of an arbitrary equation of state into three classes (genome $\vec{g} = (g_i)$, phenotype $\vec{z} = (z_i)$, and environment $\vec{y} = (y_i)$). Formally, then, we can consider the quantities of the form $\partial z_k/\partial g_i$, which measure the effect of genome on phenotype; these then measure <u>genetic information</u>. Likewise, quantities of the form $\partial z_k/\partial y_j$ measure <u>environmental information</u>, and quantities of the form $\partial z_k/\partial z_m$ measure <u>somatic information</u>. We assert that these three kinds of "information" are <u>all different</u>, and in general not interchangeable or inter-translatable.

We get a deeper insight into these measures of "information" by looking at a more limited class of equations of state. In this class some of the environmental quantities can be regarded as the velocities of some of the others. Such an equation of state can generally be written in the form

$$d\vec{x}/dt = \Phi_{\vec{g}}(\vec{x}, \vec{u}) \quad ,$$

in which the environmental magnitudes $\vec{x}$ whose velocities satisfy the equation are segregated out as _state variables_, and the other environmental magnitudes are regarded as _controls_. In these circumstances, we will have further informational measures, which can be written in the symbolic form

$$\frac{\partial}{\partial g_j}\left(\frac{dx_i}{dt}\right) \quad ; \quad \frac{\partial}{\partial u_j}\left(\frac{dx_i}{dt}\right) \quad ; \quad \frac{\partial}{\partial x_j}\left(\frac{dx_i}{dt}\right)$$

Expressions of this kind are well known in the study of dynamical systems, both in abstract stability studies and in numerous areas of application. In the latter, they are frankly associated with "information"; indeed, Higgins (1967) long ago proposed that the signs of these quantities could be interpreted in terms of _activation_ and _inhibition_.

There is no reason to stop at this stage; we can further consider expressions of the form

$$\frac{\partial}{\partial x_k}\left(\frac{\partial}{\partial x_j}\left(\frac{dx_i}{dt}\right)\right) \quad .$$

(to use purely somatic variables). These too have immediate "informational" interpretations; if such a quantity is positive in a state, we may say that x_k is an _agonist_ of x_j; otherwise, an _antagonist_.

Iterating these expressions, we obtain an infinite array which I propose to call an _influence structure_. Given a particular set of rate laws, the influence structure is completely determined thereby. However, if we attempt to _posit_ such an influence structure initially, we will find that the structure can almost never be derived from a single set of rate laws. There are good independent reasons for calling systems whose influence structures cannot be derived from a set of rate laws _complex systems_; those which can, _simple systems_. But this is a long and involved story, which we cannot enter upon here.

Let us instead make one simple observation, pertaining to the effect of introducing a single layer of deep genome into the simple polarity-generating morphogenetic network of the preceding section. Let us in particular look at what we have done to the influence structure. In the original network, in which only the epigenome was present, and in which our pattern is generated entirely epigenetically, we have

$$\frac{\partial}{\partial x_1}\left(\frac{dx_i}{dt}\right) = D - a \ .$$

On the other hand, when we have incorporated the deep genome, the counterpart of the above expression is

$$\frac{\partial}{\partial x_1}\left(\frac{dx_1}{dt}\right) = (D - a) + x_1 \frac{\partial}{\partial x_1}(D - a) - x_2 \frac{\partial D}{\partial x_1}$$

and analogously for all other quantities. In a nutshell, then, we have not only added more genome (i.e. more identity); we have enormously enlarged the influence structure, and with it, the amount of "information" in the system.

The next step in this sequence of ideas is to entirely replace the local state transition rules in the morphogenetic networks by the more general notion of influence structures. The effect of this step will be to pass from a study of pattern generation in simple systems to the corresponding study of pattern generation in complex systems. There is good reason to believe that these new complex metaphors will produce some surprises, and at the same time will be more directly interpretable in the context of real biological morphogenesis.

LITERATURE

1. Goldschmidt, R. B. 1958. Theoretical Genetics. University of California Press, Berkeley, California.

2. Higgins, J. 1967. Industrial and Engineering Chemistry 59, 18-62.

3. Nanney, D. L. 1958. Proceedings of the National Academy of Science 44, 712-715.

4. Rosen, R. 1981. Progress in Theoretical Biology 6, 161-210.

PATTERNS OF STARVATION IN A DISTRIBUTED PREDATOR-PREY SYSTEM

F. Rothe
Lehrstuhl für Biomathematik
Universität Tübingen

Auf der Morgenstelle 28,
D 74 Tübingen

ABSTRACT We investigate the asymptotic behavior of a predator-prey system with one diffusing and one sedentary species. Exactly one of the following three cases (1)(2)(3) occurs: (1)the diffusing species dies out; (2)the diffusing species converges to a positive equilibrium; (3)the diffusing species grows infinitely. In any case the nondiffusing species converges to equilibrium. It depends on the region Ω_0 where the nondiffusing component vanishes initially which one of the three possibilities (1)(2)(3) actually occurs and which equilibrium $(\bar{u},\bar{v})$ is approached in case (2). We can decide by looking at the signs of some principal eigenvalues which alternative occurs. Case (3) can only occur for diffusing predator or if the nondiffusing species is absent. Finally we give a compatibility condition which characterizes the asymptotic equilibrium in case (2) in terms of the region Ω_o.

THE MODEL AND THE MAIN RESULTS

Simple predator-prey systems can be studied quite completely by analytical tools.Yet they illustrate some interesting phenomena which may occur in more realistic systems, too. In this note we study the formation of spatial patterns of extinguished and surviving prey (or predator) in a spatially distributed predator-prey system. We consider a two-species model with one sedentary and one mobile species. The mobile species may be the prey or the predator. For illustration the reader may take the latter case and imagine e.g. cows feeding from grass. The motion of the mobile species is modelled by diffusion in a bounded domain Ω. For simplicity we take impermeable boundaries, which are described by homogeneous Neumann boundary conditions on $\partial\Omega$. Furtheron we exclude any autocatalytic effects for both species. Thus the growth rates of both species are decreasing functions of the density of the same species. (As pointed out e.g. by Mimura [2], autocatalytic effects can cause a diffusion induced instability.) The interaction of prey and predator is assumed to be given by the mass-action law. All parameters may depend on the space coordinate, although this is not necessary to get the patches of extinction we want to study.

We come to a more formal statement.

Let $\Omega \subset R^n$ be a bounded domain with smooth boundary $\partial\Omega$ of class $C^{2+\nu}$ for some $\nu \in (0,1)$. In the following, $x \in \Omega$ is the space coordinate and $t \in [0,\infty)$ denotes the time. Furtheron $u \in [0,\infty)$ and $v \in [0,\infty)$ denote the densities of the diffusing and nondiffusing species, respectively. The biologically more realistic case of diffusing predator and sedentary prey corresponds to the upper signs in the formulas below.

The dynamics of the two species living in the domain Ω are governed by the following differential equations (1a)(1b), the boundary condition (2) and the initial conditions (3a)(3b) for the functions $u = u(x,t)$ and $v = v(x,t)$:

$$u_t - \Delta u = u[-f(x,u) \pm a(x)v] \quad (1a)$$

$$\text{for all } x \in \Omega,\ t > 0;$$

$$v_t = v[d(x) - c(x)v \mp b(x)u] \quad (1b)$$

$$\partial u/\partial n = 0 \qquad \text{for all } x \in \partial\Omega,\ t > 0; \quad (2)$$

$$u(x,0) = u_o(x) \quad (3a)$$

$$\text{for all } x \in \Omega.$$

$$v(x,0) = v_o(x) \quad (3b)$$

Here we assume for the initial data

$$u_o,\ v_o \in L_\infty(\Omega)\ ;\quad u_o(x),\ v_o(x) \geqq 0 \quad \text{for almost all } x \in \Omega; \quad (4)$$

$$\int_\Omega u_o(x)dx > 0. \quad (5)$$

The set of initial data satisfying (4)(5) will be denoted by $\mathcal{J}$.

For the functions f,a,b,c,d occurring in the differential equations (1) we assume that for some $\alpha \in (0,1)$:

$f: (x,u) \in \bar{\Omega} \times [0,\infty) \rightarrow f(x,u) \in R$ is α-Hölder continuous; (6)

the partial derivative f_u exists and is continuous on $\bar{\Omega} \times [0,\infty)$; (7)

$f_u(x,u) > 0$ for all $(x,u) \in \bar{\Omega} \times [0,\infty)$. (8)

$a,b,c,d: x \in \Omega \rightarrow a(x), b(x), c(x), d(x) \in R$ are α-Hölder continuous; (9)

$a(x) \geqq 0$, $b(x) > 0$, $c(x) > 0$ for all $x \in \bar{\Omega}$. (10)

Since spatial discontinuities occur naturally for the nondiffusing component v, it is not sufficient to consider smooth classical solutions of the initial-boundary value problem (1)(2)(3). One is forced to include generalized solutions of some type. We will consider <u>mild solutions</u> as they are used in [3] p.45 or [4] p.111. These are measurable functions $(u,v): (x,t) \in \Omega \times [0,T) \to (u,v)(x,t) \in R^2$ such that $u(.,t),\ v(.,t) \in L_\infty(\Omega)$ for all $t \in (0,T)$ and an integral equation heuristically equivalent to (1)(2)(3) holds. Existence of mild solutions of this type is well-known, for a detailed proof see e.g. [4] p.111, Theorem 1. The mild solution can be extended to a maximal interval $[0,T_{max})$. If $T_{max} < \infty$, the L_∞-norm blows up for $t \to T_{max}$.

We are mainly interested in the asymptotic behavior of the solution (u,v) of the initial-boundary value problem (1)(2)(3). Crucial in this context are the equilibria. By an <u>equilibrium</u> we mean a pair of nonnegative functions $(\bar{u},\bar{v}) \in L_\infty(\Omega) \times L_\infty(\Omega)$ which is a time-independent mild solution of the initial-boundary value problem (1)(2)(3) for initial data $(u_o,v_o) = (\bar{u},\bar{v})$. The following notions are important for the characterization of equilibria:

$$\Omega_p = \{x \in \Omega \mid \bar{v}(x) = 0\}; \tag{11}$$

$$\Omega_p^+ = \{x \in \Omega \mid \bar{v}(x) = 0 \quad \text{and} \quad d(x) \mp b(x)\bar{u}(x) > 0\}. \tag{12}$$

Furtheron for given initial data (u_o,v_o) we define $\Omega_o \subset \Omega$ and $\chi_o \in L_\infty(\Omega)$:

$$\Omega_o = \{x \in \Omega \mid v_o(x) = 0\}; \tag{13}$$

$$\chi_o = \begin{cases} 1 & \text{if} \quad v_o(x) > 0 \\ 0 & \text{if} \quad v_o(x) = 0. \end{cases} \tag{14}$$

We shall say that the initial data $(u_o,v_o) \in \mathcal{J}$ and the equilibrium $(\bar{u},\bar{v})$ are <u>compatible</u> if and only if (after neglecting null sets)

$$\Omega_p^+ \subset \Omega_o \subset \Omega_p. \tag{15}$$

Roughly spoken, the compatibility condition (15) means that no species v can be present at places where it was not present at the beginning and that the species v can get extinguished only if the zero state is locally stable (if one considers the equation (1b) locally at a definite space point).

Now we can formulate our first result:

<u>THEOREM 1</u> *Suppose that for some nonnegative initial data* (u_o,v_o) *satisfying (4)(5) there exists an equilibrium* $(\bar{u},\bar{v})$ *such that* (u_o,v_o) *and* $(\bar{u},\bar{v})$ *are compatible, i.e. (15) holds.*

Then there exists a global mild solution
$(u,v)\colon (x,t)\in\Omega\times(0,\infty) \rightarrow (u,v)(x,t)\in(0,\infty)\times[0,\infty)$
of the initial-boundary value problem (1)(2)(3).

This solution satisfies

$$\sup_{0<t<\infty} [\|u(.,t)\|_\infty + \|v(.,t)\|_\infty] < \infty; \tag{16}$$

$$\lim_{t\to\infty} \|u(.,t) - \bar{u}\|_{C^\nu(\bar{\Omega})} = 0 \quad \text{for all } \nu\in[0,2); \tag{17}$$

$$\lim_{t\to\infty} \|v_o(v(.,t) - \bar{v})\|_\infty + \|[v(.,t) - \bar{v}]_+\|_\infty = 0. \tag{18}$$

<u>COROLLARY</u> *For all initial data* $(u_o,v_o) \in J$ *there exists at most one equilibrium* $(\bar{u},\bar{v})$ *such that* (u_o,v_o) *and* $(\bar{u},\bar{v})$ *are compatible.*

Next we ask whether there exists an equilibrium which is compatible to some given initial data $(u_o,v_o) \in J$. By definition the equilibrium satisfies in the mild sense

$$0 = \Delta\bar{u} + \bar{u}[-f(x,\bar{u}) \pm a(x)\bar{v}] \quad \text{in } \Omega; \quad \partial\bar{u}/\partial n = 0 \quad \text{on } \partial\Omega; \tag{19a}$$

$$0 = \bar{v}[d(x) - c(x)\bar{v} \mp b(x)\bar{u}] \text{ for almost all } x \in \Omega. \tag{19b}$$

One can check that (19b) together with the compatibility assumption (15) is equivalent to

$$\bar{v}(x) = [\chi_o/c](x)[d(x) \mp b(x)\bar{u}(x)]_+ = \tag{20}$$

$$= \begin{cases} [d(x) \mp b(x)\bar{u}(x)]/c(x) & \text{if } v_o(x) > 0 \text{ and } d(x)\mp b(x)\bar{u}(x) > 0, \\ 0 & \text{otherwise.} \end{cases}$$

Inserting (20) in (19a) yields the nonlinear elliptic problem

$$0 = \Delta\bar{u} + \bar{u}\, h_\pm(x,\bar{u}) \quad \text{in } \Omega\,; \quad \partial\bar{u}/\partial n = 0 \quad \text{on } \partial\Omega \quad \text{in the mild sense}$$

$$\text{with} \qquad h_\pm(x,\bar{u}) = -f(x,\bar{u}) \pm (a\chi_o/c)(x)[d(x)\mp b(x)\bar{u}]_+ \tag{21$\pm$}$$

Here and in the following we use the subscripts $\pm$ to distinguish the cases of diffusing predator and diffusing prey.

The problem of existence and uniqueness of solutions of (21) can be handled quite straightforward by the method of upper and lower solutions. We use the subscripts u and l to denote upper and lower solutions. For a review about nonlinear elliptic problems see e.g. [1].

Let $U \in [0,\infty)$ be arbitrary. By $\lambda(U)$ and $\varphi(x,U)$ we denote the principal eigenvalue and eigenfunction of the linear elliptic eigenvalue problem

$$\lambda_{\pm}(U)\varphi_{\pm}(x,U) = \Delta\varphi_{\pm}(x,U) + \varphi_{\pm}(x,U)\, h_{\pm}(x,U) \quad \text{in } \Omega \ ; \quad \partial\varphi_{\pm}/\partial n = 0 \quad \text{on } \partial\Omega \quad \text{in the mild sense} \tag{22$\pm$}$$

$$\lambda_{o}(U)\varphi_{o}(x,U) = \Delta\varphi_{o}(x,U) - \varphi_{o}(x,U) f(x,U) \quad \text{in } \Omega \ ; \quad \partial\varphi_{o}/\partial n = 0 \quad \text{on } \partial\Omega \tag{23}$$

From the variational principle we get the monotonities

$$\lambda_{-}(U) \leqq \lambda_{o}(U) \leqq \lambda_{+}(U) \quad \text{for all } U \in [0,\infty); \tag{24}$$

$$0 \leqq U_1 < U_2 \quad \text{implies} \quad \lambda_{o}(U_1) > \lambda_{o}(U_2) \ ; \quad \lambda_{\pm}(U_1) > \lambda_{\pm}(U_2). \tag{25}$$

Finally define $\lambda_{+}(\infty) = \lim_{U\to\infty} \lambda_{+}(U) \in \mathbb{R}\cup\{-\infty\}$. (26)

We are now ready to investigate the existence of equilibria and to formulate our main result:

THEOREM 2 (Main result)

Let nonnegative initial data (u_o,v_o) *satisfying (4)(5) be given. As indicated in the Abstract, we have to distinguish the following cases:*

	for diffusing predator:	for diffusing prey:
(1)	$0 \geqq \lambda_{+}(0)$	$0 \geqq \lambda_{-}(0)$
(2)	$\lambda_{+}(0) > 0 > \lambda_{+}(\infty)$	$\lambda_{-}(0) > 0 > \lambda_{-}(\infty)$
(3)	$\lambda_{+}(\infty) \geqq 0$	$\lambda_{-}(\infty) \geqq 0$

In case (1) the unique equilibrium state (u_1,v_1) *compatible with the initial data* $(u_o,v_o) \in J$ *is given by*

$$u_1 \equiv 0 \ ; \quad v_1(x) = (d_{+}\chi_{o}/c)(x) = \begin{cases} (d/c)(x) & \text{if } d(x)>0 \text{ and } v_o(x)>0, \\ 0 & \text{otherwise.} \end{cases}$$

The nonlinear elliptic problem (21) has no positive mild solution.

In case (2) exists a unique equilibrium $(\bar{u},\bar{v})$ *compatible with the initial data* $(u_o,v_o) \in J$*, too. Here the function* $\bar{u}$ *is a mild solution of (21) and satisfies* $\bar{u}(x) > 0$ *for all* $x \in \bar{\Omega}$ *and* $\bar{u} \in C^{\nu}(\bar{\Omega})$ *for all* $\nu \in [0,2)$.

In case (3) there exists no equilibrium compatible with the initial data.

In any case there exists a global mild solution
$(u,v): (x,t) \in \Omega \times (0,\infty) \to (u,v)(x,t) \in (0,\infty) \times [0,\infty)$
of the initial-boundary value problem (1)(2)(3).

The asymptotic behavior of (u,v) *is given by the following relations in the cases (1)(2)(3), respectively:*

Case (1):
$$\sup_{0<t<\infty} [\|u(.,t)\|_\infty + \|v(.,t)\|_\infty] < \infty; \tag{27}$$

$$\lim_{t\to\infty} e^{\lambda t}\|u(.,t)\|_{C^\nu(\bar\Omega)} = 0 \quad \text{for all } \nu \in [0,2),\ \lambda \in \{0\} \cup (0,-\lambda_{\pm}(0)); \tag{28a}$$

$$\lim_{t\to\infty} \|v_o[v(.,t) - v_1]\|_\infty = 0. \tag{28b}$$

Case (2):
$$\sup_{0<t<\infty} [\|u(.,t)\|_\infty + \|v(.,t)\|_\infty] < \infty; \tag{29}$$

$$\lim_{t\to\infty} \|u(.,t) - \bar u\|_{C^\nu(\bar\Omega)} = 0 \quad \text{for all } \nu \in [0,2); \tag{30a}$$

$$\lim_{t\to\infty} \|v_o[v(.,t) - \bar v]\|_\infty = 0. \tag{30b}$$

Case (3):
$$\lim_{t\to\infty} e^{-\lambda t} \min_{x\in\bar\Omega} u(x,t) = \infty \quad \text{for all } \lambda \in \{0\} \cup (0,\lambda_{\pm}(\infty)). \tag{31}$$

At this point we have to distinguish the model with diffusing predator (upper signs) and the model with diffusing prey (lower signs).

For the model with diffusing predator we have

$$\lim_{t\to\infty} \|v(.,t)\|_\infty = 0. \tag{32}$$

For the model with diffusing prey, case (3) is very exceptionel and can only occur if $(a\chi_o)(x) = 0$ *for almost all* $x \in \Omega$. *This means that equation (1a) is decoupled from (1b), hence* $u_t - \Delta u = -uf(x,u)$.
For the second component we get

$$\lim_{t\to\infty} v(x,t) = \infty \quad \text{for all } x \in \Omega \text{ with } v_o(x) > 0. \tag{33}$$

INDICATIONS OF PROOFS

Since the main part of the proof of Theorem 1 is already contained in [3], we only give some hints. The Lyapunov functional

$$\Lambda(t) = \int_\Omega \bar u(u - \bar u - \bar u \log(u/\bar u))\,dx + \int_{\Omega\setminus\Omega_p} (\bar u a/b)(v - \bar v - \bar v \log(v/\bar v))dx + \int_{\Omega_p} (\bar u a/b) v\,dx$$

has the time derivative

$$d\Lambda/dt = -(\Sigma_u + \Sigma_v + \Sigma_p + \Sigma_\nabla) \quad \text{with}$$

$$\Sigma_u = \int_\Omega \bar{u}(u - \bar{u})(f(x,u)-f(x,\bar{u}))\,dx \geq 0;$$

$$\Sigma_v = \int_\Omega (\bar{u}ac/b)(v - \bar{v})^2\,dx \geq 0;$$

$$\Sigma_p = -\int_{\Omega_p} (\bar{u}a/b)v(d \mp b\bar{u})\,dx \geq 0 \quad \text{by (12)(15);}$$

$$\Sigma_\nabla = \int_\Omega (u\,\nabla(\bar{u}/u))^2\,dx \geq 0.$$

To exclude a blow-up in finite time and exploit the Lyapunov functional, one has to prove the L_∞-bound (16). An L_1-bound uniform in time is easy to get from the Lyapunov functional. In the case of diffusing predator (upper signs) an L_∞-bound for the component v is straightforward. Hence one gets an L_∞-bound for the diffusing component u by the bootstrap arguments of [3] or [4] in both cases. Finally this implies an L_∞-bound of v in the case of diffusing prey (lower signs), too. Now one uses compactness - for the diffusing component in $C^\nu(\bar{\Omega})$ and for the nondiffusing component in the weak topology of $L_\infty(\Omega)$ - to define a nonvoid ω-limit set of the trajectory. As shown in [3], one exploits the Lyapunov functional to prove $u(.,t) \to \bar{u}$ in $C^\nu(\bar{\Omega})$ and $v(.,t) \to \bar{v}$ in the weak topology of $L_\infty(\Omega)$.

It remains to prove the strong convergence (18) of the component v. The differential equation (1b) for v can be written

$$v_t = -cv(v - (\bar{v}+r)) - p$$

where $r = \pm b(\bar{u} - u)/c$ satisfies $\|r(.,t)\|_\infty \to 0$ for $t \to \infty$ and $p = -v(d - c\bar{v} \mp b\bar{u})$ satisfies $p \geq 0$ by (12)(15).

For any $\varepsilon > 0$ there exists $T(\varepsilon)$ such that

$$|r(x,t)| < \varepsilon \quad \text{for all } (x,t) \in \bar{\Omega}\times[T(\varepsilon),\infty).$$

At first, we construct an upper bound for v. The comparison principle for the solutions of ordinary differential equations implies

$$v(x,t) \leq v_u(x,t) \quad \text{for all } (x,t) \in \bar{\Omega}\times[T(\varepsilon),\infty)$$

where v_u is the solution of the initial value problem

$$(v_u)_t = -cv_u(v_u - (\bar{v}+\varepsilon)) \geq -cv_u(v_u - (\bar{v}+r)) \;; \quad v_u(x,T(\varepsilon)) = v(x,T(\varepsilon)).$$

Since $v_u \geq \bar{v}+\varepsilon$ implies $-c(v_u - (\bar{v}+\varepsilon))^2 \geq -cv_u(v_u - (\bar{v}+\varepsilon))$, we compute

$$v(x,t) \leq v_u(x,t) \leq \bar{v}(x) + \varepsilon + [c(t-T(\varepsilon)]^{-1} \quad \text{for all } (x,t)\in\bar{\Omega}\times[T(\varepsilon),\infty)$$

which yields an upper bound converging to $\tilde{v}(x)$ uniformly in x.

We construct a lower bound for v. The comparison principle implies

$$v_1(x,t) \leq v(x,t) \quad \text{for all } (x,t) \in \bar{\Omega}\times[T(\varepsilon),\infty)$$

where v_1 is the solution of the initial value problem

$$(v_1)_t = -cv_1(v_1 - (\bar{v}-\varepsilon)) \; ; \quad v_1(x,T(\varepsilon)) = v(x,T(\varepsilon)).$$

An explicit computation yields

$$v_1 = \bar{v}-\varepsilon - \frac{(\bar{v}-\varepsilon)(\bar{v}-\varepsilon-v_T)\exp[-(\bar{v}-\varepsilon)c(t-T(\varepsilon))]}{v_T \qquad (1 + (\bar{v}-\varepsilon-v_T)v_T^{-1}\exp[-(\bar{v}-\varepsilon)c(t-T(\varepsilon))])}$$

with $v_T = v(.,T(\varepsilon))$. Since $v_o(x) = Kv(x,T(\varepsilon))$ for all $x \in \bar{\Omega}$, we get (18).

We sketch the proof of Theorem 2.

By a variational argument one shows that assumption (2) is a necessary condition for the existence of a positive mild solution of the nonlinear elliptic problem (21). By use of upper and lower solutions one proves that assumption (2) is a sufficent condition, too.

Next we treat the asymptotic behavior of the solution (u,v).

In case (1) we use the Lyapunov functional

$$\Lambda_1(t) = \int_\Omega \varphi_1 u \, dx + \int_{\Omega\setminus\Omega_1} (\varphi_1 a/b)(v - v_1 - v_1 \log(v/v_1))\, dx + + \int_{\Omega_1} (\varphi_1 a/b) v \, dx$$

where $v_1 = d_+\chi_o/c$, $\Omega_1 = \{x \in \Omega | \; v_1 = 0\}$ and $\varphi_1(x) = \varphi_\pm(x,0)$ is the principlal eigenfunction of problem (22$\pm$) with U = 0.
The time derivative of the functional Λ_1 is

$$d\Lambda_1/dt = \lambda_\pm(0)\int_\Omega \varphi_1 u \, dx + \int_\Omega \varphi_1 u(f(x,0) - f(x,u))\, dx - - \int_\Omega (\varphi_1 ac/b)(v - v_1)^2 \, dx + \int_{\Omega_1} (\varphi_1 a/b) dv \, dx \leq 0.$$

Now the arguments to prove (27), (28a) with $\lambda = 0$ and (28b) are similar to those in the proof of Theorem 1.
It remains to check the exponential convergence in (28a). The differential equation (1a) for the component u can be written

$$u_t - \Delta u = u\, h_\pm(x,0) + u[f(x,0) - f(x,u) \pm a(v-v_1)]$$

For $\lambda \in (0,-\lambda_\pm(0))$ given, we can choose T large enough such that

$$|a(x)[v(x,t) - v_1(x)]| \leq -\lambda_\pm(0) - \lambda \quad \text{for all } (x,t) \in \bar{\Omega}\times[T,\infty).$$

Hence $u_t - \Delta u + \lambda u \leq u[h_{\pm}(x,0) - \lambda_{\pm}(0)]$ for all $(x,t) \in \bar{\Omega}\times[T,\infty)$.

Application of the comparison principle to the functions $u(x,t)$ and $e^{-\lambda t}\varphi_{\pm}(x)$ yields

$$u(x,t) \leq Ke^{-\lambda t}\varphi_{\pm}(x) \quad \text{for all } (x,t) \in \bar{\Omega}\times[T,\infty)$$

which proves (28a).

Case (2) repeats the results of Theorem 1.

We turn to case (3). We consider only the more meaningful case of diffusing predator and leave the other case to the reader.
Let $w = w(x,t)$ be the solution of the system without prey. Hence

$$w_t - \Delta w = -wf(x,w) \quad \text{in } \Omega; \ \partial w/\partial n = 0 \quad \text{on } \partial\Omega; \ w(x,0) = u_o(x) \quad \text{in } \Omega.$$

By the comparison principle we get

$$u(x,t) \geq w(x,t) \quad \text{for all } (x,t) \in \bar{\Omega} \times [0,\infty).$$

Since $\lambda_+(U) = \lambda_o(U)$ for U large enough, a further application of the comparison principle to the functions $w(x,t)$ and $e^{\lambda t}\varphi_o(x,U)$ with $\lambda \in \{0\} \cup (0,\lambda_+(\infty))$ yields (31). Now (32) follows easily.

DISCUSSION

This note presents a simple quite standard predator-prey system. The main feature of our model is that only one species can diffuse. This hypothesis makes the model at the same time more treatable and more interesting. The strong maximum principle does not spoil the game: we get places - we call them "patches" - where the sedentary species is absent from the beginning or becomes extinct for large time. The pattern of these patches can be prescribed to a large extend by the choice of initial data. Indeed, it turns out that there exists a whole continuum of equilibrium states. For given initial data (u_o,v_o) and equilibrium $(\bar{u},\bar{v})$, the compatibility assumption (15) is a necessary and sufficient condition that the equilibrium $(\bar{u},\bar{v})$ is approached for large time starting from the initial data (u_o,v_o). Roughly spoken, the compatibility condition tells that the sedentary species cannot be present at places where it is absent in the beginning and that it becomes extinct where one expects this by linearizing the differential equation for the sedentary species locally in space.

Existence of equilibria can be proved and disproved by reducing the system together with the compatibility condition to a scalar nonlinear elliptic boundary value problem and applying comparison methods. Convergence to equilibrium is proved via a Lyapunov functional of [5].A main peculiarity of this functional ís that it can only be written down if one already knows the asymptotic equilibrium. The convergence proof contains many mathematical difficulties: bootstrapping, weak compactness, an asymptotic version of the differential equations etc. The reader interested in details will find them in [3].

We should remark that the strong logistic saturation term introduced into the equation for the sedentary species leads to some simplifications. It bounds the sedentary species, prevents the system to oscillate (which may indeed happen in a rather special case) and allows to prove even strong convergence of the sedentary species.

REFERENCES

[1] P.I. Lions, On the existence of positive solutions of semilinear elliptic equations, SIAM Review, 24(1982) 441-467.

[2] M. Mimura and J.D. Murray, On a diffusive prey-predator model which exhibits patchiness, J. Th. Biol., 75(1979)249-262.

[3] F. Rothe, Asymptotic behavior of a nonhomogeneous predator-prey system with one diffusing and one sedentary species, Math. Meth. Appl. Sci., 5(1983)40-67.

[4] F. Rothe, A priori estimates, global existence and asymptotic behavior for weakly coupled systems of reaction-diffusion equations, Habilitationsschrift, Universität Tübingen 1983.

[5] K. Kawasaki and E. Teramoto, Spatial pattern formation for prey-predator populations, J. Math. Biol., 8(1979)33-46.

GLOBAL BRANCHES OF ONE DIMENSIONAL STATIONARY SOLUTIONS TO CHEMOTAXIS SYSTEMS AND STABILITY

Renate Schaaf
Universität Heidelberg
Sonderforschungsbereich 123
Im Neuenheimer Feld 293
D-6900 Heidelberg 1

1. This article will show some bifurcation results for stationary solutions of the Keller-Segel-model for slime mold aggregation ([4]) in one space dimension:

$$\begin{aligned} u_t &= \left(\mu(u,v)u_x - \chi(u,v)v_x\right)_x \\ v_t &= \nu_o v_{xx} + k(u,v) \end{aligned} \tag{1}$$

Here $u=u(x,t)$, $x\in[0,1]$, $t\in\mathbb{R}^+$ is a cell density, $v=v(x,t)$ is the density of some chemoattractant whose production and degradation depends on u and v. Neglecting birth and death of cells the time evolution of the cell density is due to a flux consisting of a random part $-\mu(u,v)u_x$ and a chemotaxis part $\chi(u,v)v_x$ up the gradient of v. In the v-equation ν_o is a positive diffusion coefficient, k is a reaction term modelling production and degradation of the chemoattractant. For a detailed derivation of (1) see [4].

We assume the whole process to take place on a bounded interval in $\mathbb{R}^1$, normalized to [0,1]. No flux boundary conditions reduce to homogeneous Neumann conditions:

$$u_x(0) = u_x(1) = v_x(0) = v_x(1) = 0. \tag{2}$$

Conditions on μ and k are

$\mu:\mathbb{R}^+ \times \mathbb{R}^+ \to \mathbb{R}^+$ is twice continuously differentiable,
$k:\mathbb{R}^+ \times \mathbb{R}^+ \to \mathbb{R}^+$ is three times continuously differentiable,
$k^{-1}(\{0\}) \neq \emptyset$, $\partial_1 k(r,s) \geq 0$, $\partial_2 k(r,s) < 0$ for all $r,s\in\mathbb{R}^+$.

(Here ∂_1 (∂_2) denotes differentiation with respect to the first (second) variable.)

The chemotaxis coefficient χ we will specify to the following situation: Suppose that by some receptor mechanism cells do not measure the gradient of v but of some $\phi(v)$ with a sensitivity function $\phi:\mathbb{R}^+\to\mathbb{R}^+$, $\phi'>0$ (see [5]). The speed of a single cell moving up the gradient of $\phi(v)$ is assumed to be proportional to the random motility coefficient $\mu(u,v)$, such that the chemotactic flux is

$$\chi_0 \mu(u,v) u(\phi(v))_x ,$$

that is

$$\chi(u,v) = \chi_0 \mu(u,v) u \phi'(v) . \tag{3}$$

For a derivation of (1) and (3) as a diffusion approximation to a biased random walk model see [1] and [2].
Several normalized forms of ϕ have been suggested:

$$\phi(v) = v \text{ (direct measurement)} \tag{4}$$

$$\phi(v) = \ln(c+v) \quad \text{(logarithmic measurement, } c \geq 0\text{)} \tag{5}$$

$$\phi(v) = \frac{v}{1+cv} \quad \text{(Michaelis-Menten kinetics, } c>0\text{)} \tag{6}$$

$$\phi(v) = \frac{v^2}{1+cv^2} \quad \text{(cooperative binding, } c>0\text{)} . \tag{7}$$

We make the assumptions

$$\chi_0 \in \mathbb{R}^+ , \quad \phi : \mathbb{R}^+ \to \mathbb{R}^+ \text{ three times continuously differentiable.}$$

The stationary problem then becomes equivalent to

$$\begin{aligned} & u' - \chi_0 u \phi'(v) v' = 0 \\ & \nu_0 v'' + k(u,v) = 0 \\ & u'(0) = u'(1) = v'(0) = v'(1) = 0 \end{aligned} \tag{8}$$

If the production of v can equilibrate degradation ($k^{-1}(\{0\}) \neq \emptyset$) then the system admits a one parameter family of spatially homogeneous solutions. As is argued in [4] the onset of aggregation then may occur whenever those homogeneous solutions become unstable at some critical parameter value. For the time evolution of the system beyond such a critical point so far S. Childress and J.K. Percus ([3]) have pointed out that this depends critically on the number of space variables considered. For the case $\phi(v)=v$, $k(u,v)=\alpha u-\beta v$, they give an L^2-a priori bound in one space dimension, whereas e.g. in dimension ≥ 3 there may be convergence to a δ-distribution in finite time.
The purpose here is to study bifurcation of one dimensional steady states which have a spatial pattern and eventually become stable when the homogeneous solutions loose their stability. In more generality this has been done in [7]. Here we will only consider case (3) with ϕ as in (4), (5) and (6), since under those conditions we can use analytical techniques

from [6] and [8] in order to get the whole qualitative global bifurcation behaviour of (8).

2. (8) can be reduced to a parameter dependent scalar problem in the following way: $(u,v)\in C^2([0,1],\mathbb{R}^+)$ solves the first equation of (8) iff

$$u(x) = \varphi(v(x),\lambda) := \lambda e^{\chi_0 \phi(v(x))} \tag{9}$$

for some $\lambda\in\mathbb{R}^+$. Thus by (9) the stationary system becomes equivalent to

$$\begin{aligned} &v''+f(v,\lambda) = 0 \\ &v'(0) = v'(1) = 0 \end{aligned} \tag{10}$$

with

$$f(s,\lambda) = \frac{1}{\nu_0} k(\varphi(s,\lambda),s) \quad , \qquad \lambda\in\mathbb{R}^+ \quad .$$

We have to study bifurcation from the set

$$M:=\{(m,\lambda)\in\mathbb{R}^+ \times \mathbb{R}^+ \mid f(m,\lambda)=0\}$$

which corresponds to the constant solution set of (8). M is a one dimensional C^2-manifold because of the assumptions on k and ϕ.

THEOREM 1

a) A necessary condition for $(m,\lambda)\in M$ *to be a bifurction point of (10) is*

$$\partial_1 f(m,\lambda) = j^2\pi^2>0 \quad \textit{for some } j\in\mathbb{N}$$

b) Let $A_0\subset\mathbb{R}^+$ *be an open* λ*-interval over which* M *is a graph, i.e. there is a* C^2*-map* $\lambda\mapsto m(\lambda)$ *with*

$$(m(\lambda),\lambda)\in M,\ \partial_1 f(m(\lambda),\lambda)>0 \quad \textit{for all } \lambda\in A_0.$$

Also assume for each fixed $\lambda\in A_0$ *that* $\liminf_{s\to 0} f(s,\lambda)\geq 0$ *and*

$$\begin{aligned} &5\partial_1^2 f(s,\lambda)-3\partial_1^3 f(s,\lambda)\,\partial_1 f(s,\lambda)>0 \\ &\textit{for all } s\in\mathbb{R}^+ \textit{ with } \partial_1 f(s,\lambda)\geq 0 \end{aligned} \tag{11}$$

and

$$\begin{aligned} &\textit{if } \partial_1 f(s,\lambda)<0, \quad s<m(\lambda) \quad \textit{then } \partial_1^2 f(s,\lambda)>0, \\ &\textit{if } \partial_1 f(s,\lambda)<0, \quad s>m(\lambda) \quad \textit{then } \partial_1^2 f(s,\lambda)<0\ . \end{aligned} \tag{12}$$

Define sets $A_j\subset A_0$ *by*

$$A_j = \{\lambda \in A_o \mid \partial_1 f(m(\lambda),\lambda) > j^2\pi^2\}$$

Then for each j *with* $A_j \neq \emptyset$ *there are two distinct* C^1*-solution curves*

$$V_j^+ : A_j \to C^2([0,1],\mathbb{R}^+)$$
$$V_j^- : A_j \to C^2([0,1],\mathbb{R}^+)$$

such that $(V_j^{+(-)}(\lambda),\lambda)$ *solves (10).*
All solutions $V_j^{+(-)}(\lambda)$, $\lambda \in A_j$ *have the same spatial pattern, i.e. the number of maxima and minima on* $[0,1]$ *is* $j+1$.
If $\lambda_j \in \partial A_j$ *with* $\partial_1 f(m(\lambda_j),\lambda_j) = j^2\pi^2$ *then*

$$V_j^{+(-)}(\lambda) \to m(\lambda_j) \quad \textit{for } \lambda \to \lambda_j$$

that is $(m(\lambda_j),\lambda_j)$ *is a bifurcation point.*

This theorem is an application of results in [6] and [8] on the so called time map (see [9]) $T(a,\lambda)$ for parameter dependent Neumann problems. (10) is equivalent to a series of problems $T(a,\lambda)=\frac{1}{j}$, $j \in \mathbb{N}$, $(a,\lambda) \in D(T) \subset \mathbb{R}^+ \times \mathbb{R}^+$. In [8] (11) and (12) are shown to be a criterion for $\partial_1 T(a,\lambda)$ to be positive which gives the parametrized global solution branches $(V_j^{+(-)}(\lambda),\lambda)$ via the implicit function theorem. (See [6] thm. 4.2)

3. In this section we will apply theorem 1 to (10) with $f(s,\lambda)=\frac{1}{\nu_o} k(\varphi(s,\lambda),s)$, $\varphi(s,\lambda)=\lambda e^{\chi_o \phi(s)}$ and ϕ as in (4), (5) and (6). For k we will assume the most simple reaction mechanisms, that is

$$k(r,s) = \nu_o(\alpha r - \beta s) \quad \text{with } \alpha,\beta > 0$$

$$\text{or} \quad k(r,s) = \nu_o\left(\alpha r - \frac{\beta s}{1+\gamma s}\right) \text{ with } \alpha,\beta,\gamma > 0.$$

3.1 $\phi(s)=s$, $k(r,s)=\nu_o(\alpha r-\beta s)$

We have $f(s,\lambda) = \alpha\lambda e^{\chi_o s} - \beta s$.
It is easy to show (11) and (12) for f. We will illustrate the application of theorem 1 in figures 1 and 2 below.The marked region of M consisting of points with $\partial_1 f(m,\lambda)>0$ can be parametrized over A_o by some $\lambda \to m(\lambda)$. The graph $\lambda \to \partial_1 f(m(\lambda),\lambda)$ is strictly decreasing with $\partial_1 f(m(\lambda),\lambda) \to \infty$ for $\lambda \to 0$. Thus we have infinitely many branches parametrized over $A_j=\{\lambda \in A_o \mid \partial_1 f(m(\lambda),\lambda) > j^2\pi^2\}$. The resulting qualitative bifurcation diagram for the system (8) is given in figure 2. As a projection of the (u,v)-space in $\mathbb{R}^2$ we use parameters $\bar{u}=\int_0^1 u(x)dx$ and $v(0)$. The pattern of solutions within the branches is indicated.

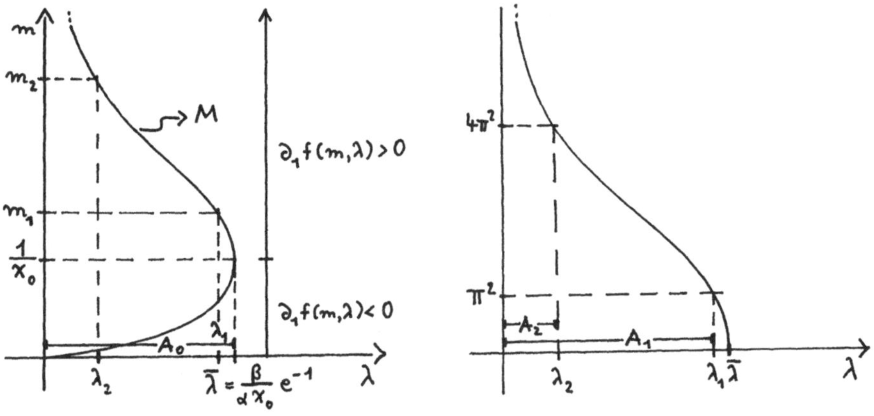

Figure 1: $M=\{(m,\lambda)\,|\,f(m,\lambda)=0\}$ $\qquad$ $\lambda \mapsto \partial_1 f(m(\lambda))$

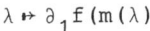

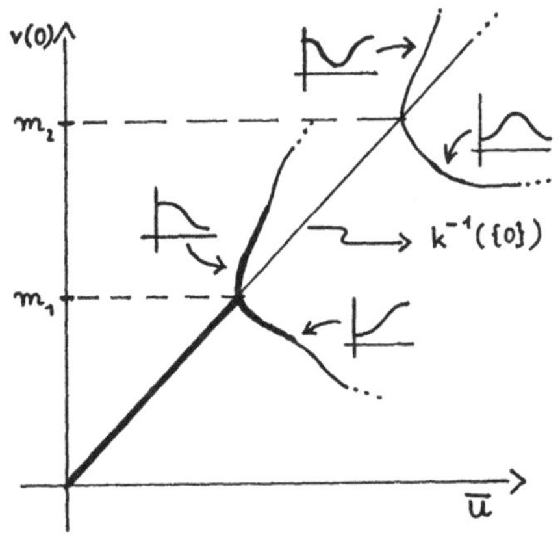

Figure 2: Bifurcation diagram (3.1: $\beta<2\pi^2$)

3.2 $\phi(s)=\ln s, \quad k(r,s)=\nu_o(\alpha r-\frac{\beta s}{1+\gamma s})$

(The case $\phi(s)=\ln(c+s)$ with $c>0$ can be handled in the same way. Here

$$f(s) = \alpha\lambda s^{\chi_o} - \frac{\beta s}{1+\gamma s}\ .$$

We consider the case $1\leq\chi_o<2$ for which we have $\partial_1^3 f(s,\lambda)<0$, such that (11) and (12) are trivially satisfied.

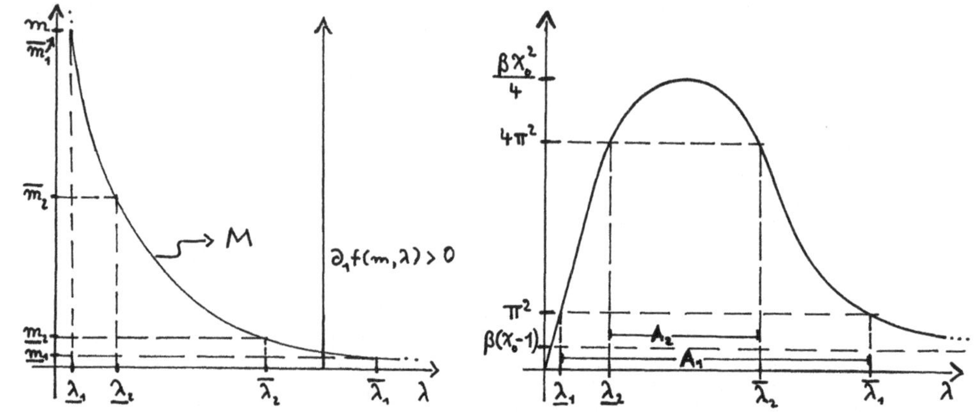

$M=\{(m,\lambda)\,|\,f(m,\lambda)=0\}$ $\qquad\qquad$ $\lambda \mapsto \partial_1 f(m(\lambda),\lambda)$

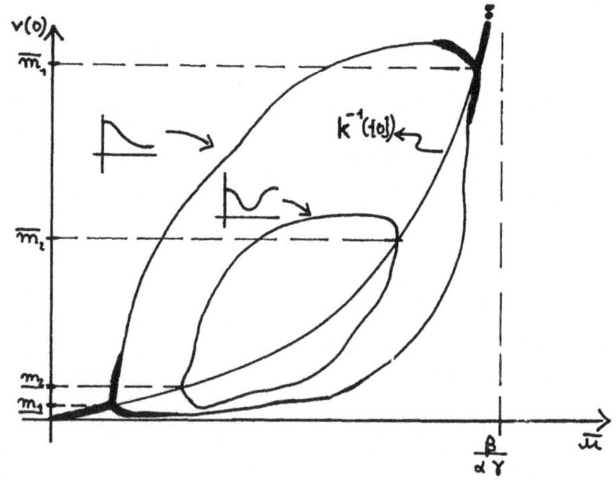

Figure 3: Bifurcation diagram (3.2: $1<\chi_0<2$, $\beta(\chi_0-1)<\pi^2$)

From figure 3 we see that the number of bifurcating branches and also the amount of spatial structure in the branches depends on $\beta\chi_0$. $\nu_0\beta$ is measuring a sort of amplifying effect on the gradient, since it is the degradation rate for small chemical densities; ν_0 is the diffusion coefficient of the chemoattractant. Thus it makes sense that the amount of spatial structure increases as $\nu_0\beta$ increases for fixed ν_0,χ_0, or as ν_0 decreases for fixed $\nu_0\beta,\chi_0$. The situation for $\chi_0=1$ looks similar. If $\beta(\chi_0-1)>\pi^2$ one or more branches split off and become unbounded.

3.3 $\quad \phi(s) = \dfrac{s}{1+cs}$, $c>0$, $k(r,s) = \nu_0(\alpha r-\beta s)$

For $\chi_0>\psi c$ we have the following situation:

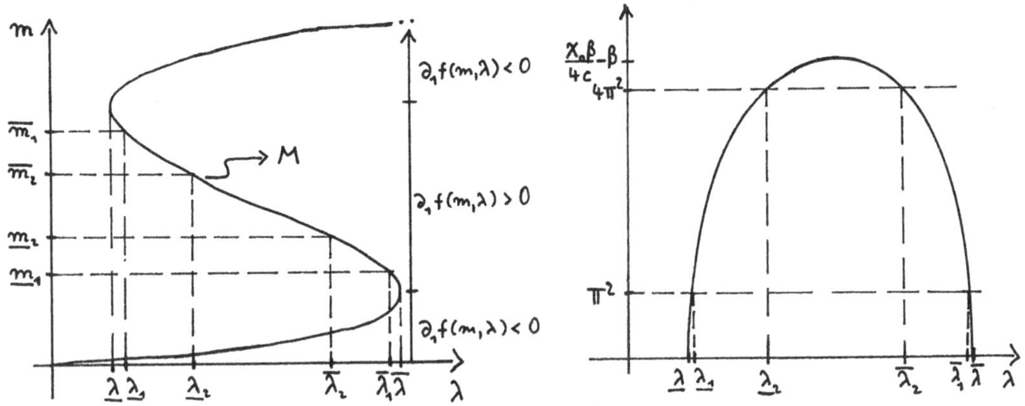

Figure 4: $M=\{(m,\lambda)\,|\,f(m,\lambda)=0\}$ $\lambda \mapsto \partial_1 f(m(\lambda),\lambda)$

Calculations to show (11) and (12) become more lengthly but are straightforward. So we have the following bifurcation diagram:

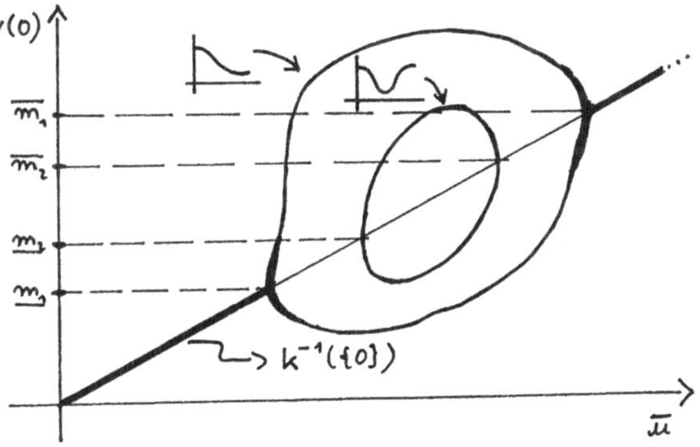

Figure 5: Bifurcation diagram (3.3)

4. Finally we will discuss the stability of the bifurcating solutions of (8). Proofs for the statements in this section can be found in [7]. Since from (1) we get that $\int_0^1 u(x,t)\,dx=\bar{u}$ is a constant, asymptotic stability with respect to small perturbations (u,v) with $\int_0^1 u(x)\,dx=0$ is studied. The stability of spatially homogeneous solutions in $k^{-1}(\{0\})$ can be decided by a criterion in terms of f, $f(s,\lambda)=\frac{1}{\nu_0}\,k(\varphi(s,\lambda),s)$:

LEMMA

A constant solution $(u,v)\equiv(n,m)$ *with* $k(n,m)=0$ *is stable if* $n=\varphi(m,\lambda)$ *and*

$$\partial_1 f(m,\lambda) < \pi^2.$$

If

$$\partial_1 f(m,\lambda) > \pi^2$$

then $(n,m)=(\varphi(m,\lambda),m)$ *is unstable.*

Thus for exchange of stability one has to consider bifurcation points $(n,m)=(\varphi(m,\lambda),m)$ with $\partial_1 f(m,\lambda)=\pi^2$. It turns out that stability of bifurcating solutions near the bifurcation point depends on the direction of the branch with respect to the parameter $\bar{u}$, as is shown in figure 6:

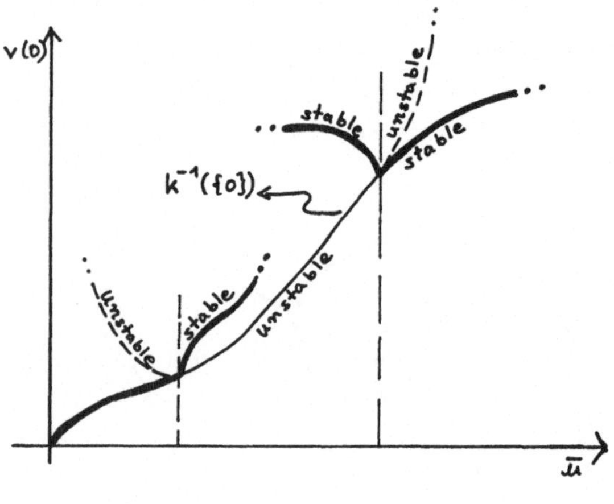

Figure 6: Exchange of (in-)stability

By Taylor expansion up to third order we can calculate the direction of the branch from an expression involving derivatives up to third order of f an φ at the point (m,λ) which is a bifurcation point for (10) corresponding to a bifurcation point $(\varphi(m,\lambda),m)$ for (8):

THEOREM 2

Let $(n,m)\in k^{-1}(\{0\})$ *be a bifurcation point of (8) with* $n=\varphi(m,\lambda)$, $\partial_1 f(m,\lambda)=\pi^2$. *Define*

$$J(m,\lambda):=[(\partial_1^2\varphi\partial_1 f-\partial_1\varphi\partial_1^2 f)(m,\lambda)]\cdot[(\partial_1\partial_2 f-\frac{\partial_1^2 f\partial_2 f}{\partial_1 f})(m,\lambda)]$$
$$+[(\partial_2\varphi-\frac{\partial_1\varphi\partial_2 f}{\partial_1 f})(m,\lambda)]\cdot[(\tfrac{5}{6}(\partial_1^2 f)^2-\tfrac{1}{2}\partial_1 f\partial_1^3 f)(m,\lambda)] \quad .$$

Then the bifurcating solutions of (8) near the bifurcation point are stable (supercritical bifurcation) *if* $J(m,\lambda) < 0$,

unstable (subcritical bifurcation) *if* $J(m,\lambda) > 0$.

We can apply theorem 2 to examples 3.1-3.3 and find (see [7]):

3.1 Bifurcation is

supercritical for $\beta < 2\pi^2$

subcritical for $\beta > 2\pi^2$

(This result has also been obtained in [3].)

3.2 Bifurcation is supercritical for

$1 \leq \chi_0 < 2, \quad \beta\chi_0^2 > 4\pi^2, \quad \beta(\chi_0 - 1) < \pi^2.$

3.3 Bifurcation is supercritical for

$\chi_0 > 4c \ , \quad \beta\left(\frac{\chi_0}{4c} - 1\right) > \pi^2, \quad \beta > \frac{\pi^2}{2(1+\sqrt{2})}$

Regions of stability have been marked in the bifurcation diagrams of section 3. Figure 2 is showing the case $\beta < 2\pi^2$.

REFERENCES

[1] W. Alt: Biased random walk models for chemotaxis and related diffusion approximations.J.Math.Bio. 9, 147-177 (1980)
[2] W. Alt: Orientation of cells migrating in a chemotactic gradient. Biological Growth and Spread, Proceedings, Heidelberg 1979, Springer Lecture Notes in Biomathematics 38, 353-366 (1980)
[3] S. Childress, J.K. Percus: Nonlinear aspects of chemotaxis. Math. Biosci. 56, 217-237 (1981)
[4] E.F. Keller, L.A. Segel: Initiation of slime mold aggregation viewed as an instability. J.Theor.Bio. 26, 399-415 (1970)
[5] I.R. Lapidus, M. Levandowsky: Modeling chemosensory responses of swimming eukaryotes. Biological Growth and Spread, Proceedings, Heidelberg 1979, Springer Lecture Notes in Biomathematics 38, 388-396 (1980)
[6] R. Schaaf: Global behaviour of solution branches for some Neumann problems depending on one or several parameters. Sonderforschungsbereich 123, preprint Nr. 162. To appear in J.reine angew.Math.
[7] R. Schaaf: Stationary solutions of chemotaxis systems. Preprint SFB 123, to appear
[8] R. Schaaf: Global solution branches via the time map. In preparation
[9] J. Smoller, A. Wasserman: Global bifurcation of steady state solutions. J. of Diff.Equ. 39, 269-290 (1981)

ANALYTICAL AND TOPOLOGICAL METHODS FOR REACTION-DIFFUSION EQUATIONS

J.A. Smoller* and J.S. Shi
The University of Michigan
Department of Mathematics
Ann Arbor, Michigan 48109
U.S.A.

§1. We shall discuss some mathematical methods in the theory of reaction-diffusion equations and show how they can be sometimes used to find patterns. The patterns we find will be dynamic patterns, as opposed to stationary patterns, in the sense that we shall (qualitatively) find the complete solution space. The success of our method relies on the fact that we can take a certain parameter (a specific diffusion coefficient) to be large. As this parameter decreases toward zero, the solution space undergoes radical changes and we have included some computational results concerning the asymptotic behavior of solutions in this case.

Our methods will involve both analytical and topological techniques. In order that the topological methods can be applied, it is necessary that there be a certain amount of compactness available; this is achieved by having (bounded) global attracting regions in (u-z)-space, and such regions can be located by using standard "invariant region" techniques [2]. Once this is established, one can apply the Conley index theory to the stationary solutions, in order to calculate their unstable manifolds. This usually requires some "gradient structure," and is often difficult to establish. We get around this problem by showing that for certain parameter values, the equations are "effectively" gradient-like in the sense that all solutions of (1), (2) tend to stationary solutions as $t \to +\infty$.

Our equations will vary smoothly with the parameters, and as the parameters change, the space of solutions undergoes changes in such a way that a relevant homotopy invariant, the Conley index, doesn't change. This means that we can find algebraic invariants, (exact sequences of cohomology groups), which are "discrete" objects and can be computed since they remain constant under small perturbations. That is, we can "continue" our equations to simpler ones in which these invariants can be explicitly computed. These invariants are useful in obtaining information on solutions. Detailed proofs can be found in [1].

*Research supported in part by NSF contract #MCS-80-02337.

§2. The setting is provided by the following system of reaction-diffusion equations in one space dimension:

$$u_t = \alpha u_{xx} + a(u)f(z)$$
$$z_t = \gamma z_{xx} + b(u)g(z) + \phi(z) \ , \tag{1}$$

where $|x| < L$, $t > 0$, and both α and γ are positive constants. We assume that u and z satisfy the following homogeneous boundary conditions:

$$(u_x(\pm L,t) \ , \ z(\pm L,t)) = (0,0) \ , \ t > 0 \ . \tag{2}$$

The functions a , f , b , g and ϕ , which appear in (1) are as depicted in figure 1 below. Namely,

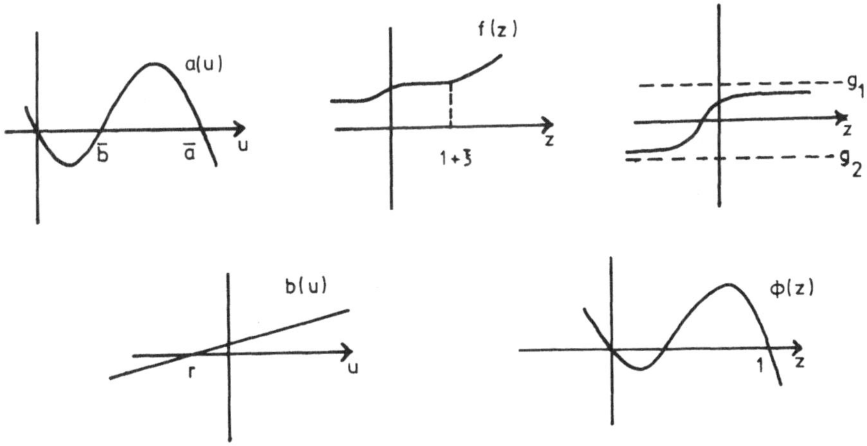

Figure 1

ϕ and a are cubic-like functions, $|g_1-g_2|$ is small, r is near zero, b has small slope, and f is constant on $0 \leq z \leq 1 + \xi$, $\xi > 0$.

If $\varepsilon > 0$ is a small positive number, then if $|g_1-g_2|$ is small, and b has small slope, it is fairly straightforward to show that the set

$$R = \{(u,z): -\varepsilon \leq u \leq \bar{a} \ , \ \theta \leq z \leq 1 + \varepsilon\}$$

is a global attracting region for all solutions of (1), (2). It follows that all stationary solutions of (1), (2) lie in R .

§3. We now consider the problem of finding all the stationary solutions of our problem. Note that if $a(\bar{u}) = 0$, then $(\bar{u}, z(x))$ is such a solution, provided that z satisfies

$$\begin{aligned} \gamma z'' + \phi(z) + b(\bar{u})g(z) &= 0 \qquad |x| < L \\ z(\pm L) &= 0 \quad . \end{aligned} \tag{3}$$

The graph of $\Phi(z) \equiv \phi(z) = b(\bar{u})g(z)$ is depicted in Figure 2, together with the corresponding phase plane for (3). Note that there are three solutions

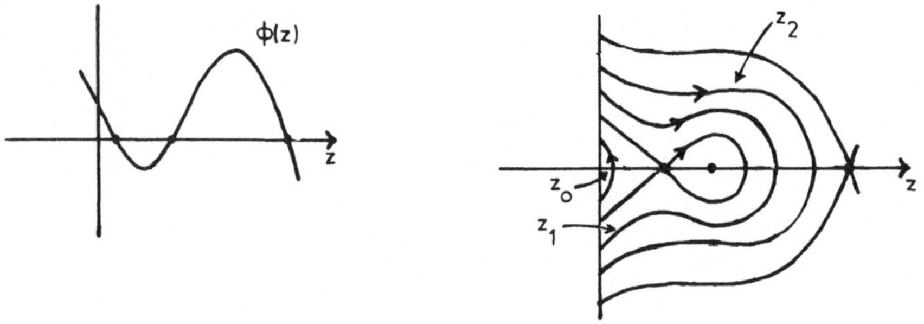

Figure 2

z_0 , z_1 and z_2 , as depicted, where z_1 and z_2 exist only if $L >> 1$ or $\gamma << 1$. From the results in [2], the Conley index, $(h(\cdot))$ of these solutions are $h(z_0) = h(z_2) = \Sigma^0$, $h(z_1) = \Sigma^1$, where Σ^k denotes the pointed k-sphere. That is, z_0 and z_2 are attractors for the associated time dependent problem

$$z_t = \gamma z_{xx} + \Phi(z) \; , \; z(\pm L, t) = 0 \; , \quad |x| < L \; , \; t > 0 \; , \tag{4}$$

while z_1 has a one-dimensional unstable manifold. In addition, there exist solutions of (4) which connect z_1 to both z_0 and z_1. It follows easily from this that the corresponding rest points (ie, stationary solutions) of (1), (2) are connected by orbits of (1), (2); namely, there are solutions v and w of (4) such that

$$(\bar{u},v(x,t)) \begin{cases} \nearrow (\bar{u},z_1(x)) & \text{as} \quad t \to -\infty \\ \searrow (\bar{u},z_0(x)) & \text{as} \quad t \to \infty \end{cases}$$

$$(\bar{u},w(x,t)) \begin{cases} \nearrow (\bar{u},z_1(x)) & \text{as} \quad t \to -\infty \\ \searrow (\bar{u},z_2(x)) & \text{as} \quad t \to +\infty \end{cases},$$

uniformly for $|x| < L$.

If we assume that $L \gg 1$ or $\gamma \ll 1$, then for each root 0, $\bar{b}$ or $\bar{a}$ of **a**, we have functions and phase planes as in figure 2, and corresponding connecting orbits. We thus have the following 9 stationary solutions:

$$P_0 = (0,z_0(x)) \ , \ P_1(0,z_1(x)) \ , \ P_2(0,z_2(x))$$

$$Q_0 = (\bar{b},\tilde{z}_0(x)) \ , \ Q_1(\bar{b},\tilde{z}_1(x)) \ , \ Q_2(\bar{b},\tilde{z}_2(x))$$

$$R_0 = (\bar{a},\bar{z}_0(x)) \ , \ R_1(\bar{a},\bar{z}_1(x)) \ , \ R_2(\bar{a},\bar{z}_2(x)) \ ,$$

together with the corresponding (six) connecting orbits, as depicted in figure 3.

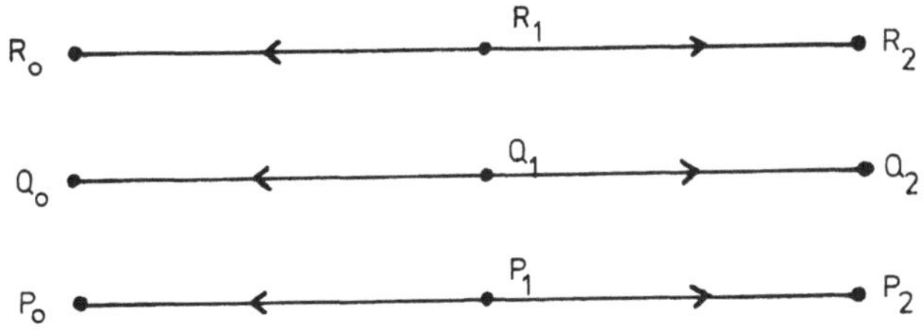

Figure 3

Now fix $L > 0$, and take γ so small that the above 9 stationary solutions exist.

Lemma 1. If α is sufficiently large, then u tends to its spatial average as $t \to \infty$, and the above 9 solutions are the only stationary ones.

The proof consists of showing that

$$\int_{-2}^{2} |u(x,t) - v(t)|^2 \, dx \leq \int_{-2}^{2} u_x^2(x,t)\,dx \leq ce^{-\sigma t} \, , \quad \sigma > 0 \, ,$$

where v is the spatial average of u.

Having obtained all of the stationary solutions, we now want to calculate the dimensions of their unstable manifolds. We can thus linearize (1), (2) about these solutions and we easily obtain the following lemma.

Lemma 2. $h(P_0) = h(P_2) = h(R_0) = h(R_2) = \Sigma^0$;
$h(P_1) = h(R_1) = \Sigma^1$; $h(Q_0) = \Sigma^k$, $h(Q_1) = \Sigma^m$
$h(Q_2) = \Sigma^n$, where k , m , n are all positive integers.

Note that linearization techniques do not allow us to obtain $h(Q_i)$ precisely. We need topological techniques to get this information (Lemma 4, below).

We next show that our equations are "effectively" gradient-like.

Lemma 3. If α is sufficiently large, then all solutions tend to stationary solutions.

To prove this, we note that u tends to its spatial average v, as $t \to \infty$, but also, it is not hard to show that v tends to a root σ of $a(u)$. Thus, u tends to a root σ of $a(u)$. In order to complete the proof, we must show that z tends to a solution of

$$\gamma z'' + b(\sigma)g(z) + \phi(z) = 0 \, , \quad |x| < L \; ; \quad z(\pm L) = 0 \, .$$

This is done by a comparison theory argument for scalar equations.

§4. Having obtained the compactness and the gradient-like structure, we can use the Conley index in order to determine the structure of the solution set, when $\gamma << 1 << \alpha$.

Let I_0 (resp $I_{\bar{a}}$) denote the set consisting of the rest points P_0, P_1 and P_2 (resp R_0, R_1 and R_2) , together with their connecting orbits. If

$$\psi(t) = \int_{-L}^{L} u(x,t)\,dx \ ,$$

then

$$\psi'(t) = \int_{-L}^{L} f(z(x,t))\,a(u(x,t))\,dx$$

so $\psi' < 0$ if $u(x,t) < \bar{b}$, $|x| < L$. Since small neighborhoods of I_0 lie in $u < \bar{b}$, the invariant sets in such a neighborhood tend to one of the rest points P_0, P_1 or P_2 . Hence I_0 (and similarly $I_{\bar{a}}$) is an isolated invariant set.

To compute the Conley index of I_0 , we deform $a(u)$ as depicted in figure 4. Under this deformation, I_0 continues to the maximal

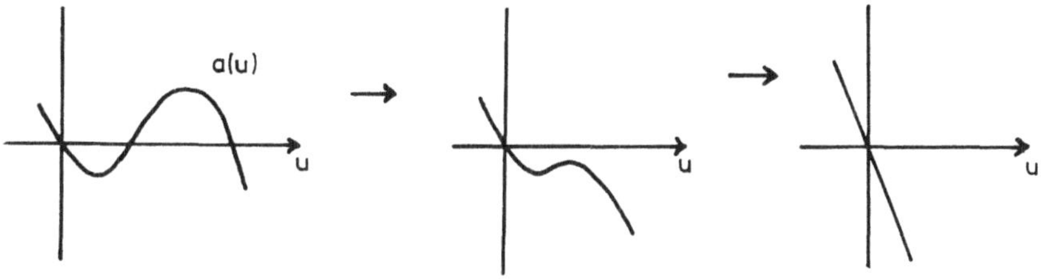

Figure 4

invariant set in R ; hence $h(I_0) = \Sigma^0$ by the continuation theorem [2]. Similarly, $h(I_{\bar{a}}) = \Sigma^0$. We can use these facts to compute the index of the Q_i's .

Lemma 4. $h(Q_0) = h(Q_2) = \Sigma^1$; $h(Q_1) = \Sigma^2$.

Proof. If $I(R)$ denotes the maximal invariant set in R, then $h(I(R)) = \Sigma^0$, since R is an attracting region. Now deform ϕ by "pushing down the hill" (see figure 4). Then R_1 cancels against R_2, Q_1 against Q_2 and P_1 against P_2. This leaves us with R_0, Q_0 and P_0 as the only rest points. Since $h(R_0) = h(P_0) = \Sigma^0$, there must be orbits which connect Q_0 to both P_0 and R_0 ([2] Ch 24, §E). If M is the isolated invariant set consisting of Q_0, P_0 and the orbits connecting them, then $h(M) = \bar{0}$, (the homotopy class of a point), since Q_0 and P_0 cancel when we deform a by "pulling up the valley" (see figure 5).

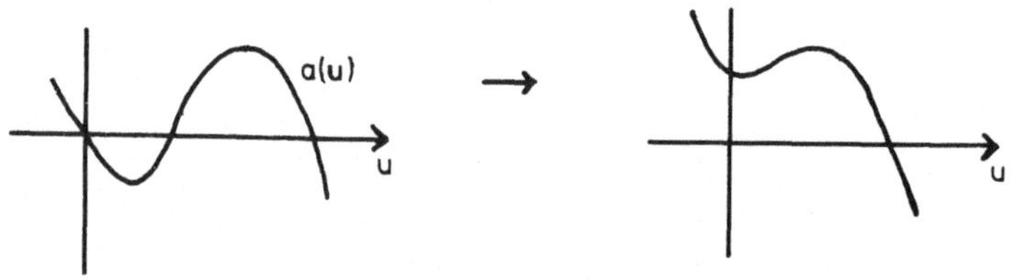

Figure 5

Now we use the following exact sequence of cohomology groups ([2] Ch 23, §D)

$$\ldots \to H^{n-1}(h(M)) \to H^{n-1}(h(P_0)) \to H^n(h(Q_0)) \to H^n(h(M)) \to \ldots ,$$

or

$$\ldots \to 0 \to H^{n-1}(\Sigma^0) \to H^n(\Sigma^k) \to 0 \to \ldots .$$

Thus for each n, $H^{n-1}(\Sigma^0) = H^n(\Sigma^k)$, so that $k = 1$. Similarly, we can show that $h(Q_2) = \Sigma^1$. Since Q_1 and Q_2 cancel when we push down the valley in ϕ (c.f. figure 4), an argument similar to the one just given shows $h(Q_1) = \Sigma^2$. This completes the proof.

Now by using these algebraic and continuation methods, we can show that there are orbits running from Q_i to both R_i and P_i, $i = 0, 1, 2$. Thus the complete solution space can be depicted as in figure 6. Namely, there is an invariant two-dimensional manifold in R consisting of orbits connecting rest points as depicted. All other solutions tend to one of these rest points as $t \to +\infty$. Figure 6 can be considered as a "dynamic" pattern formation.

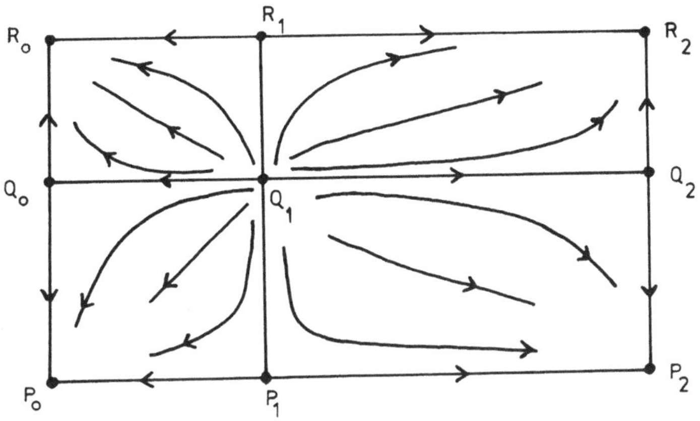

Figure 6

§5. It is reasonable to inquire as to how the solutions of (1),(2) behave as we relax the assumption $\alpha >> 1$. We have not considered this problem analytically, but instead, we have run some computer experiments, and it is this which we wish to discuss here.

We consider the following finite-difference scheme:

$$\frac{u_i^{j+1} - u_i^j}{\Delta t} = \alpha \frac{u_{i+1}^j - 2u_i^j + u_{i-1}^j}{(\Delta x)^2} + \frac{1}{2}\left\{a(u_{i+1}^j)f(z_{i+1}^j) + a(u_{i-1}^j)f(z_{i-1}^j)\right\}$$

$$\frac{z_i^{j+1} - z_i^j}{\Delta t} = \gamma \frac{z_{i+1}^j - 2z_i^j + z_{i-1}^j}{(\Delta x)^2} + \frac{1}{2}\left\{b(u_{i+1}^j)g(z_{i+1}^j) + b(u_{i-1}^j)g(z_{i-1}^j) + \phi(z_{i+1}^j) + \phi(z_{i-1}^j)\right\}$$

where $\frac{\Delta t}{(\Delta x)^2} \cdot \max(\alpha,\gamma) \leq \frac{1}{2}$.

As initial data, we take the following four functions (on $|x| < L$) :

A. $(u(x,0),z(x,0)) = (\cos \frac{\pi x}{L}, \sin \frac{\pi x}{L})$

B. $(u(x,0),z(x,0)) = (2 \sin \frac{\pi x}{L}, \frac{3}{2} \sin \frac{\pi x}{L})$

C. $(u(x,0), z(x,0)) = (2 \sin \frac{2\pi x}{L} , \frac{3}{2} \sin \frac{2\pi x}{L})$

D. $(u(x,0), z(x,0)) = (2 \cos \frac{3\pi x}{L} , \frac{3}{2} \sin \frac{3\pi x}{L})$.

The corresponding solutions are given in the figures below. These (and many other calculations) indicate that for large t , i) z is always near zero for all data if α is sufficiently small, and ii) u approaches $u(x,0)$ as $t \to \infty$ and $\alpha \to 0$. In particular, the solutions do not go to steady-state solutions. It would be interesting to rigorously investigate the case where $0 < \alpha << 1$.

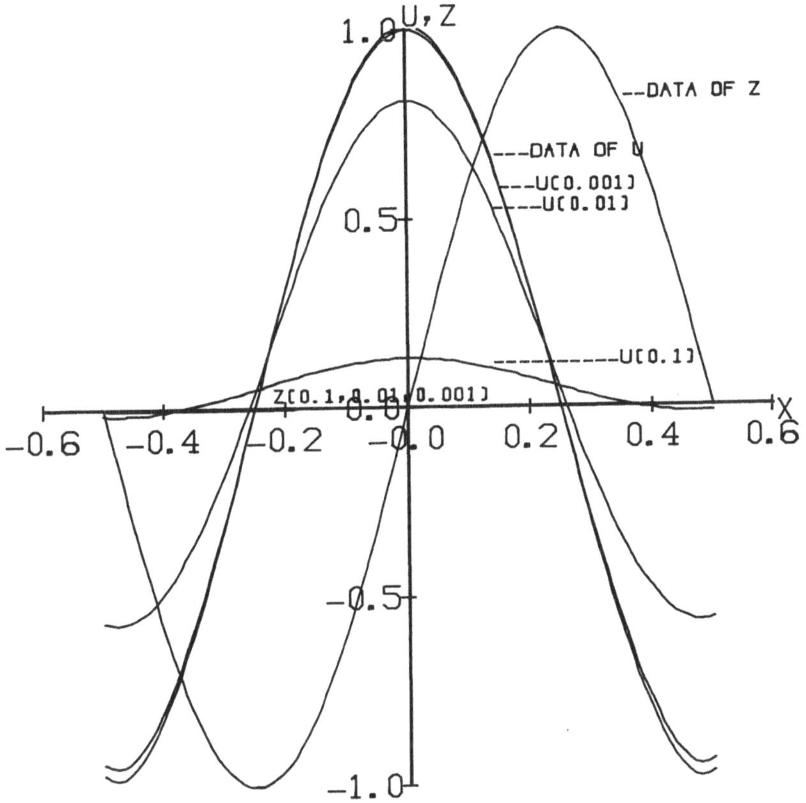

Figure 7 (case A)

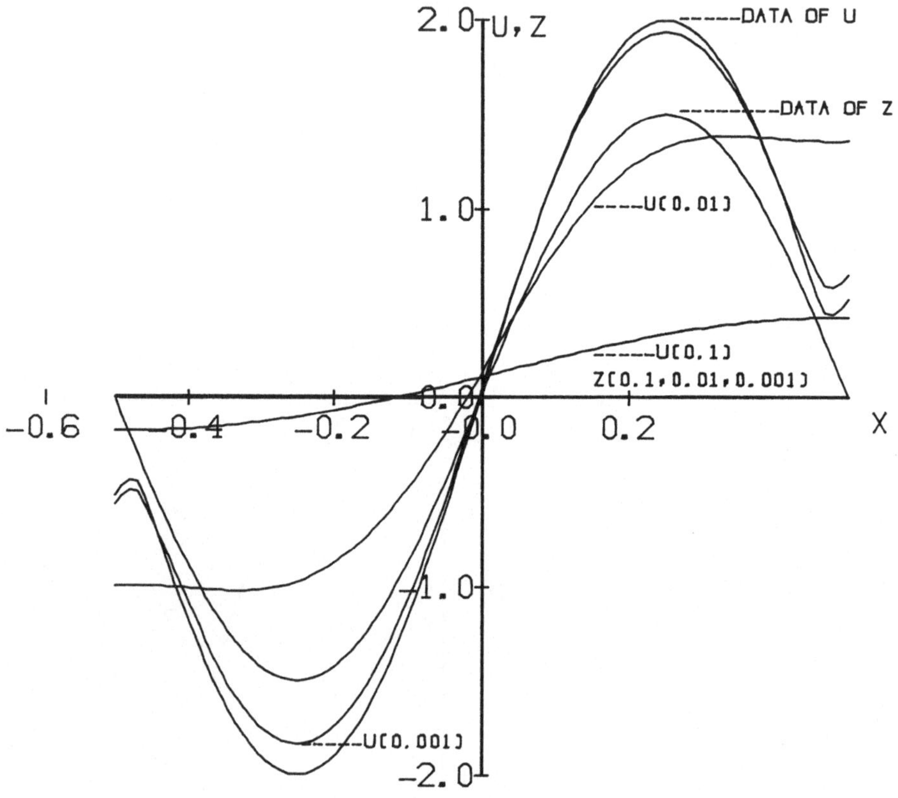

Figure 8 (case B)

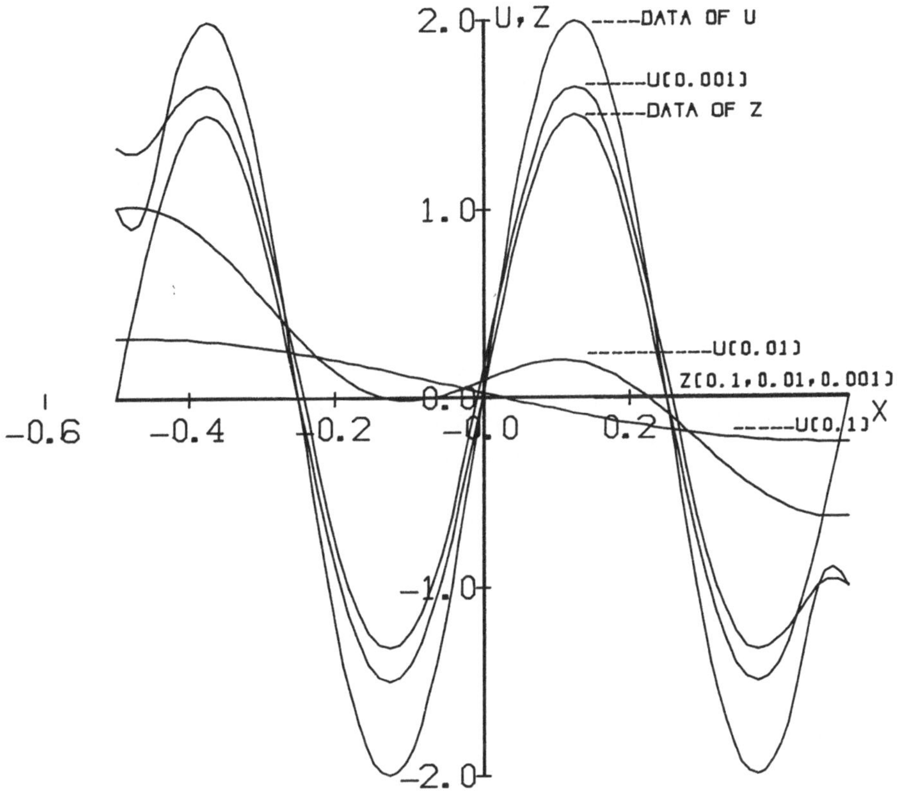

Figure 9 (case C)

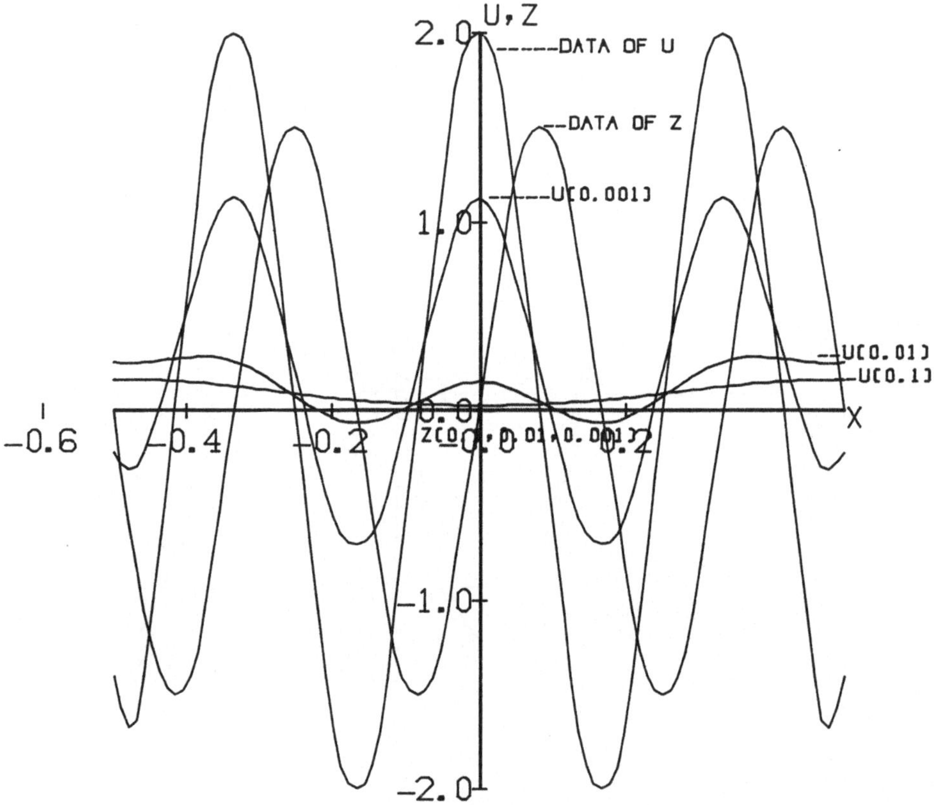

Figure 10 (case D)

BIBLIOGRAPHY

1. Smoller, J., The complete solution space for a system of reaction-diffusion equations, Proc. UAB International Conf. on Diff. Eqns., Spring 1983, to appear.

2. Smoller, J., Shock Waves and Reaction-Diffusion Equations, Springer-Verlag, New York, 1983.

BRANCHING PROCESSES WITH INTERACTION AS MODELS OF CELLULAR PATTERN FORMATION

P.Tautu
Institute for Documentation,Information and Statistics
Department of Mathematical Models
German Cancer Research Center
Heidelberg

A mathematician,like a painter or a poet,
is a maker of patterns.
G.H.Hardy

O.Introduction

The branching processes with interference (BPI) have been introduced by R.A.Holley and T.M.Liggett(1975)as dual(adjoint)processes to certain stochastic infinite systems of interacting particles (IIP systems,for short). Their generalization as 'branching processes with interaction' is due to K.Schürger(1980,1981)who studied the asymptotic geometrical behaviour of these Hunt processes. The arguments for this approach are the investigation of ergodicity conditions and the existence of stationary measures for different classes of IIP systems,e.g., the proximity processes. The ergodic properties of the latter should correspond to long-time behaviour of BPI and,analogically,questions about the existence of stationary measures can be turned into questions about harmonic functions for BPI. For example,it has been proved (Liggett,1977,Th.3.2.1)that,if ξ_t is a proximity process and $\hat{\xi}_t$ is its dual, a BPI in this case,ξ_t is ergodic iff $P^{\Lambda}\{\hat{\xi}_t \neq \emptyset\} \to 0$ as $t\to\infty$ for all finite sets Λ of a countable set S. Also,by introducing a harmonic function for the dual $\hat{\xi}_t$,i.e.,$h(\Lambda) = \lim_{t\to\infty} P^{\Lambda}\{\hat{\xi}_t \neq \emptyset\}$,one can define the ergodicity conditions for some IIP systems,like contact processes. Notwithstanding the fact that in no case do the results appear to even come near to best possible,there is one class of processes where the use of the duality approach gives complete results. These processes belong to the class of 'voter models',in fact a subclass of proximity processes(see Liggett,1980).

The present paper has the following objectives:

(i)The definition of BPI as duals of IIP systems by using the theory of duality for Markov processes introduced by G.A.Hunt(1957) with the refinements by M.A.Garcia Alvarez and P.A.Meyer(1973). Duality is expressed in terms of resolvents of semigroups. The idea at the back of it is that one can construct a BPI (as a Hunt process) if one knows the resolvent of an IIP system and the corresponding duality relationship (2.3). It must be pointed out that there are several notions of duality(see,e.g.,Getoor and Sharpe,1981,for dual density). As a matter of fact,the duality approach,which employs a 'percolation substructure' (Harris,1978),is frequently used but,in spite of some interesting results (see for biological processes Bramson and Griffeath, 1981;Donnelly,1983),it will not be discussed here.

(ii)The characterization of stochastic semigroups with branching and interaction properties. Such mathematical objects suggest the construction of strong Markov processes 'with creation and interaction' analogous to the Markov processes 'with creation and annihilation' (Helms,1967,1970;Silverstein,1968;Nagasawa,1969). The latter are generalized Markov processes in which both the starting time and the terminal time are random variables (see Leviatan,1973).

(iii)The possibility of approximating BPI by some measure-valued processes.

The underlying biological motive of this paper is the following: It is suggested that IIP systems may be adequate stochastic models for pattern formation processes that include positional information,growth and differentiation. Because of the complexity of such 'stochastic morphogenetic systems',their dual processes (if they exist) could be more tractable mathematically.

1.Stochastic models for pattern formation processes

Intuitively,what is assumed to happen in a BPI is that if there is a particle located on a lattice point $x\in\mathbb{Z}^d$,$d\geq 1$,is that this particle can split into $j>1$ identical particles which will move to the adjacent sites of x. The regularity rule of exclusion of multiple occupancy does not permit more than one cell to be located at one site at a time so that the progeny must search for sites. If the neighbouring sites are not vacant,the new particles must interact with the old occupants ; the result of this interaction may be either annihilation (both colliding particles die) or coalescence (the two interacting particles fuse into a single one). Of course,the interaction with vacant sites must also be taken into consideration. In a BPI the split rate depends on the neighbourhood structure so that crowding effects can be controlled. Moreover, if a changing state mechanism exists,it is also dependent on the actual states of its nearest neighbours. The characteristic example is the "voter" behaviour : At exponential times the voter at a point x switches his mind. The rate for the voter change is equal to the proportion of his neighbours who hold an opinion opposite to his. (Note that the voter consensus is dimension-dependent,following the laws known in the theory of random walks.)

Instead of coalescence one perceives in many biological situations the selected survival of a single particle at a point be it the newcomer or the old occupant[1)]. The competition for a site is a game of survival. One can further assume that the defeated cell dies (is "killed") or is shifted to the next vacant site. For instance,in renewing cell populations the "cell loss" is a common process so that the assumption that the old occupant at x vanishes might be biologically plausible. In this case,any new occupant generates a new BPI ; the existence of a long path of occupied sites by the offspring (which illustrates the expansion of a 'descendant clone' generated by a single ancestor) would be the consequence of a weak competition or of a strong biological advantage of the newcomers,respectively. By 'descendant clone' is meant any group of clonally related cells irrespective of whether they have remained contiguous throughout development. Descendant clones reflect the cell lineage relationship within a differentiated tissue(Nesbitt,1974; West,1978). The size of descendant clone is dependent only on the number of surviving mitotic progeny. In contrast,a 'coherent clone' reflects the spatial relationship of clonally related cells at a time, and its size is a function of the degree of "cell mixing"(by interaction).

---

1)Cellular hybridization - which should correspond to coalescence - is not a frequent phenomenon occurring in vivo everywhere and in every generation of cells(see Ringertz and Savage,1976). The term 'coalescence' has been used in a biological context by P.Geddes(1880)and H.V. Wilson(1907)in their studies on the coalescence of amoeboid cells into plasmodia and on coalescence and regeneration in sponges.

An interesting biological application is suggested by the complex process of morphogenesis. The mathematical work carried out by M.Eden (1958,1961)is fundamental and can be compared with the previous work of the Cambridge mathematician A.Turing(1952)who suggested that the formation of spatial patterns during morphogenesis could be explained in terms of the instability of solutions of a reaction-diffusion system. The most important assumption in Eden's model is that the cell growth can be described by a stochastic branching process which takes place on a 2-dimensional square lattice. In Eden's own terms(1961),"we may view the successive division of cells of a colony as a branching process and examine the sequence of configurations through which the colony passes in its growth. If this sequence of configurations can be described by a recursive relation,then a generating process[1] can be formulated for the branching process,and the probability of certain subsets of the set of possible configurations may be computed." In this model,cells are assumed to be immortal - which amounts to the assumption that an occupied site will always remain occupied[2].

Clearly,this is a simple spatial growth model that can describe the expansion of mosaic systems and of systems that regulate itself by epimorphosis,i.e.,systems showing size dependence of pattern. An identical observation has already been made for Turing's model(Bard and Lauder, 1974;Othmer and Pate,1980). As the considered spatial growth model is a simple one,there are no assumptions on cell differentiation or migration,cell size variation as well as on preferential orientation,except that it is peripheral - "a condition due most likely to the fact that diffusion of nutrient is too slow to permit any large number of cell divisions in the interior of the growth"(Eden,1958). In fact,the periph-

---

1)This is a misprint : the problem was the construction of a generating <u>function</u>,a very useful tool in the algebra of sequences (in this case,sequences of configurations). Eden's 2-dimensional morphogenetic problem is similar to the 2-dimensional Ising problem,that is,the deduction of a generating function for the number of configurations (resp. labeled admissible subgraphs) with a given number of connected cells (resp.edges). See for the Ising problem,e.g.,M.Kac and J.C.Ward(1952). R.C.Read presented in 1962 a "cell-growth problem" as the expansion of connected square-celled 'animals',asking how many connected animals A_m with m cells there are,up to isomorphism. The problem of recursivity for growth patterns has been studied by S.Ulam(1962)within the framework of cell automata theory (see also Schrandt and Ulam,1970). For combinatorial analysis the reader is referred to W.K.Whitten(1978)who studied 1-,2-,and 3-dimensional random mosaics.

2)This kind of formal 'immortality' must be distinguished from the immortality assumed to be characteristic for stem cells and malignant cells (in vitro). In the last case,cellular immortality might be a result of recessive disfunctions or alterations in the genetic program that limits the division of normal cells(Pereira-Smith and Smith,1983).

eral growth is a consequence of model's technical assumptions. If the cells are immortal,the only change that can occur is in the boundary of the occupied and vacant sites.

Eden's model actually deals with the acquisition of unoccupied sites of the 2-dimensional square lattice. His computations show that the configurations with the largest number of connected cells are most probable,while configurations with many short branches are more probable than those with a few long branches. Large configurations will be essentially circular in outline[1)]; they have a high density,i.e.,include very few "holes" and short "tentacles". Yet,even in more complicated IIP systems with many types of particles - e.g.,carcinogenetic systems - the developed heterogeneous 'tissue' tends to be "round" and "solid" for large t (Schürger and Tautu,1976,Conjecture 3.2).

In a seminal paper,D.Richardson(1973)studied the problem of the asymptotic shape for a special class of stochastic spatial growth models in $\mathbb{R}^d$. The main feature of these models is,roughly speaking,that they describe the change of 'white' cells (interpreted as 'normal'cells) into 'black' cells (interpreted as 'transformed'cells) whenever the white cells have black neighbours. In other words,a growth process of this type realizes a configuration B of occupied white-black lattice points if B is accessible from a preceding configuration A. The condition of accessibility clearly implies that a black cell in B has already been black in A or had a black neighbour in A.

The simplest example of this class of models is Richardson's model Gp : At each time unit a white cell becomes a black one,with probability p,if there is at least one black neighbour. For this model.D.Richardson proved that the asymptotic shape is circular. As a function of p and t,the shape changes from a diamond to a circle (see also the computer experiments done by Kindermann and Snell,1980,pp.88-89);as p decreases,the roughness of the boundary seems to increase. Formally, D.Richardson showed that there exists a norm $N(\cdot)$ on $\mathbb{R}^d$ such that for any $0<\varepsilon<1$,the configuration of occupied sites at time t,A_t,contains an N-ball of radius $(1-\varepsilon)t$ and is itself contained in an N-ball of radius $(1+\varepsilon)t$. The probability of this double circular inclusion tends to 1 as $t\to\infty$. In other words,if A_t is the configuration at time t,then the set A_t/t of sites will approach a circle whose radius increases at a linear rate. (Versions of Richardson's main theorem can be found in H.Kesten

---

1)In his combinatorial approach,R.Ransom(1977)suggests that his computer model with non-random division essentially simulates radial cell division,because all cells attempt to divide in the direction of the nearest free edge.

(1973)and J.Hammersley(1974,Note 7);this is "the strong law of large configurations" by K.Schürger(1979,1981).)

In order to point up the convergence of different mathematical approaches (and also some confusion in the nomenclature !),one must take notice of the fact that in the class of models considered by D.Richardson there are included the Eden model above as well as the Williams-Bjerknes model(1972)for the spread of a malignant clone in the epithelium. The rough description of the latter is as follows : The $\mathbb{Z}^2$ lattice contains white cells which can split with exponential rate 1. At t=0 only one cell(located at site x=(0,0))is black,but this cell and its descendants will have a certain birth advantage,dividing themselves with rate $\kappa > 1$. The case $\kappa = \infty$ is identified with the Eden model for morphogenesis and also with the Morgan-Welsh model(1965)for Poissonian spread of an infection. For a generalized Eden model,K.Schürger(1980, Th.4.2;1981,Th.5.1)found that the growth radii are between $(1-\varepsilon)n^{1/d}$ and $(1+\varepsilon)n^{1/d}$,$d\geq 2$; also,the asymptotic shape for the Williams-Bjerknes model has been proved by M.Bramson and D.Griffeath(1980 a,b;1981).

Now,we must notice that Richardson's Gp model can be identified with a simple,discrete time IIP system;for its analysis one can use the existent relationships between three mathematical theories for spatial growth,namely percolation,contact processes(a class of IIP systems)and branching random walks(see Durrett and Liggett,1981). In this example the dual process has been called 'coalescing random walk with nearest neighbour births' ; actually,it belongs to the BPI family. By analogy with the above,the Williams-Bjerknes model is identified with a 'biased voter model'(Schwartz,1977). As a matter of fact,it is contained in a large class of spatial cell growth models - that includes multivariate cases and competition - which have been considered in 1976 by K.Schürger and P.Tautu. (The asymptotic shape has been speculated in Conjecture 3.2 (1976) and Theorem 17.1 (1977) by the same authors.) Throughout this paper I will use the suggestive name of 'carcinogenetic system' for this model,in (biological) contrast with the hereafter discussed 'morphogenetic systems'.

It is my purpose to suggest that certain IIP systems and their duals BPI may be adequate stochastic models to describe and analyse the pattern formation process as it is conceived by theoretical biologists. Pattern formation is in their interpretation the process of specifying the spatial pattern of cellular differentiation. "In very general terms, the pattern problem may be stated as follows:given an ensemble of more or less identical cells,how can states be assigned to these cells such that,when they undergo molecular differentiation,the cells will form a well defined spatial pattern...A suggested solution to the pattern problem is that the cells are assigned positional information which effectively gives them their position in a coordinate system,and this positional information is used to determine the cell's molecular or cytodifferentiation"(Wolpert,1971).

A mindful re-reading of the above description of BPI behaviour will suggest the contribution of this type of stochastic models to highlighting some conceptual problems in the theory of spatial cellular differentiation.

(I)Firstly,BPI(as well as the associated IIP systems)may be regarded as models of the spatial organization of multiplicative biological systems. Although cell division,growth and size are closely linked to pattern formation,the growth process has a place apart in the conception of many embryologists who consider that "during the evolutionary change from morphallaxis to epimorphosis,the same basic mechanisms of pattern formation were preserved,and growth was just added on top as a complicating factor. Growth did not,therefore,replace already existing mechanisms of pattern formation"(Maden,1981). Or,in the stochastic models considered,growth and interaction between progenies represent the pattern generators. In the univariate case,i.e.,no differentiation,as in the Eden model,growth is essential for the creation of a form. Such IIP systems are called 'birth-death systems'(Liggett,1980;Cocozza and Kipnis,1980)or,simply,'growth models'(Durrett and Griffeath,1982). Incontestably,in some stages of morphogenesis there is cell death(see the survey by Hincliffe,1981)and cell growth of invasive type(e.g.,the invasion of mesenchyme into 3-dimensional collagen gels:Runyan and Markwald,1983). If growth is not assumed,one may use another class of IIP systems,i.e.,the exclusion processes : they are particle motion processes in which the basic motion of each particle is that of a continuous time Markov chain on the lattice[1].

(II)Secondly,in IIP systems and BPI the assumption of neighbourhood relationships is decisive for growth and pattern formation. Although this assumption can be simply evidenced by the existence of gap junctions(that appear in the embryonal eight-cell stage !),it has not primarily been considered as a form of positional information because

1)A deterministic analogue can be found in a paper by G.Rogers and N.S.Goel(1978)on cell-sorting,cellular migration through a mass of cells and contact inhibition.

of the gradient paradigm[1]. However,in some new theoretical models there are assumptions having resemblance to those existing in the voter model: the 'spatial averaging property' (Slack,1980) expresses the tendency of a subgroup of embryonic cells to adopt the same state as in the majority of neighbouring territories. Thus,the construction of cell contact models - where the transmission of information is the result of local or at least short-range events - can be performed by the application of 'contact processes',particularly the basic (nearest neighbour)contact processes on the 3-dimensional lattice. They exhibit critical phenomena, an important property that presumably is common to all morphogenetic systems. I speculate that the formation of compartments (and presumably that of islands in carcinogenetic systems) should result from such phenomena[2]. As it is known,compartments define a group of cells(polyclone), usually in small numbers,whose progeny never subsequently give rise to cells in adjacent compartments(Crick and Lawrence,1975). Taking into account the cell competition within compartments(Simpson,1979,1981),one can imagine that the construction of (multivariate) BPI with random birth and death times would be adequate. Yet,the basic contact processes are self-dual,that is,the BPI have exactly the same probabilistic structure as the original process(Griffeath,1981;see also Holley and Liggett,1981;Griffeath and Liggett,1982),so that such BPI may have the power of a local description.

(III)Finally,I will consider the problem of pattern formation in a three-dimensional space. It seems that the occurrence of simultaneous determinations of positions in three dimensions is,for biologists,an open question(McMahon and West,1976). As S.V.Bryant(1982)states,"the reason for even considering the possibility of two dimensional positional information in three dimensional structures is that in the development of an embryo,the interaction between sheets of cells organized in two dimensions where one sheet transfers information to the other,is a major recurring theme". Apparently,space for embryologists is 2-dimensional(e.g.,proximo-distal axis,antero-posterior polarity,the graph of

---

1)Although L.Wolpert envisaged "a mechanism of cell counting"(1969) and states of the cell membrane(Wolpert and Gingell,1969),he still assumed that the interpretation of cell position is unaffected by the behaviour of adjacent cells. He confessed,however,that this is "a rather extreme and perhaps simplistic view"(1977). Moreover,"from the point of view of cell-to-cell communication it is very important to realize that models of positional information can be based on the movement of small molecules between cells. The only interactions required are those necessary to set up the system of positional values and there need be no interaction between the differentiating regions as such"(1978).

2)Recent results on gap-junctional communication compartments(Weir and Lo,1984)suggest the hypothesis that metabolic(but not morphologic) instabilities may be correlated with the differentiation of cell types.

morphogen concentration,etc.). The tendency is to use other coordinate systems(i.e.,the polar coordinates)and to refine them by using topological arguments(Lewis,1981). Also,the search for a different metric(Gierer,1981)has been suggested. There are,however,old and new mathematical results corroborating the idea that a cell system growing in a 3-dimensional space behaves itself unlikely than in a two-dimensional space (see,e.g.,Tautu,1978).

Remark : IIP systems may be models for some other multiplicative processes with pattern formation like virus plaque formation(Schwöbel et al.,1966;see also Koch,1964)or hemolytic plaque formation(DeLisi and Bell,1974). A membrane model called 'the dynamic receptor pattern generation model'(Koch et al.,1979)can also be represented by a configuration process.

2.The background material for the duality problem

It is known that the argument leading to the investigation of dual processes is that each Markov process can be considered in two directions of time. The "traditional pair"$\{\xi_t,\hat{\xi}_t\}$ of Markov processes in duality can be thought of as the ancestors of a single stationary two-sided Markov process with random birth and death times. The transition probability function $(P_t)_{t\geq 0}$ of the original process will give the transitions forward in time and the function $(\hat{P}_t)$ of the dual process the transitions backward in time. The processes ξ_t and $\hat{\xi}_t$ are in duality relative to a σ-finite measure ν if their stationary transition functions are connected by the relation

$$\nu(dx)P_t(x,dy) = \nu(dy)\hat{P}_t(y,dx). \tag{2.1}$$

In general,no relation between the paths of ξ_t and $\hat{\xi}_t$ is requested (Dynkin,1978)but the path connection can be established with the aid of time reversion. This is the most fundamental symmetry operation of dynamics. However,it has been remarked that by reversing time one ignores if the process $\hat{\xi}$ (that retains the Markov property) admits a transition probability ; if $(\hat{P}_t)$ exists,it is non-homogeneous in time. Moreover, $(\hat{P}_t)$ will not depend only on (P_t) but also on the entrance law of ξ_t.

The starting point in the case of IIP systems is the following. Suppose,for example,that$\{\xi_t\}_{t\geq 0}$ and $\{\hat{\xi}_t\}_{t\geq 0}$ are two Markov processes with state spaces $\Xi = X^S$ and,respectively $\Xi_o = \{F\subset S:F \text{ is finite}\}$, S countable, X finite sets ; let $f(\xi,\hat{\xi})$ be a real-valued bounded measurable function on $\Xi\times\Xi_o$. Then ξ_t and $\hat{\xi}_t$ are said to be dual to one another with respect to f if

$$E_A[f(\xi_t,F)] = E_F[f(A,\hat{\xi}_t)] \ , \ A\in\Xi, F\in\Xi_o, t\geq 0 \tag{2.2}$$

(Liggett,1977). Thus,if $\{f_F(\cdot)\}$ is a determining class of functions for the process ξ_t,then $\hat{\xi}_t$ can be used to study ξ_t via (2.2). The suggestion for using dual processes can be found in (Spitzer,1970 - for symmetric simple exclusion processes);T.Harris(1976)used the term of 'associate' processes. Now,if one takes into account that the original process develops on a geometrical structure - as in the case of IIP on $\mathbb{Z}^d$ - then the dual process will develop on the dual lattice by an operation of space-time reversion. Also,if one takes into account the usual definition of duality in terms of resolvents,the dual BPI is a Hunt process which can be constructed given its resolvent or semigroup. The modus operandi to follow - if one builds up a Hunt process from its resolvent- may be found in (Stoica,1983) but the complete construction procedure with respect to a certain fine topology has been introduced by M.G.Shur (1977,Th.1).

Let us introduce the following essential notation and terminology which partially follows Blumenthal and Getoor(1968)-referred as[BG].

(I)S denotes the state space of the original process ξ (all the objects specifying the dual process ξ will carry the symbol "^"). S is assumed to be a locally compact space with a countable basis(LCCB).

(II)$\tilde{S}$ = S$\cup\{\Delta\}$,where Δ is a point isolated from S (Δ is called 'cemetery','graveyard' or 'coffin'). If S is a locally compact Hausdorff space,$\tilde{S}$ is a compact Hausdorff space(the Alexandroff one-point compactification). A topological space is a Lusin(resp.U-)space if it is homeomorphic to a universally(resp.Borel)subspace of a compact metric space.

(III)$\mathcal{S}=\mathcal{B}(S)$,the set of the minimal σ-field of sets of S containing the open(or closed)sets of S(i.e.,the Borel algebra in S).

(IV)$\mathcal{M}(S)$,the family of all probability measures on S. $\mathcal{M}_R(S)$ is the family of Radon measures on $(S,\mathcal{S})$,i.e.,all measures ν such that $\nu(K)<\infty$ for every compact K. Every Radon measure is continuous,σ-finite and regular,that is,for $A\in\mathcal{S}$,

$$\nu(A) = \sup\{\nu(K):K\subset A,K \text{ compact}\}= \inf\{\nu(B):B\supset A,B \text{ open}\}.$$

The space of Radon measures on S(assumed to be LCCB)will be endowed with the vague(weak)topology. It is Polish.

(V)$\mathcal{S}^*=\bigcap_{\mu\in\mathcal{M}}\mathcal{S}^\mu$ is the σ-algebra of universally measurable sets over $(S,\mathcal{S})$. $\mathcal{S}^\mu$ is the completion of $\mathcal{S}$ with respect to a measure μ on $(S,\mathcal{S})$; it consists of all sets $A\subset S$ for which there exist $B_1,B_2\in\mathcal{S}$ such that $B_1\subset A\subset B_2$ and $\mu(B_2-B_1)=0$. The set A is called "nearly Borel";the σ-field of nearly Borel sets will be denoted by $\mathcal{S}^\circ$. We have $\mathcal{S}\subset\mathcal{S}^\circ\subset\mathcal{S}^*$.

(VI)$C(S)$ is the vector space of real valued continuous(rvc)functions on a topological space S, which are bounded; we let $\|f\| = \sup\{|f(x)|: x\in S\}$. Then C becomes a Banach space and its norm is called the supremum (uniform, convergence) norm. Since $\tilde{S}$ is compact, every f on $\tilde{S}$ is bounded. $C_o(S)$ is the space of rvc functions which tend to 0 at infinity :

$$C_o(S)=\{f\in C(S), \forall \varepsilon>0, \exists\ K \text{ compact}, |f(x)|<\varepsilon, \quad x\in S\setminus K\}.$$

Further, $C_k(S)$ will denote the class of rvc functions with compact support (vanishing outside some compact K). Clearly, $C_k\subset C_o\subset C$. Each of C, C_o and C_k are algebras and, also, lattices ordered by $f\geq g$.

(VII) $\Omega = D([0,\infty), S)$ is the set of all mappings $\omega: \mathbb{R}^+\to S$ which are right-continuous, have left limit and possess a lifetime Θ, namely

$$\omega(t) \neq \Delta \text{ for } t<\Theta(\omega) \quad \text{and } \omega(t) = \Delta \text{ for } t\geq\Theta(\omega).$$

Ω is endowed with the usual Skorohod topology which makes Ω a separable metric space (e.g., Polish). F is a σ-field in Ω.

(VIII) Operators and generators. By a semigroup of operators on a Banach space $\mathbb{B}$ it is generally meant a collection $(R_t)_{t\geq 0}$ of linear operators on $\mathbb{B}$ which satisfy the following conditions : (1) $R_o=I$, (2) $R_{s+t}= =R_sR_t$ for all $s\geq 0, t\geq 0$, (3) $\|R_t\|\leq 1$, (4) $R_tf\to f$ as $t\downarrow 0, f\in\mathbb{B}$. The commonly used Banach spaces are $L^1(S,\mathcal{S},\mu)$, the space of real valued, μ-integrable functions and $L^\infty(S,\mathcal{S},\mu)$, the space of real valued, μ-essentially bounded functions, with their respective usual norms. A linear mapping $R:L^1\to L^1$ will be called a Markov operator if $Rf\geq 0$ and $\|Rf\|=\|f\|$, for $0\leq f\in L^1$.

(a) $(P_t)_{t\geq 0}$ is a strongly continuous Markov semigroup of transition operators on L^1, associated with a standard transition probability matrix function $\underset{\sim}{P}(t)$ with elements $P_t(\cdot,\cdot)$.

(b) With $P(\cdot,\cdot)$ we may associate an operator T defined by

$$\int_A Tf d\mu = \int_S fP(\cdot,A)d\mu \ , \ f\in L^1, \ A\in\mathcal{S}$$

which is a positive linear operator of norm 1 on L^1 provided only that the measure $\int\mu(dx)P(x,\cdot)$ is absolutely continuous with respect to μ. T_t is called a contraction semigroup if for all $t\geq 0$, $\|T_tf\| \leq \|f\|$, $f\in L^1$.

(c) Let $(T_t)_{t\geq 0}$ be a uniformly continuous semigroup. Then there exists a bounded operator A such that $T_t=\exp\{tA\}$ for $t\geq 0$. The operator A is given by the formula $A = \lim_{h\downarrow 0}(T_h-I)/h$. It is called the differential operator or the infinitesimal generator of the semigroup T_t.

(d) The resolvent U^α, $\alpha>0$, is a bounded linear operator of norm $\|U^\alpha\| \leq\alpha^{-1}$, defined by

$$U^\alpha f = \int_0^\infty e^{-\alpha t} \int f(y)P_t(x,dy)dt = \int_0^\infty e^{-\alpha t}T_tfdt$$

Moreover, $U^\alpha=(\alpha I-A)^{-1}$.

The resolvent has as property the celebrated "resolvent equation" $U^\alpha-U^\beta=(\beta-\alpha)U^\alpha U^\beta$, for every $\alpha,\beta>0$.

(e) The operator $Vf=s\cdot\lim_{\alpha\downarrow 0}U^\alpha f$ is called the potential (or cogenerator) of the semigroup T_t iff the limit exists for f in a dense subset of

$\mathbb{B}$. Also, $V = -A^{-1}$.

(IX) A function $f \in S, 0 \le f \le \infty$, is called α-superaveraging (supermedian) relative to (P_t) iff $e^{-\alpha t} P_t f \le f, t \ge 0$, and it is α-excessive iff in addition $f = \lim_{t \downarrow 0} e^{-\alpha t} P_f f, \alpha > 0$. Then f is α-excessive for the resolvent $U^\alpha f$ if $\lim_{\alpha \to \infty} \alpha U^\alpha f = f$.

(X) Exceptional sets. A set $A \subset S$ is thin if there exists a set $E \in S^\circ$ such that $A \subset E$ and $P^x\{T_E = 0\} = 0$ for all $x \in S$, where $T_E = \inf\{t > 0, \xi_t \in E\}$ is the first hitting time. A set $A \subset S$ is semipolar if it is contained in a countable union of thin sets. A semipolar set A in S^* is of potential zero : $U^\alpha(x, A) = 0, x \in S$. A set $A \subset S$ is polar if there exists a set $E \in S^\circ$ such that $A \subset E$ and $P^x\{T_E < \infty\} = 0$, for all $x \in S$ (see [BG, p.79]). Example: If A is semipolar, then $\xi_t \in A$ a.s. for only countably many values of t.

Let us now assume that there are two standard processes ξ and $\hat{\xi}$ (relative to Δ) with semigroups (P_t) and $(\hat{P}_t)$, respectively, defined canonically on Ω. It is assumed that they have the same state space S which may be Hausdorff or Lusin. Also, U^α and $\hat{U}^\alpha$ are their resolvents. The two processes ξ and $\hat{\xi}$ are in duality relative to a reference measure $\nu \in M_R(S)$, if, for each $\alpha > 0$, and all $f, g \in S^*$ the following two conditions are satisfied ([BG, p.254]):

(i) The measures $U^\alpha(x, \cdot)$ and $\hat{U}^\alpha(\cdot, x)$ are absolutely continuous with respect to measure ν for each $\alpha > 0$ and $x \in S$.

$$\text{(ii)} \int_S f(x) U^\alpha g(x) \nu(dx) = \int_S f \hat{U}^\alpha(x) g(x) \nu(dx). \tag{2.3}$$

Condition (i) represents Hypothesis (L) by P.A.Meyer (1962); if it is not considered, then ξ and $\hat{\xi}$ are said to be in weak duality. The left and right hand sides of (ii) are, respectively, the Laplace transforms of $\int f(x) P_t g(x) \nu(dx)$ and $\int g(x) \hat{P}_t f(x) \nu(dx), f, g \in S^*$. Since f and g are continuous and ν-integrable, both transforms are equal and can be written as

$$E_\nu[f(\xi_o) g(\xi_t)] = \hat{E}_\nu[g(\hat{\xi}_o) f(\hat{\xi}_t)], \quad f, g \in S^*, \ t \ge 0 \tag{2.4}$$

where $E_\nu[\eta] = \int E_x[\eta] \nu(dx)$. Condition (2.3) implies that the semigroups (P_t) and $(\hat{P}_t)$ are also in (weak) duality. An immediate consequence of (2.4) is that the reference measure ν is also excessive (resp. co-excessive), i.e., $\int P_t(x, A) \nu(dx) \le \nu(A)$, for all $t \ge 0$ and $A \in S$ (Walsh, 1972; Mitro, 1979). If ν is assumed to be invariant under (P_t), one has (2.1) : the study of dual processes relative to an <u>invariant</u> reference measure originates in A.N.Kolmogorov 1936 paper.

Following a theorem given by M.A.Garcia-Alvarez and P.A.Meyer (1973), the transition semigroup $(\hat{P}_t)$ of the dual process $\hat{\xi}$ on a Lusin space has the following four properties :

(a) The process $\{\hat{\xi}_t\}_{t \ge 0}$ is moderate Markovian with respect to $(\hat{F}_t)$, [i.e.,almost all its paths have left limits(Smythe and Walsh,1973)].

(b)For each $x\in S$ there exists a Markov process which is left continuous and moderate Markovian admitting $(\hat{P}_t)$ as transition semigroup and $(\varepsilon_x\hat{P}_t)_{t>0}$ as entry law.

(c)The resolvent $\hat{U}^{\alpha}$ has a dual U^{α} with respect to the reference measure $\nu = \alpha U$ (see Condition (2.3) above).

(d)For any bounded positive function f and $\alpha>0$, $\hat{U}^{\alpha}f$ is finely continuous outside a polar set [i.e.,this set is not 'charged' by ν],and $\lim_{\beta} \beta U^{\beta}\hat{U}^{\alpha}f$ exists.

Thus $(\hat{P}_t)$ is chosen taking into account that U^{α} and $\hat{U}^{\alpha}$ must be compact with base ν ,and a polar set A must exist such that all functions $\hat{U}^{\alpha}f$ (f positive bounded,$\alpha>0$) be finely continuous outside A.

If we assume that U^{α} is a sub-Markov resolvent,then a Hunt process ξ_t associated to U^{α} exists if the following conditions are satisfied (Stoica,1983,Th.4.3):

(i) $U^{\alpha}C_k(S)\subset C(S)$, for each $\alpha>0$,

(ii) $\lim_{\alpha\to\infty} U^{\alpha}f(x) = f(x)$, for each $f\in C_k(S)$,$x\in S$,

(iii) $U^{\alpha}f\in \mathcal{D}(E_{\alpha})$, for each $f\in C_k(S)$,$\alpha>0$,

where E_{α} is the cone of all α-excessive functions. (For each excessive function $f\in E$,the process $\{f(\xi_t)\}_{t\geq 0}$ is a supermartingale with respect to the family $(F_t)_{t\geq 0}$.) This allows the construction of BPI as Hunt processes starting from their resolvents.

3.Growth systems and their dual pattern-forming processes

In this section I am going to discuss the following points:

(A)the construction of some IIP systems capable of having dual processes,

(B)the definition of BPI and their resolvents.

Let $\mathbb{Z}^d=\{x|x=(x^1,\dots,x^d),x^i \text{ integer},1\leq i\leq d\}$ denote the d-dimensional square lattice on which a site $x\in Z^d$ may or may be not occupied by a particle(:cell). In most cases the infinite set of points of $\mathbb{Z}^d$ will be identified with the arbitrary(infinitely)countable set S defined in (2.I). Let N_x denote the set of the nearest neighbours of $x\in\mathbb{Z}^d$.

The elements of a finite set X will indicate the situation at each site. In the simplest case X is a two-element set,X={0,1},designating the states in which a site may be : 0 if the site is vacant and 1 if the site is occupied (with the additional characteristic of the occupant:infected,excited,malignant,etc.).

A configuration ξ is defined as the mapping $\xi:\mathbb{Z}^d\to X$. Then $\xi(x)$ denotes the situation at $x\in\mathbb{Z}^d$, e.g., $\xi(x)=1$ if $x\in\mathbb{Z}^d$ is occupied. Put $\Xi=X^{\mathbb{Z}^d}$ for the set of all configurations and provide it with the usual product topology. Remark that between Z, the set of all subsets of $\mathbb{Z}^d$ and Ξ there is a 1-1 correspondence if $|X|=2$ and denote by Ξ_o the set of finite subsets of $\mathbb{Z}^d$. The evolution in time of configurations is represented by the stochastic process $\{\xi_t\}_{t\geq 0}$.

Of course, the changes in ξ_t are influenced by the behaviour of particles located on $\mathbb{Z}^d$. Let $\xi\in\Xi$ define the initial configuration. Each particle waits a random (holding) time which is exponentially distributed with mean 1. Since the holding times have a continuous distribution only one particle moves at a time. At the end of the holding time the particle attempts to jump of its site (say, x) to the site y with probability $p(x,y)$. This is the transition function of an irreducible Markov chain (in many cases a simple symmetric random walk with $p(0,y-x)$), with $p(x,y)\geq 0$, $\sum p(x,y)\leq 1$, and $\sup_y \sum_x p(x,y)<\infty$. If $\|x-y\|=1$ and equiprobable choice of $y\in N_x$, $p(x,y)=1/2d$. The change at each site $x\in\mathbb{Z}^d$ happens with rate $c(x,\xi)$, that is, $c(x,\xi)h + o(h)$ is the probability that if at a certain instant the IIP system is in state ξ, a particle at $x\in\mathbb{Z}^d$ will attempt a transition during the small interval h. The evolution of the considered system can be described by

$$P_\xi\{\xi_{t+h}(x) \neq \xi_t(x)\,|\,\xi_t=\xi\} = c(x,\xi)h+o(h)$$
$$P_\xi\{\xi_{t+h}(x) \neq \xi_t(x), \xi_{t+h}(y)\neq \xi_t(y)\} = o(h), \text{for } x\neq y$$

as $h\downarrow 0$ and $x\in\mathbb{Z}^d$.

The configuration changes are denoted as follows :

$$\xi_{y,z}(x)=\begin{cases}\xi(x) & \text{if } x\neq y,z\\ \xi(y) & \text{if } x=z\\ \xi(z) & \text{if } x=y\end{cases} \qquad \xi_y(x)=\begin{cases}\xi(x) & \text{if } x\neq y\\ 1-\xi(x) & \text{if } x=y\end{cases}$$

if $X=\{0,1\}$. Then $\xi_{y,z}=\xi$ if y and z are both vacant or both occupied. Otherwise $\xi_{y,z}$ represents a transition from one site to another. For more complicated X, in a multivariate case, the reader is referred to (Schürger and Tautu, 1976).

I will call 'growth systems' (syn.: growth model (Durrett and Griffeath, 1982), birth-death system (Liggett, 1981)) the particle systems defined in

Definition 1. Let $\{c(x)\}_{x\in\mathbb{Z}^d}$ be a collection of uniformly bounded, Borel measurable functions, $c : \Xi^2 \to [0,\infty)$, generally called speed functions, Define an operator A with rate $c(x)$ by

$$Af(\xi) = \sum_x c(x,\xi)\Delta_x f(\xi), \tag{3.1}$$

for all $f\in C(\Xi)$, where $\Delta_x f(\xi)=f(\xi^{(x)})-f(\xi)$, $\xi^{(x)}$ being the configuration with changed state at site $x\in\mathbb{Z}^d$. A growth system with speed rates $\{c(x)\}$

and initial configuration ξ ($\xi=\phi$ is an absorbing state) is a Ξ-valued strong Markov process $(\Omega,F,P,\xi_t,F_o^t,t[0,\infty))$ satisfying

(i) $P\{\xi_o = \xi\} = 1$,

(ii) $f(\xi_t) - \int_o^t Af(\xi\ )ds$ is a P_ξ-martingale

for all $\xi\in\Xi, f\in C_o$.

A probability measure P on (Ω,F) is said to be a solution to the martingale problem for c (or for A) having initial configuration ξ. It is proved(Holley and Stroock,1976)that if there is a unique solution P_ξ to the martingale problem for c,then $\{\xi_t\}_{t\geq 0}$ is a strong Feller process with respect to P_ξ. (The existence of a unique solution depends on the conditions imposed to $c(x,\xi)$ in Theorem 1.2 by T.Liggett(1972) or relation(4.6)deduced by R.Holley and D.Stroock(1976).)

Such a growth system has the following properties :

(a)it is translation invariant : $c(x,\xi)=c(0,\xi-x)$,for $\xi-x=\{y|x+y\in\xi\}$;

(b)it is attractive : if $A\subset B\in\Xi$, ξ_t with initial configuration A(i.e., $\xi_o^A=A$)and the system ξ_t with initial configuration B(i.e., $\xi_o^B=B$)can be defined on the same (Ω,F,P) in such a way that $\xi^A\subset\xi_t^B$ for all $t\geq 0$. This property is called "set-monotonicity property" and,in fact,follows from the property of additivity,

$$\xi_t^{A\cup B} = \xi_t^A \cup \xi_t^B \ , \ A,B\in\Xi,\ t\geq 0 \qquad (3.2)$$

(c)it is local : $\exists\, r<\infty$, $c(x,\xi)=c(x,\xi\ \cap\{y|x-r\leq y\leq x+r\})$. This is called "the finite range property".

The models briefly discussed in Section 1,e.g.,the Eden models,the Richardson Gp model,the Williams-Bjerknes model and some other simple contact processes belong to this group of growth systems. A special mention must be made for a new model called 'diffusion-limited aggregation' (DLA : Witten and Sander,1983)which has some similarities with the Eden model but also some different features.

Remarks:(1)Following Theorem 5.1 in (Spitzer,1971),the equilibrium state Π of a growth system is,under certain conditions,an isotropic Markov random field(or,equivalently,an isotropic nearest neighbour Gibbs state). Also,the stationary Markov process corresponding to the interaction process is time-reversible. (2)A very interesting approach can be suggested by associating two random fields with a growth system with symmetric transition density : the free(Gaussian)random field,with mean zero,and the occupation field which may give us the lifespan of a particle at each site $x\in\mathbb{Z}^d$(see Dynkin,1984). (3)The solution to the martingale problem for A can be represented as a multiple random time change of the solution for A_o, the operator for which all the speed functions are constant (Holley and Stroock, 1976; see also Kurtz, 1980).

I will give now the conditions which ensure that a growth system has a dual.

Let $\{\xi_t\}_{t\geq 0}$ be a Ξ_o-valued standard process and let (T_t) be a Markov semigroup on $C(\Xi_o)$. Then (T_t) has a dual $(\hat{T}_t)$ iff

(i) $AC_o \subset C_o$,

(ii) $\psi A g_F(\{G\}) \geq 0$, for each $F, G \in \Xi_o$, $F \neq G$.

In the inequalities (ii) there are introduced functions $\psi: C(\Xi_o) \to M(\Xi_o)$, and $g \in C_o$, defined by

$$\psi f(\{F\}) = \sum_{E \subset F} (-1)^{|E|+1} f(F^c \cup E), \quad f \in C, F \in \Xi_o$$

$$g_F(G) = \begin{matrix} 1 \text{ if } F \text{ 'touches' } G : F \cap G = \emptyset \\ 0 \text{ otherwise} \end{matrix}$$

The functions $\{g_F, F \neq \emptyset\}$ are a basis for C_o and

$$f = \sum_{F \neq \emptyset} \psi f(\{F\}) g_F , \quad f \in C_o.$$

Moreover, $\psi A g_F(\{G\})$ will be the transition intensities for $F \to G, F \neq G$, in the dual process $\hat{\xi}_t$ (Harris, 1976, Th.4.4; Schwartz, 1977).

A Markov semigroup $(\hat{T}_t)$ is called to be in duality with (T_t) if

$$T_t g_F(G) = \hat{T}_t g_G(F) , \quad F, G \in \Xi_o \tag{3.3}$$

that corresponds to the known duality formula

$$P_F\{\xi_t \cap G = \emptyset\} = \hat{P}_G\{\hat{\xi}_t \cap F = \emptyset\}, \tag{3.4}$$

analogue to (2.2). The Feller property extends (3.3) to the situation where only one of the configurations F or G is finite.

If a BPI exists as a dual of a growth system, it presumably must have some of the properties above. Indeed, the branching processes with interaction are translation invariant and also point-symmetric. All branching processes are additive in the sense that there exists the independence of lines of descent (Athreya and Kaplan, 1978). Similar with (3.2), if we start at x+y then the resulting process is equivalent to the sum of two independent branching processes, one starting at x and the other at y. If competition on the lattice is assumed, independent lines of descent does not exist (e.g., as in the case of branching processes in random environment which are examples of random processes with complete connections).

Assume $\mathbb{Z}^d = S$. Then the state space of the growth system is $\Xi = \{0,1\}^S$.

Let $\tilde{\Xi}_o=\{F|F$ is a finite subset of $S\cup\{\infty\}\}$. If at time t the dual process is in state $F\in\tilde{\Xi}_o$, then each of the particles in the configuration ξ_t attempts to spread forth to a neighbouring set G, independently of all other particles, with the intensity (3.6). If the particle at $x\in F$ is the first to split, it (or its offspring) can choose G as location set, then the sites of $G\cap(F\backslash x)^c$ will be added to the already occupied set $F\backslash x$ with probability (3.7). (For the sake of simplicity I write x instead of $\{x\}$.) Yet, at sites $G\cap(F\backslash x)$ there is competition; if y is one of these sites, only one of the interacting particles will survive with probabilities (3.8). The competition at $y\in G\cap(F\backslash x)$ takes place independently at each of the neighbouring sites. After each branching and competition the process runs again.

The approach follows R.Holley and D.Stroock (1979) and is based on the use of speed functions with absolutely convergent series. Let consider Ξ as a compact Abelian group which is the direct product of (cyclic) groups $\{0,1\}_x, x\in S$. The dual group is then identifiable with Ξ_o, with symmetric difference Δ as group operation in Ξ_o, and the identification

$$F \to \gamma_F(\xi) = \prod_{x\in F}\xi(x) \ , \quad \xi\in\Xi \ , \ F\in\Xi_o$$

between the elements of Ξ_o and the characters γ of the Abelian group Ξ. It must be mentioned that there is a strong relation between the characters of Ξ and its characteristic function, χ, that is,

$$\chi(F) = E_\mu[\gamma_F(\xi)] = \int_\Xi \gamma_F(\xi)\mu(d\xi) \ ,\xi\in\Xi, F\in\Xi_o$$

μ being a probability measure on Ξ. Moreover, the action of two different generators on the system of characters $\{\gamma_F(\cdot)\}$ is exactly the same so that the duality relationship (2.2) can be written as

$$E^{P_\xi}[\gamma_F(\xi_t)] = E^{\hat{P}_F}[\gamma_{\hat{\xi}_t}(\xi)] \ ,\xi\in\Xi \ ,F\in\Xi_o, t\geq 0 \ . \tag{3.5}$$

Recalling the verbal description above about the BPI behaviour, we have to define the branching intensity of a particle at x by

$$q(x) = v_x\sum_F p(x,F) \ , \tag{3.6}$$

where $v=\{v_x, x\in S\}\subset(0,\infty)$, $\sum p(x,F)\leq 1$ and additional assumption

$$\sup v_x[\sum p(x,F)(|F|+1)] = C < \infty \ .$$

Further,

$$\frac{v_x}{q(x)}\, p(x,G) \tag{3.7}$$

will define the probability of occupation of a neighbouring set G. Also, following competition, the survival probabilities at site $y\in G\cap(F\backslash x)$ will be

$$r_1 = \frac{2a_y}{1+a_y}\ , \text{ for a particle of type 1}$$
$$r_o = \frac{1-a_y}{1+a_y}\ , \text{ for a particle of type 0} \tag{3.8}$$

where $a=\{a_x, x\in S\}\subset[0,1]$.

The speed functions of interest,

$$c(x,\xi) = v_x[(1-\xi(x))+(2\xi(x)-1)\sum_F p(x,F)\gamma_F(\xi)], \tag{3.9}$$

are the speed functions of a proximity process. Then, using a similar computation,

$$A\gamma_G(\xi) = \sum_{x\in G} v_x \sum_F p(x,F)[\gamma_{(G\setminus x)\cup F}(\xi)-\gamma_G(\xi)]. \tag{3.10}$$

For each $F\in\Xi_o$ there is exactly only one solution P_F to the martingale problem for A starting from state(configuration)F. Then for all $\alpha>0$

$$U^\alpha f = \int_0^\infty e^{-\alpha t} E^{P_F}[\ f(\xi_t)]dt\ ,\ F\in\Xi_o$$

(Lemma 7.2, Holley and Stroock, 1976). As it is known, each strongly continuous resolvent of positive linear operators defined in a Banach space is connected with a unique strongly continuous semigroup of positive linear operators (see, e.g., Th.X.13, Meyer, 1966). The constructed BPI is a Hunt process on the state space $\tilde{\Xi}_o$ governed by a Markov family $(\hat{P}_F)$, $F\in\tilde{\Xi}_o$, whose Q-matrix $\hat{A}$ is derived from functions $c(x,\xi)$ as above. The BPI is assumed to be minimal.

4. Final remarks

It is already known that if a branching process has a complex state space, one needs more information in order to specify it, e.g., the branching kernel, the nonbranching process, the lifetime distribution, etc. The reader is referred to the fundamental papers by N. Ikeda, M. Nagasawa and S. Watanabe (1968-1969) for a discussion of these ideas.

I will deal in this section with two questions concerning the branching semigroup:

(A) Are there conditions for a semigroup $(T_t)_{t\geq 0}$ to experience both the branching and the interaction property ?

(B) Are there conditions for joining together a branching semigroup of operators with a spatial motion/interaction semigroup ?

Both questions serve to prepare the construction of a BPI or, at least, its first approximation. The answer to the first question is : A branching semigroup satisfies the interaction property iff it is <u>degenerated</u> (Takahashi, 1971). Let S be the LCCB Hausdorff space and denote $S^{(n)}$, the n-fold direct product of S with product topology, and

$S^{*}=\bigcup_{n\geq 1} S^{(n)}$; $S^{(n)}$ is compact and S^{*} is locally compact.

$S_{(n)}$, the n-fold symmetric product of S, and $S_{*}=\bigcup_{n\geq 1} S_{(n)}$.

R_n, the restriction map from $C(S^{*})$ onto $C(S^{(n)})$.

I_n, the identity operator from $C(S^{(n)})$ into $C(S^{*})$, i.e., $I_n f=f$ on $S^{(n)}$ for $f\in C(S^{(n)})$.

Y_n, the symmetrizing operator on $C(S^{(n)})$, i.e.,

$$Y_n f(x_1,\ldots,x_n)=\frac{1}{n!}\sum_{\alpha} f(x_{\alpha(1)},\ldots,x_{\alpha(n)}),$$

where the summation is taken over all permutations α of $\{1,\ldots,n\}$, $x_i\in C(S^{(n)})$, $1\leq i\leq n$.

M, the multiplicative operator from $C(S)$ into $C(S^{*})$, i.e.,

$$Mf(x_1,\ldots,x_n)=f(x_1)\ldots f(x_n) \text{ on } S_{(n)},\ n\geq 1.$$

When the specification is not stringent, one writes $\mathbb{C}$ for spaces $C(S_{*})$, $C(S^{*})$, $C_o(S_{*})$. Note that $C(S_{*})$ is endowed with the structure of commutative algebra (without unit) with multiplication $*$ defined as follows:

$$\phi*\psi=\sum I_n Y_n[\sum_{i+j=n}(R_i\phi)\otimes(R_j\psi)],\quad \phi,\psi\in C(S_{*})$$

where $\otimes$ denotes the tensor product.

A semigroup $(T_t)_{t\geq 0}$ of linear operators on C is called a <u>branching</u> semigroup if it satisfies the property

$$(1)\quad T_t Mf = MR_1 T_t Mf\ ,\ t\geq 0,\ Mf\in\mathbb{C},\ f\in C(S).$$

It also satisfies the <u>interaction</u> property

$$(2)\quad T_t(\phi*\psi) = T_t\phi * T_t\psi,\ t\geq 0,\ \phi,\psi,(\phi*\psi)\in\mathbb{C}$$

iff it is <u>degenerated</u>, that is, if

$$(3)\quad T_t[(I_1f_1)*\ldots*(I_1f_n)] = I_n\,[(U_tf_1)*\ldots*(U_tf_n)],\ t\geq 0$$

where $(U_t)_{t\geq 0}$ is a semigroup of linear operators on $C(S)$ and $f_i\in C(S)$, $1\leq i\leq n$.

It is proved (Takahashi, 1971, Th.4.1) that if the dual semigroup $(\hat{T}_t)$ is a branching Markov semigroup, then the original process ξ_t is a Hunt process with interaction. Conversely, if $\hat{\xi}_t$ is a branching process, then (T_t) is an interaction Markov semigroup. If $(\hat{T}_t)$ is degenerated (Condition 3 above), then $\hat{\xi}_t$ is equivalent to the n independent copies of a branching process with state space S on each $S_{(n)}$.

The answer to the second question leads to the construction of a class of measure-valued processes. A first approach has been suggested by N.Ikeda et al. (1969) who considered that a Markov branching process ξ_t has a nonbranching subprocess X_t^o. The latter can take different forms so that by joining the branching part with a specified nonbranching part one obtains a special branching process which can be again a Markov process (under some conditions imposed to its multiplicative functional) or no more Markovian.

By using a nonlinear version of Trotter's formula(1959)for the product of semigroups of operators,D.Dawson(1975,Th.6.3)introduced

$$U_t = \lim_{u\to\infty}(T_{t/u}S_{t/u})^u \ , \ t\geq 0$$

as the semigroup of nonlinear operators on $C_k(\mathbb{R}^d)$,which defines a special Markov process living on the family of translation-invariant measures $\mathcal{M}_I(\mathbb{R}^d)$. In his approach,D.Dawson assumes that (T_t) is the semigroup of a multiplicative Brownian process with state space $\mathcal{M}(\mathbb{R}^d)$ and that (S_t) is the semigroup associated with a conservative Markov process living in $\mathbb{R}^d$(e.g.,a spatial motion process). Moreover,$(S_{t/u})$ may be a symmetric stable process in $\mathbb{R}^d$ with index α,$0<\alpha\leq 2$.

The process ξ_t corresponding to (U_t) is called "the critical multiplicative measure diffusion process" (Dawson,1977) and is uniquely determined by the characteristic functional of the initial distribution ξ_o and by the Laplace transition functional

$$L_{t,\xi_o}(f) = \exp\{i\int U_t f(x)\xi_o(dx)\},\ \xi_o\in\mathcal{M}_I(\mathbb{R}^d),f\in C_k(\mathbb{R}^d).$$

Because $L_{t,\xi_o}(f)$ is also the characteristic functional of an infinitely divisible random measure(for each t>0),the above processes are related to the multiplicative Markov point processes(e.g.,branching random fields:Dawson and Ivanoff,1978;Ivanoff,1980).

References

Athreya,K.B.,Kaplan,N.(1978) The additive property and its applications in branching processes. In:Branching Processes(A.Joffe,P.Ney eds.), pp.27-60. New York:M.Dekker

Bard,J.,Lauder,I.(1974) How well does Turing's theory of morphogenesis work ? J.Theor.Biol.45,501-531

Blumenthal,R.M.,Getoor,R.K.(1968) Markov Processes and Potential Theory. New York:Academic Press

Bramson,M.,Griffeath,D.(1980-1981) On the Williams-Bjerknes tumour growth model. I.Ann.Probability 9,173-185; II.Math.Proc.Camb.Phil.Soc.,88, 339-357

Bramson,M.,Griffeath,D.(1980) The asymptotic behavior of a probabilistic model for tumor growth. Lecture Notes in Biomath.,Vol.38,pp. 165-172. Berlin-Heidelberg-New York:Springer

Bryant,S.V.(1982) Introduction to the Symposium:Principles and problems of pattern formation in animals. Amer.Zool.22,3-5

Clifford,P.,Sudbury,A.(1973) A model of spatial conflict. Biometrika, 60,581-588

Cocozza,C.,Kipnis,C.(1980) Processus de vie et de mort sur ℝ avec interaction selon les particles les plus proches. Z.Wahrscheinlichkeitstheorie verw.Gebiete 51,123-132

Crick,F.H.C.,Lawrence,P.A.(1975) Compartments and polyclones in insect development. Science 189,340-347

Dawson,D.A.(1975) Stochastic evolution equations and related measure processes. J.Multivar.Anal.5,1-52

Dawson,D.A.(1977) The critical measure diffusion process. Z.Wahrscheinlichkeitstheorie verw.Gebiete 40,125-145

Dawson,D.A.,Ivanoff,G.(1978) Branching diffusions and random measures. In:Branching Processes(A.Joffe,P.Ney eds.),pp.61-103. New York: M.Dekker

DeLisi,C.P.,Bell,G.I.(1974) The kinetics of hemolytic plaque formation. Proc.Natl.Acad.Sci.USA 71,16-20

Donnelly,P.(1983) The transient behaviour of some models in population genetics. In press,Math.Proc.Camb.Phil.Soc.

Durrett,R.,Griffeath,D.(1982) Contact processes in several dimensions. Z.Wahrscheinlichkeitstheorie verw.Gebiete 59,535-552

Durrett,R.,Liggett,T.M.(1981) The shape of the limit set in Richardson's growth model. Ann.Probability 9,186-193

Dynkin,E.B.(1978) On duality for Markov processes. In:Stochastic Analysis(A.Friedman,M.Pinsky eds.),pp.63-77. New York:Academic Press

Dynkin,E.B.(1984) Gaussian and non-Gaussian random fields associated with Markov processes. J.Funct.Anal.,55,344-376

Eden,M.(1958) A probabilistic model for morphogenesis. In:Symposium on Information Theory in Biology(H.P.Yockey ed.),pp.359-370. New York: Pergamon Press

Eden,M.(1961) A two-dimensional growth process. Proc.4th Berkeley Symp. Math.Statist.Probability,Vol.IV,pp.223-239. Berkeley:Univ.California Press

Garcia,A.M.A.,Meyer,P.A.(1973) Une théorie de la dualité à ensemble polaire près. Ann.Probability 1,207-222

Getoor,R.K.,Sharpe,M.J.(1981) Two results on dual excursions. In:Seminar on Stochastic Processes,1981(E.Çinlar,K.L.Chung,R.K.Getoor eds.), pp.31-52. Boston:Birkhäuser

Gierer,A.(1981) Generation of biological patterns and form:Some physical, mathematical,and logical aspects. Prog.Biophys.Molec.Biol.37,1-47

Griffeath,D.(1981) The basic contact process. Stoch.Process.Appl.,11, 151-185

Griffeath,D.,Liggett,T.M.(1982) Critical phenomena for Spitzer's reversible nearest particle systems. Ann.Probability 10,881-895

Hammersley,J.M.(1974) Postulates for subadditive processes. Ann.Probability 2,652-680

Harris,T.E.(1976) On a class of set-valued Markov processes. Ann.Probability 4,175-194

Harris,T.E.(1978) Additive set-valued Markov processes and graphical methods. Ann.Probability 6,355-378

Helms,L.L.(1967,1970) Markov processes with creation of mass.I,II. Z.Wahrscheinlichkeitstheorie verw.Gebiete 7,225-234;15,208-218

Hincliffe,J.R.(1981) Cell death in embryogenesis. In:Cell Death in Biology and Pathology(J.D.Bowen,R.A.Lockshin eds.),pp.35-78. London: Chapman and Hall

Holley,R.A.,Liggett,T.M.(1975) Ergodic theorems for weakly interacting infinite systems and the voter model. Ann.Probability 3,643-663

Holley,R.,Liggett,T.M.(1981) Generalized potlatch and smoothing processes. Z.Wahrscheinlichkeitstheorie verw.Gebiete 55,165-195

Holley,R.A.,Stroock,D.W.(1976) A martingale approach to infinite systems of interacting processes. Ann.Probability 4,195-228

Holley,R.A.,Stroock,D.W.(1979) Dual processes and their application to infinite interacting systems. Advances in Math.,32,149-174

Hunt,G.A.(1957,1958) Markoff processes and potentials.I-III. Illinois J.Math.,1,44-93;316-369;2,151-213

Ikeda,N.,Nagasawa,M.,Watanabe,S.(1968,1969) Branching Markov processes. I-III. J.Math.Kyoto Univ.,8,233-278;365-410;9,95-160

Ivanoff,G.(1980) The branching random field. Advances Appl.Probability, 12,825-847

Kac,M.,Ward,J.C.(1952) A combinatorial solution of the two-dimensional Ising model. Phys.Rev.,88,1332-1337

Kesten,H.(1973) Discussion on J.F.C.Kingman's paper on subadditive ergodic theory. Ann.Probability 1,9o3

Kindermann,R.,Snell,J.L.(1980) Markov Random Fields and Their Applications. Providence:Amer.Math.Soc.

Koch,A.L.(1964) The growth of viral plaques during the enlargement phase. J.Theor.Biol.,6,413-431

Koch,A.S.,Fehér,G.,Lukovits,I.(1979) A simple model of dynamic receptor pattern generation. Biol.Cybernetics 32,125-138

Kurtz,T.G.(1980) Representations of Markov processes as multiparameter time changes. Ann.Probability 8,682-715

Leviatan,T.(1973) On Markov processes with random starting time. Ann. Probability 1,223-230

Lewis,J.(1981) Simpler rules for epimorphic regeneration:The polar-coordinate model without polar coordinates. J.Theor.Biol.,88,371-392

Liggett,T.M.(1972) Existence theorems for infinite particle systems. Trans.Amer.Math.Soc.,165,471-481

Liggett,T.M.(1977) The stochastic evolution of infinite systems of interacting particles. Lecture Notes in Math.,Vol.598,pp.187-248.

Liggett,T.M.(1980) Interacting Markov processes. Lecture Notes in Biomath.,Vol.38,pp.145-156. Berlin-Heidelberg-New York:Springer

Maden,M.(1981) Morphallaxis in an epimorphic system:size,growth control and pattern formation during amphibian limb regeneration. J.Embryol.Exp.Morph.,65(Suppl.),151-167

McMahon,D.,West,C.(1976) Transduction of positional information during development. In:The Cell Surface in Animal Embryogenesis and Development(G.Poste,G.L.Nicholson eds.),pp.449-493. Amsterdam:North Holland

Meyer,P.A.(1962) Fonctionnelles multiplicatives et additives de Markov. Ann.Inst.Fourier 12,125-230

Meyer,P.A.(1966) Probability and Potentials. Waltham:Blaisdell

Mitro,J.B.(1979) Dual Markov processes:Construction of a useful auxiliary process. Z.Wahrscheinlichkeitstheorie verw.Gebiete 47,139-156

Morgan,R.W.,Welsh,D.J.A.(1965) A two-dimensional Poisson growth process. J.Roy.Statist.Soc.Ser.B,27,497-504

Nagasawa,M.(1969) Markov processes with creation and annihilation. Z. Wahrscheinlichkeitstheorie verw.Gebiete 14,49-60

Nesbitt,M.N.(1974) Chimaeras vs.X-inactivation mosaics:Significance of difference in pigment distribution. Devel.Biol.,38,202-207

Othmer,H.G.,Pate,E.(1980) Scale-invariance in reaction-diffusion models of spatial pattern formation. Proc.Natl.Acad.Sci.USA,77,4180-4184

Pereira-Smith,O.M.,Smith,J.R.(1983) Evidence for the recessive nature of cellular immortality. Science 221,964-966

Ransom,R.(1977) Computer analysis of cell division in Drosophila imaginal discs:Model revision and extension to simulate leg disc growth. J.Theor.Biol.,66,361-377

Read,R.C.(1962) Contributions to the cell-growth problem. Canad.J.Math., 14,1-20

Richardson,D.(1973) Random growth in a tesselation. Proc.Camb.Phil.Soc., 74,515-528

Ringertz,N.R.,Savage,R.E.(1976) Cell Hybrids. New York:Academic Press

Rogers,G.,Goel,N.S.(1978) Computer simulation of cellular movements: Cell-sorting,cellular migration through a mass of cells and contact inhibition. J.Theor.Biol.,71,141-166

Runyan,R.B.,Markwald,R.R.(1983) Invasion of mesenchyme into three-dimensional collagen gels:A regional and temporal analysis of interaction in embryonic heart tissue. Devel.Biol.,95,108-114

Schrandt,R.G.,Ulam,S.M.(1970) On recursively defined geometrical objects and patterns of growth. In:Essays on Cellular Automata(A.W.Burks ed.),pp.232-243. Urbana:Univ.Illinois Press

Schürger,K.(1979) On the asymptotic geometrical behaviour of a class of contact interaction processes with a monotone infection rate. Z.Wahrscheinlichkeitstheorie verw.Gebiete 48,35-48

Schürger,K.(1980) On a class of branching processes on a lattice with interactions. Lecture Notes in Biomath.,Vol.38,pp.157-164. Berlin--Heidelberg-New York:Springer

Schürger,K.(1981) On a class of branching processes on a lattice with interactions. Advances Appl.Probability 13,14-39

Schürger,K.,Tautu,P.(1976) A Markovian configuration model for carcinogenesis. Lecture Notes in Biomath.,Vol.11,pp.92-108. Berlin-Heidelberg-New York:Springer

Schürger,K.,Tautu,P.(1977) A spatial stochastic model for carcinogenesis:A Markov configuration model. Invited paper,10th European Meeting of Statisticians (Manuscript)

Schwartz,D.L.(1977) Applications of duality to a class of Markov processes. Ann.Probability 5,522-532

Schwöbel,W.,Geidel,H.,Lorenz,R.J.(1966) Ein Modell der Plaquebildung. Z.Naturforsch.,21,953-959

Shur,M.G.(1977) On dual Markov processes. Theor.Probability Appl.,22, 257-270

Silverstein,M.L.(1968) Markov processes with creation of particles. Z. Wahrscheinlichkeitstheorie verw.Gebiete 9,235-257

Simpson,P.(1979) Parameters of cell competition in the compartments of the wing disc of Drosophila. Devel.Biol.,69,182-193

Simpson,P.(1981) Growth and cell competition in Drosophila. J.Embryol. Exp.Morph.,65(Suppl.),77-88

Slack,J.M.W.(1980) A serial threshold theory of regeneration. J.Theor. Biol.,82,105-140

Smythe,R.T.,Walsh,J.B.(1973) The existence of dual processes. Invent. Math.,19,113-148

Spitzer,F.(1970) Interaction of Markov processes. Advances in Math.,5, 246-290

Spitzer,F.(1971) Random Fields and Interacting Particle Systems. MAA Summer Seminar. Math.Ass.America

Stoica,L.(1983) On the construction of Hunt processes from resolvents. Z.Wahrscheinlichkeitstheorie verw.Gebiete 64,167-179

Takahashi,Y.(1971) Markov semigroups with simplest interaction.I,II. Proc.Japan Acad.,47(Suppl.II),974-978;1019-1024

Tautu,P.(1978) Blackening a d-dimensional lattice. Rev.Roumaine Math. Pures et Appl.,23,141-152

Trotter,H.(1959) On the product of semi-groups of operators. Proc.Amer. Math.Soc.,10,545-551

Turing,A.M.(1952) The chemical basis of morphogenesis. Phil.Trans.Roy. Soc.,Ser.B,237,37-72

Ulam,S.M.(1962) On some mathematical problems connected with patterns of growth of figures. Proc.Symp.Appl.Math.,14,215-224

Walsh,J.B.(1972) Transition functions of Markov processes. Sémin.de Probabilités VI,pp.215-232. Berlin-Heidelberg-New York:Springer

Weir,M.P.,Lo,C.W.(1984)Gap-junctional communication compartments in the Drosophila wing imaginal disk. Devel.Biol.,102,130-146

West,J.D.(1978) Clonal growth versus cell mingling. In:Genetic Mosaics and Chimeras in Mammals(L.B.Russell ed.),pp.435-444. New York: Plenum Press

Whitten,W.K.(1978) Combinatorial and computer analysis of random mosaics. In:Genetic Mosaics and Chimeras in Mammals(L.B.Russell ed.) pp.445-463. New York:Plenum Press

Williams,T.,Bjerknes,R.(1972) Stochastic model for abnormal clone spread through epithelial basal layer. Nature 236,19-21

Witten,T.A.,Sander,L.M.(1983) Diffusion-limited aggregation. Phys.Rev. B,27,5686-5697

Wolpert,L.(1969) Positional information and the spatial pattern of cellular differentiation. J.Theor.Biol.,25,1-47

Wolpert,L.(1971) Positional information and pattern formation. Current Topics Devel.Biol.(A.A.Moscona,A.Monroy eds.),Vol.6,pp.183-224. New York:Academic Press

Wolpert,L.(1977) Positional information and morphogenetic signals:An introduction. In:Cell Interaction in Differentiation(M.Karkinen--Jääskeläinen,L.Saxén,L.Weiss eds.),pp.79-87. London:Academic Press

Wolpert,L.(1978) Gap junctions:Channels for communication in development. In:Intercellular Junctions and Synapses(J.Feldman,N.B.Gilula, J.D.Pitts eds.),pp.83-96. London:Chapman & Hall

Wolpert,L.,Gingell,D.(1969) The cell membrane and contact control. In: Homeostatic Regulators(G.Wolstenholme,J.Knight eds.),pp.241-259 London:Churchill

PERIODIC GROWTH PHENOMENA IN SPATIALLY ORGANIZED MICROBIAL SYSTEMS

Julian W.T. Wimpenny, Steve Jaffe and J. Philip Coombs
Department of Microbiology
University College
Newport Road
Cardiff CF2 1TA, Wales, UK

Modern microbiology has as one of its sacred cows the pursuit of the homogeneity paradigm. This, roughly speaking, implies that respectable research uses well mixed liquid cultures preferably operating under steady state conditions. Fortunately there are signs that this attitude is beginning to change, for natural ecosystems are seldom homogeneous and operate in the short term at far from steady state conditions. Life for the "heterogeneous" microbiologist is infinitely richer and more satisfying once one starts to appreciate the regularity and order that is possible in a spatially ordered three dimensional environment.

Pattern formation is seen in a number of different circumstances. The aggregation of slime molds (Newell, 1983) and the structure of the bacterial colony (Wimpenny *et al*, 1983 for example) constitute relatively complex three dimensional patterns. There is a family of periodic structures associated with cell motility. Thus the swarming cultures of *Proteus* species form a series of concentric rings on the surface of agar petri dishes (Smith, 1972; Williams & Schwarzhoff 1978, Rowbury *et al* 1983). There have been a number of accounts of growth bands appearing in liquid cultures of motile bacteria, following the work of Adler (1966). Concentric rings can also be observed in cultures of certain fungi growing on an agar surface.

The second main class of periodic growth structures is seen in cultures which are substantially immobilised. The classic way to immobilise bacteria yet to allow their continued growth is to incorporate them into semi-solid agar (so called "shake cultures"). This technique has been widely employed for determining viable counts but has also been used to investigate oxygen sensitivity (Pringsheim, 1910; Spray, 1966; Whittenbury, 1963; de Vries & Stouthamer, 1969).

In an interesting series of papers Williams (1938a,b 1939a,b) investigated the growth of a range of bacterial species exposed to different partial pressures of oxygen and carbon dioxide in such gel-stabilised systems. He noted the appearance of multiple growth bands and described them as 'bacterial growth spectra'. Sadly this was a period when interest in the homogeneity paradigm was growing explosively. Williams gave a paper at a meeting of the American Society of Pathologists where one M.C. Sevag of Philadelphia chastised him thus: "I think it would be more advantageous to deviate from antiquated procedures and study the mechanisms of simple systems of the physiological activities of organisms." effectively closing the door for about forty years on what we now see as one of the more interesting manifestations of microbial behaviour. During this interval there were sporadic reports describing band formation. Tschapek & Giambiagi (1954) noted them this time in shake cultures of Azotobacter vinelandii under conditions when it was fixing nitrogen, a process sensitive to oxygen poisoning. Nitsch & Kutzner (1973) showed examples of band formation in shake cultures of various streptomycetes, especially when nutrients were restricted (Figure 1).

Figure 1. Band formation in nutrient restricted gels containing streptomycetes (Nitsch & Kutzner, 1973).

Particularly interesting were the observations of Hoppensteadt and Jager (1979). They described a system where a salmonella requiring the amino acid histidine, was grown as a lawn on the surface of an agar petri dish containing glucose and buffered salts. A drop of histidine solution was placed at the centre of the dish and the plate incubated. Growth took place in this system as a series of concentric rings. Hoppensteadt and Jager developed a mathematical model of this system. This will be considered in more detail later.

We noticed periodic behaviour whilst developing methods to investigate the growth of bacteria in spatially heterogeneous environments. One of our laboratory model systems used agar to stabilise growth media to determine the spatial organisation of cells and solutes in the system. The addition of the gelling agent prevents solute transport by any method other than molecular diffusion. Each 'microcosm' was established in a 250 cm3 glass beaker. A source layer containing 1.0% w/v agar, the basic growth medium and a diffusible carbon and energy source (usually glucose) was allowed to set in its base. Above this was poured an upper layer containing in the same volume the same basal medium, 0.4% w/v agar and an initially homogeneous inoculum of bacteria. Whittaker and Wimpenny (1979) first reported multiple band formation in these systems. A description of the method and some of the conditions associated with band formation was published by Wimpenny *et al* (1981). Many species demonstrated band formation which seemed to be associated more with facultatively anaerobic to microaerophilic physiologies than with the more strictly aerobic way of life although some aerobes generated a small numbers of bands near the surface. All species examined with the exception of *Lactobacillus confusus* produced surface growth as well as subsurface bands. Of all the organisms used, *Bacillus cereus* was most prolific in band production showing as many as eight discrete bands under appropriate circumstances. For this reason it was decided to continue work with this organism to try to explain the mechanism of band production.

The growth medium used contained casamino acids, yeast extract and mineral salts whilst glucose was the diffusing carbon source in all experiments. Except in a few experiments all the systems were incubated aerobically. The time course of band production was monitored using a gel scanner. The earliest band appears near the centre of the gel. After a period second and subsequent bands formed in order above the first and almost completely separated from one another by clear zones (Figures 2 & 3).
Band position was significantly affected by a number of factors. Increasing the glucose concentration in the source layer or decreasing the basal nutrient concentration throughout the whole system both

force band formation higher in the growth layer. Incubating the gel anaerobically, even if the gels were poured in the presence of oxygen abolished band formation, as did incorporating glucose throughout the gel or allowing glucose to diffuse at 4oC before the cells were allowed to grow at 30 oC. Various types and concentrations of agar

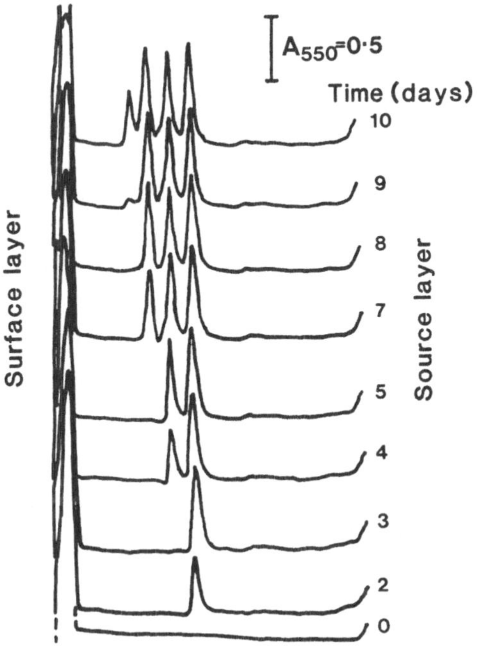

Figure 2. Periodic growth bands in 6-day gel cultures of B. cereus grown in glucose-oxygen counter gradients at 37oC.
Figure 3. Gel scans of B.cereus band formation over a period of 10 days.

were used however there was very little difference in the number or position of bands formed. This was an interesting observation and gave us evidence that cell motility was not responsible for band production since agar concentrations as high as 2% w/v did not affect band position. A number of different strains of Bacillus cereus were examined, again there were no very great changes in position or number of bands.

More information was gained by analysis of various physico-chemical factors in the gel. Needle electrodes for oxygen tension, pH and Eh were deployed using a micromanipulator. In addition gel sampling is comparitively easy though destructive to the gel, and entails removal of a gel core using a cork borer. The core is then sliced into 1 or 2 mm slices. It is possible to assay various solutes or to determine the total and actual numbers of bacteria in each slice. Steep pH gradients were noted in gel systems which showed pronounced band formation. It seemed clear that two quite distinct processes were taking place in the gel: first in the lower anaerobic parts of the growth layer cells were fermenting glucose generating acidic fermentation products which in the end took the pH down to concentrations which we ascertained in separate experiments, inhibited growth. Secondly aerobic growth by organisms in the surface layer generated alkaline products which diffused downwards into the growth layer. pH gradients from 7.7 at the surface to 5.3 in the lower parts of the growth region were sustained in some cases over a distance of about 10 mm. The junction between acid and alkali gradients was marked by a sharp inflexion in pH. By observation it was clear that growth bands appeared near the point of inflexion.

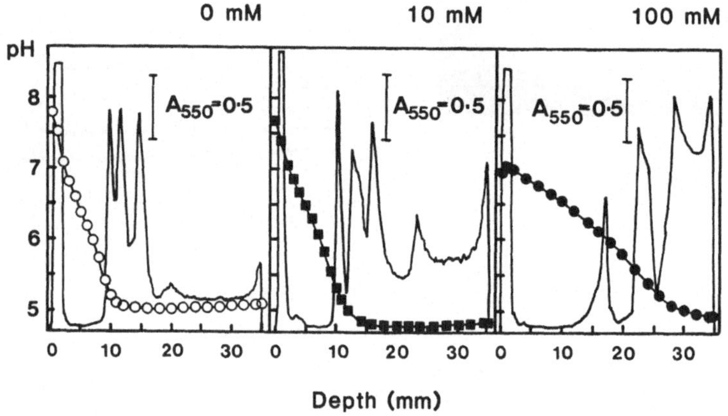

Figure 4. The effect of added buffer concentration on the pH gradient and band position in gel-stabilised cultures of B.cereus grown in glucose-oxygen counter gradients.

It seemed possible that pH changes were related to band production so that two confirmatory experiments were performed. In the first the buffer concentration in the medium was varied. pH gradients changed as a result of this and band position moved downwards as the buffer concentration increased (Figure 4). In a second experiment gels were incubated in the absence of oxygen. The control gel showed no band formation and the pH value fell to low growth inhibitory levels throughout the gel. Another gel was poured incorporating an alkaline buffer in a thin agar layer on the surface of the growth layer. Two clear bands appeared in this gel and pH gradients similar to those seen in the aerobic system developed (Figure 5).

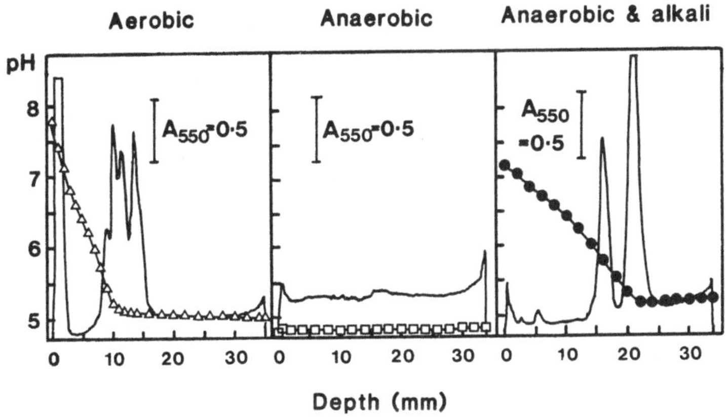

Figure 5. Band formation in gel-stabilised cultures of B.cereus. Cultures were incubated aerobically and anaerobically or anaerobically with an alkaline gel layer poured over the surface of the gel system.

An explanation for band formation on the basis of the results presented here goes something like this: as cells start to grow two metabolic processes can be discerned. In the lower parts of the gel glucose diffusing upwards from the source layer is fermented by cells growing anaerobically. Acidic fermentation products are formed which accumulate and diffuse upwards. The pH value falls in these regions to growth inhibitory levels. Meanwhile at the upper gel surface cells are growing aerobically on nitrogenous compounds present in the casamino-acid yeast-extract. The products of amino acid oxidation are generally alkaline and the pH rises here above levels in uninoculated media. These alkaline products diffuse downwards. Glucose continuing to diffuse upwards reaches a neutral zone where the down flow of alkali meets the ascending acidity. Growth now proceeds to form a distinct band. Acid products accumulate once again and inhibit further growth. Glucose diffuses through this region to another position higher up the gel where conditions are once again favourable for further band formation.

It seemed a convincing explanation of band formation at least at a superficial level. It was felt that additional insight into the whole process should be sought using the techniques of mathematical modeling.

MATHEMATICAL ASPECTS OF MICROBIAL GROWTH PATTERN FORMATION

We in common with earlier investigators (Williams, 1938a,b;1939a,b; Tschapek & Giambiagi, 1954) noticed the similarity between these microbial growth patterns and the banded precipitation patterns of reacting diffusible inorganic solutes discovered by Liesegang in 1896. Unfortunately, this hardly serves as an explanation, since, despite much investigation (cf Stern, 1967) the mechanisms for the formation of Liesegang rings remains unclear. There appear to be two leading contenders for the answer, one old and one recent. The old idea, first put forward by Ostwald (1925), is that the banding of precipitation is due to a mechanism involving supersaturation: that is, no reaction occurs until the reactants are present in excess of their equilibrium level, and the reaction proceeds very rapidly once a "supersaturation threshold" is crossed. The recent idea of Flicker & Ross (1974), is that the patterns arise from a diffusive instability, whose proximal cause is the autocatalytic nature of the reaction. We will briefly discuss the key features of these theories, remark on their similarities and differences, and describe the results of numerical solutions to models of microbial growth incorporating these ideas.

As Hedges (1932) pointed out, most of the early theories of Liesegang rings formation, even if they do not involve supersaturation, rely on a threshold or critical concentration. The important thing to note about these threshold theories is that they involve a threshold of a very special kind, which we have called <u>asymmetric</u>: no reaction occurs until a threshold is crossed, but once started, the reaction can proceed to pull the concentration of reactants down to well below the threshold. Thus it is a threshold for switching on, not for switching off.

That such a mechanism is capable of producing banded growth seems clear at a heuristic level. If there were no threshold mechanism, a narrow growth zone would proceed smoothly along the gel. For the one source system with infinitely-fast reaction in a semi-infinite gel, analytic solutions of this "moving boundary" problem can be obtained (Crank, 1975). The moving boundary solution remains valid when there is a symmetrical threshold that is determined by each reactant separately (i.e. of the form A>A*, B>B*).

Now consider what happens in the assymetric threshold model when the threshold has just been reached at some point in the gel. Call the more concentrated nutrient A, and the less concentrated one B. It is somewhat easier to picture, but not essential, to suppose that B determines the threshold, that is the threshold is reached when A exceeds some tiny level, and B exceeds a large threshold level. The situation is illustrated in Figure 6. Now at the growth zone, growth, once started, is able to deplete B entirely, and this removes B, by

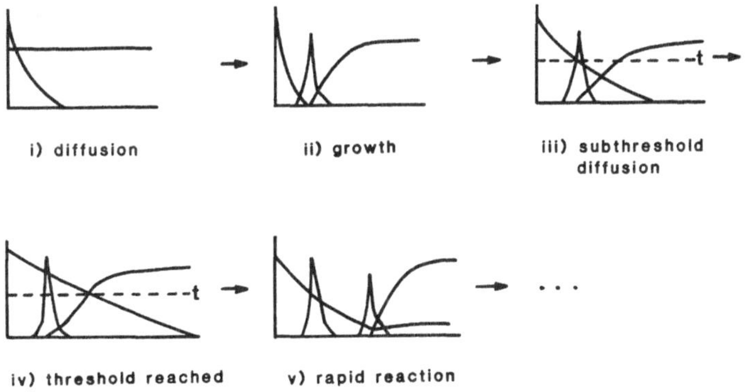

Figure 6. Asymmetric threshold model for band formation in opposing solute gradients.

diffusion, from a neighbouring region. The excess flux of A diffuses past the growth zone into a region depleted of B, and must therefore travel some distance before the threshold level of B is reached, at which point a new growth zone forms. This heuristic explanation can be made somewhat more quantitative, and semi-analytic solutions have been obtained by Wagner (1950) and by Prager (1956) under various simplifying assumptions. These results for the classical one-source system include the prediction that band spacing should increase geometrically.

An even more drastically simplified calculation can be made for our system with nutrient sources at opposite ends of the gel, so that the concentration of A is fixed at the left boundary, and B is fixed at the right. Assume that reaction, once started, is so rapid that the concentration of B at the growth zone is nearly zero, and that diffusion is rapid enough compared to the movement of the growth zone that the concentration profile of B is in a quasi-steady-state. Referring to Figure 7, if the depth of the growth zone is _d_, then the profile of B is just a straight line from B0 at the boundary to 0 at _d_. If the threshold is reached when B = Bth, and A exceeds some small value, then the next band will occur at a depth _d_x (Bth/B0), so that the spacing of bands is a decreasing geometric progression.

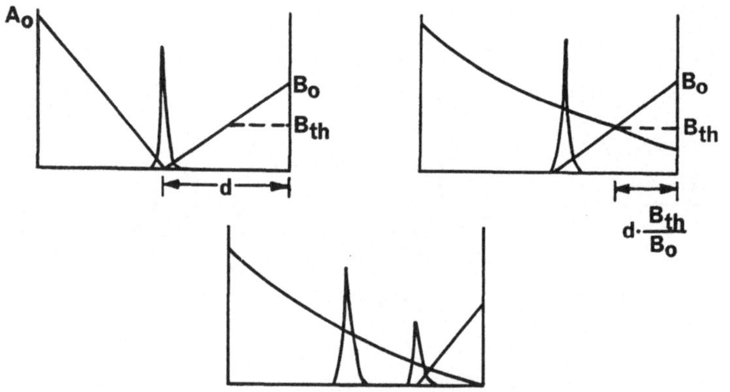

Figure 7. A simplified activation threshold model for band formation.

The asymmetric threshold can be realized in many ways, the essential point being that growth rate is not determined solely by instantaneous solute concentrations, but by the history of the reaction also. We have done numerical experiments with a model in which cells exist in two states, active and inactive. The equations are shown below:

The cell activation model is defined by the following equations :

$$\frac{\partial A}{\partial t} = D_A \Delta A - R_A \mu(A,B)\cdot C \quad \text{- - - - - - - -} \quad (1)$$

$$\frac{\partial B}{\partial t} = D_B \Delta B - R_B \mu(A,B)\cdot C \quad \text{- - - - - - - -} \quad (2)$$

$$\frac{\partial C}{\partial t} = \mu(A,B)\cdot C + \lambda(A,B)\cdot C_i \quad \text{- - - - - - -} \quad (3)$$

$$\frac{\partial C_i}{\partial t} = \lambda(A,B)\cdot C_i \quad - - - - - - - - - - - - (4)$$

Cell growth rate is given by :

$$\mu(A,B) = \mu_{max} \cdot \frac{A}{K_A + A} \cdot \frac{B}{K_B + B} - - - - (5)$$

Whilst the rate of conversion of inactive to active cells is :

$$\lambda(A,B) = \begin{cases} \lambda_{max} \frac{A}{K'_A + A} \cdot \frac{B}{K'_B + B} - \theta & \text{if} > 0 \\ 0 & \text{otherwise} \end{cases} \quad -(6)$$

A and B are the two substrates which have growth yield values R_A and R_B and diffusion coefficients D_A and D_B. C and C_i are active and inactive-cell population respectively whilst μ is the growth rate constant and λ the rate of conversion of inactive to active cells. θ is the threshold above which cells become activated. Finally K_A, K_B and K'_A, K'_B are substrate affinity constants for growth and activation respectively.

Active cells grow at a rate determined by the instantaneous substrate levels, and can grow at quite low levels of substrate. Inactive cells turn into active ones by a first-order process, but only when the substrate level exceeds some fairly high threshold. Both the growth rate-constant and the threshold are based on a product of Monod type functions, one for each substrate. The cells are initially all in the inactive state. Thus no growth occurs until the threshold is reached, then they activate fairly rapidly. Once active, cells are able to continue growing down to very low substrate levels. Numerical results show that this model is indeed capable of generating a variety of patterns of banded growth in both the one-source and the two-source systems.

Typical results are shown in Figure 8. The pattern is sensitive to changes in parameters, boundary and initial conditions in a manner consistent with experimental observations. These patterns are not produced if the same parameters are used, but there is no threshold, or if the threshold is made symmetrical, so that growth stops completely below the threshold value. Figure 9 shows the growth in the two-source system with no threshold.

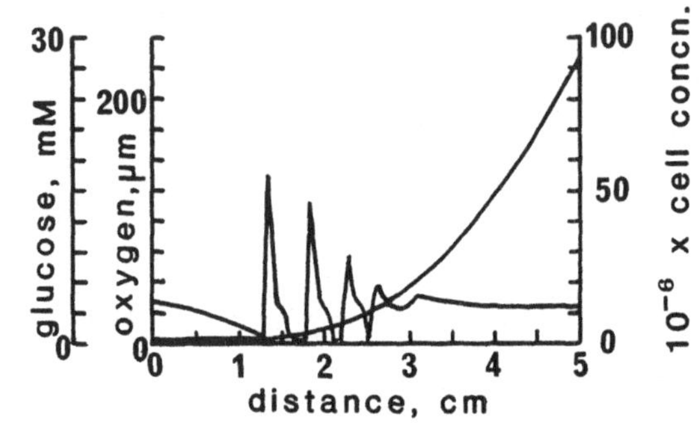

Figure 8. A numerical simulation of band formation using an asymmetric activation threshold model in a single-layer gel.

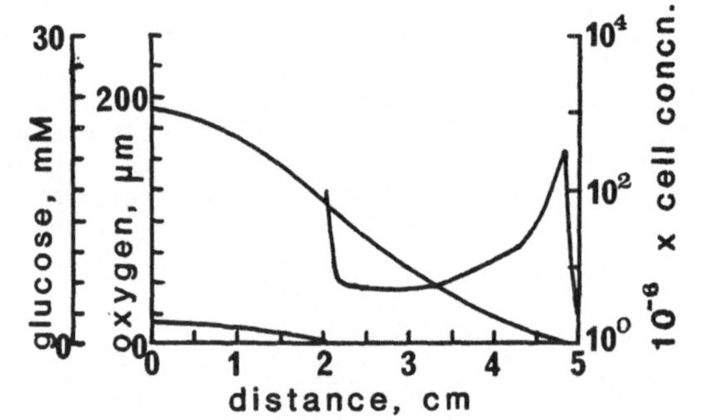

Figure 9. A numerical simulation of band formation. Two layer gel, no threshold: all other parameters as for Figure 8.

This mechanism can be modified to include a transition of cells from active to inactive at low substrate levels, and clearly this must happen to generate the initial inactive state. We have not included it in our calculations, which is equivalent to assuming that the inactivation happens only slowly and/or at such low substrate levels that it has no effect on the progressive development of the pattern. That is, we assume that each cell switches on only once, and it switches off only when it is growing so slowly that it is not affecting the pattern anyway. It is interesting to note that the Hoppensteadt:Jager model does allow for the repeated switching on and off of cell growth. It would be interesting to know to what extent this occurs in their computations.

Hoppensteadt and Jager have pointed out that what we call an asymmetric threshold can be seen as a form of hysteresis, which can be produced by a dynamical system that depends smoothly on a parameter (here substrate concentration), but which has two stable stationary states for a certain range of parameter values, and only a single stationary state otherwise. In the present case, the system would represent the cellular growth rate-constant. If the intrinsic dynamics are fast compared to changes in the substrate, then the system will always lie on the equilibrium curve, and there will be two switching points, one "on" and one "off", where the number of stationary states jumps from 3 to 1. Figure 10a shows a hypothetical equilibrium surface for specific growth rate versus substrate concentration. Thus the essence of this mechanism is fast dynamics and multiple steady states, leading to rapid switching.

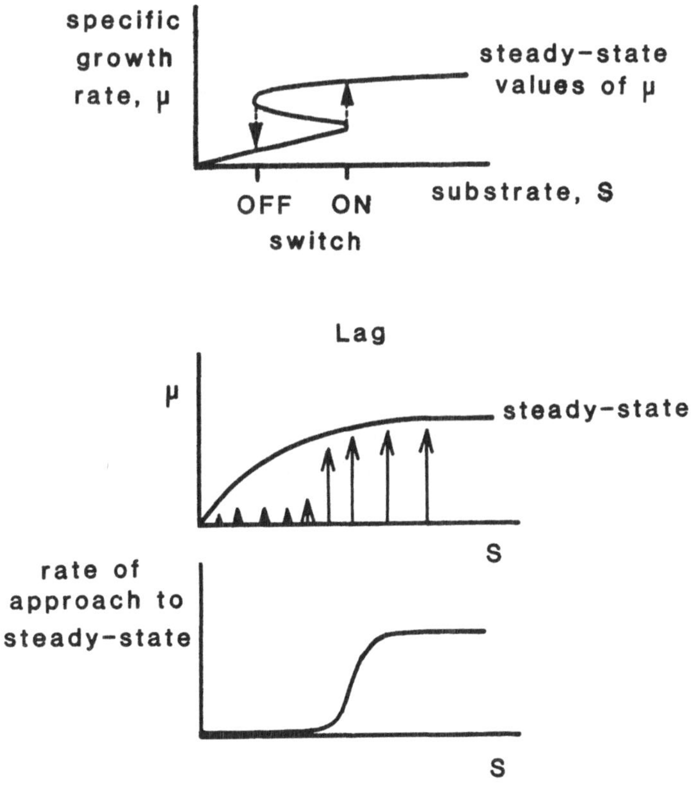

Figure 10. Two possible ways of generating asymmetric activation: A. A multiple steady state model. B. A lag model.

A very different mechanism based on time lag can produce the same asymmetric threshold behaviour. Suppose that there is a single steady state at each substrate level, but the rate of approach to this steady state, far from being rapid, also depends strongly on the substrate level. In an extreme case, there would be a threshold for the rate of approach to the steady state: at low concentrations the rate of approach would be negligeable, while above the threshold level it would be rapid. Both our cell activation model and the supersaturation model of Ostwald, fit naturally into this class, with the activation rate 0 below the threshold and finite above it. Figure 10b illustrates the dynamics of this lag model.

These two mechanisms -multiple steady states versus lags- will give the sane "hysteresis loop" in response to a cyclic variation in substrate, but they give quite different steady state behaviour.

In microbiological terms, the inactive state of our model could represent cells starved of nutrients or poisoned by toxic products, such as low pH. They would be like cells in the stationary or death phases of a batch culture. It is known that many bacteria undergo significant morphological and biochemical changes under conditions of low nutrient or high toxic agent. These changes include sporulation as an extreme example (Pirt, 1975). When such cells are then exposed to nutrient, or the toxic agents removed, growth may occur only after a lag whose duration depends strongly on both the previous and present nutrient conditions with times of the order of hours or even days (Hinshelwood, 1946). The data of Hinshelwood on lag and drug adaptation are particularly intriguing, since they show both threshold behaviour and a "protective" effect of previous drug exposure. The ability to respond rapidly to fluctuations in nutrient levels would be expected to be correlated with the natural environment to which the organism is adapted. This suggests that there may be a connection between the ability to form bands and the constancy of an organisms natural habitat.

We note that the models discussed above can be placed in the context of the general principles of <u>local activation</u> and <u>lateral inhibition</u> which have been emphasised in the investigations of Meinhardt and Gierer (1974). Growth produces a local positive feedback (by autocatalysis) as well as a dispersed negative feedback (by inhibitor production or nutrient depletion). As Meinhardt and Gierer point out, the important thing is that the positive feedback is stronger than the negative feedback (for example in some models the activator A changes at a rate proportional to A2/H where H is the inhibitor level. Thus the positive feedback is quadratic, whilst the feedback is only linear). This is of fundamental importance for the patterns we are considering, since the growth would be expected to choke itself off and be driven by the overall substrate gradients to move smoothly along the gel. The asymmetric threshold satisfies this requirement in a very strong way, by making "activated" regions relatively immune to the effects of low substrate, which inhibits nearby regions not yet activated. The same phenomenon is seen in the growth of bacteria that produce antibiotic substances against which they themselves are protected.

Recently a quite different, and altogether less obvious explanation for Liesegang rings has been proposed by Flicker & Ross (1974). They suggest that the patterns are caused by diffusional instability, a prerequisite for which is that the reaction be autocatalytic (a requirement clearly satisfied by bacterial growth). Thus whereas the Ostwald mechanism involves thresholds and hysteresis, the F-R model does not. Whereas the Ostwald mechanism is local in space and sequential in time, the F-R mechanism is global and simultaneous. This of course applies only to the growth of instability from a homogeneous initial state, and the significance of this analysis for the true nonlinear system with large gradients is uncertain. Flicker and Ross base their claims on a linear stability analysis, which produces a dispersion relation peaking at a dominant spatial frequency. The position of this peak is highly dependent on the diffusion coefficient of the product in our case cells. When this diffusion coefficient vanishes, so does the peak. Our own analysis of a simple conservative growth model gives a dispersion relation which, for no diffusion of cells, is monotonically increasing with spatial frequency, to an asymptote equal to the partial derivative of the growth rate with respect to cell concentration, that is the autocatalytic coefficient. Since bands are formed by non-motile organisms in very thick agar gels, it seems safe to discount the importance of "diffusion" of cells, and to assume that the corresponding diffusion coefficient is zero. In this case the linear stability analysis no longer predicts the emergence of a periodic pattern of a definite frequency, but there remains the possibility that pattern can arise from coupling of the linear instability with non-linear terms and boundary conditions.

Although the Flicker-Ross mechanism is global by nature, one can imagine that it might generate instabilities in a moving reaction zone, leading to the sequential production of localized patterns. We have in fact seen something like this in simulations of the growth of bacteria where the specific growth rate is given by the product of Monod growth functions. Most values of the parameters led to a smoothly moving growth front producing a monotone pattern of growth. However, a certain range of parameters led to an oscillatory pattern of growth superimposed on a uniform spread. This result was seen only in the one-source system. Figure 11 shows the evolution of bands in the one source system together with the profiles of growth rate. The final pattern of growth reflects a temporal oscillation in growth rate superimposed on a uniform movement of the growth zone. In this simulation peak growth rate oscillates whilst the position of the growth peak moves smoothly with time.

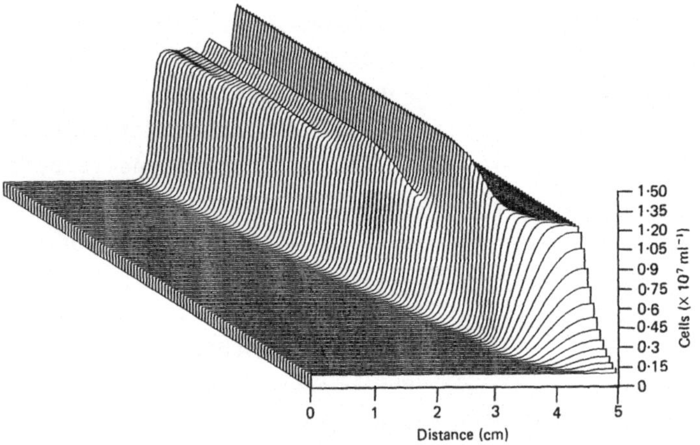

Figure 11. A numerical simulation of band formation. The "simple" model is used in a single layer gel under one specific set of conditions.

We have so far been able to produce such patterns only within a fairly narrow range of parameters. The initial concentration of the inner nutrient must be low -low enough to be depleted by growth within 2-3 hours. The level of outer nutrient should be from 1 to 10 times this (in stoichiometric units). Most important, it seems, is that the affinity constant for the inner nutrient be much higher than that for the outer nutrient. This means, as the simulations show, that the inner nutrient is generally growth-rate limiting. Periodic patterns do not form without this asymmetry in affinity constants.

There are a number of serious discrepancies between these calculations and the observed growth patterns. Both the total growth, and the peak-to-trough ratio of the growth peaks are too low; only a few bands can be formed; and bands are not formed in the two source system. It remains to be seen whether adjustments to the parameters can alleviate these problems.

It is clear from the work discussed in this paper that periodic growth phenomena are by no means uncommon in structured microbial growth systems. There are hints in the literature that they may be widely distributed in natural ecosystems, for instance aquatic sediments (Perfilev & Gabe, 1969). Clearly such structures are not only important but fascinating in their own right. A combined approach via experiment and numerical modeling should continue to yield useful results.

Acknowledgements. The authors gratefully acknowledge support from the SERC during the course of this research. They also express gratitude to the following for practical help and useful discussion: Steve Whittaker, Bob Lovitt, Hamid Abdollahi and to Professor David Hughes who first mentioned Liesegang rings to us.

REFERENCES

ADLER,J. (1966), Chemotaxis in bacteria, Science, 153, 708-716.

CRANK, J. (1975), The mathematics of diffusion, 2nd edition, Clarendon Press, Oxford.

DE VRIES,W. & STOUTHAMER,A.H. (1969), Factors determining the degree of anaerobiosis of Bifidobacterium strains, Archiv fur Mikrobiologie, 65, 275-287.

FLICKER, M. AND ROSS, J. (1974), Mechanism of chemical instability for periodic precipitation phenomena, Journal of Chemistry and Physics, 60, 3458-3465.

HEDGES, E.S. (1932), Liesegang rings and other periodic structures, Chapman and Hall, London.

HINSHELWOOD, C.N. (1946), The chemical kinetics of the bacterial cell, Clarendon Press, Oxford.

HOPPENSTEADT, F.C. AND JAGER, W. (1979), Pattern formation by bacteria, in Biological growth and spread, edited by W. Jager, H. Rost and P. Taylor, 38, 68-81, Springer-Verlag Lecture Notes in Biomathematics, Berlin,New York.

MEINHARDT, H. AND GIERER, A. (1974), Applications of a theory of biological pattern formation based on lateral inhibition, Journal of Cell Science, 15, 321-346.

NEWELL, P.C. (1983), Attraction and adhesion in the slime mold Dictyostelium, in Fungal Differentiation edited by J.E.Smith, pp43-71, Marcel Dekker, New York.

NITSCH, B. & KUTZNER, H.J. (1973), Wachstum von Streptomycetin in Schuttelagarkultur: eine neue Methode zur Feststellung des c-Quellen-Spektrums, Symposium on Technische Mikrobiologie, Berlin, 481-486.

OSTWALD, W. (1925), Kolloid Zeitung, 36, 380.

PERFIL'EV, B.V. & GABE, D.R. (1969), Capillary Methods of Investigating Microorganisms, (English translation). Oliver & Boyd, Edinburgh.

PIRT S.J. (1975), Principles of microbe and cell cultivation, Blackwell, Oxford.

PRAGER, S. (1956), Periodic precipitation, Journal of Chemistry and Physics, 25, 279-283.

PRINGSHEIM, H. (1910), Weiteres uber Verwendung von Cellulose als Energiequelle zur Assimilation des Luftstickstoffs, Centralblatt fur Bakteriologie und Parasitenkunde, Abteil II, 26, 222-227.

ROWBURY, R.J., ARMITAGE, J.P. AND KING, C. (1983), Movement, taxes and cellular interactions in the response of microorganisms to the natural environment, in Microbes in their Natural Environment, Symposium of the Society for General Microbiology, 34, 299-350.

SMITH, D.G. (1972), The Proteus swarming phenomenon, Science Progress, Oxford, 60, 487.

SPRAY, R.S. (1936), Semisolid media for the cultivation and identification of the sporulating anaerobes, Journal of Bacteriology, 32, 135-155.

STERN, K.H. (1967), Bibliography of Liesegang rings (second edition), National Bureau of Standards, Miscellaneous Publications No 292, US Government Printing Office, Washington D.C.

TSCHAPEK,M & GIAMBIAGI,N (1954), The formation of Liesegang rings by Azotobacter under oxygen inhibition. / Die Bildung von Liesegang'schen Ringen durch Azotobakter bei O2-Hemmung, Kolloid Zeitschrift, 135, 47-48.

WAGNER, C.J. (1950), Mathematical analysis of the formation of periodic precipitates, Journal of Colloid Science, 5, 85-97.

WHITTENBURY,R. (1963), The use of soft agar in the study of conditions affecting the utilization of fermentable substrates by lactic acid bacteria, Journal of General Microbiology, 32, 375-384.

WILLIAMS, F.D. & SCHWARZHOFF, R.H. (1978), Nature of the swarming phenomenon in Proteus, Annual Review of Microbiology, 32, 101-122.

WILLIAMS,J.W. (1938a), Bacterial growth "spectrum" analysis. I. Methods and application, The American Journal of Medical Technology, 4, 58-61.

WILLIAMS,J.W. (1938b), Bacterial growth "spectrums". II. Their significance in pathology and bacteriology, American Journal of Medical Technology, 14, 642-645.

WILLIAMS,J.W. (1939a), Growth of microorganisms in shake cultures under increased oxygen and carbon dioxide tensions, Growth, 3, 21-33.

WILLIAMS,J.W. (1939b), The nature of gel mediums as determined by various gas tensions and its importance in growth of microorganisms and cellular metabolism, Growth, 3, 181-196.

WIMPENNY, J.W.T., LOVITT, R.W. AND COOMBS, J.P. (1983), Laboratory model systems for the investigation of spatially and temporally organised microbial ecosystems, Symposia of the Society for General Microbiology, 34, 67-117.

WIMPENNY,J.W.T. & WHITTAKER,S. (1979), Microbial growth in gel stabilised nutrient gradients, Society for General Microbiology Quarterly, 6, 80.

Lecture Notes in Biomathematics

Vol. 1: P. Waltman, Deterministic Threshold Models in the Theory of Epidemics. V, 101 pages. 1974.

Vol. 2: Mathematical Problems in Biology, Victoria Conference 1973. Edited by P. van den Driessche. VI, 280 pages. 1974.

Vol. 3: D. Ludwig, Stochastic Population Theories. VI, 108 pages. 1974.

Vol. 4: Physics and Mathematics of the Nervous System. Edited by M. Conrad, W. Güttinger, and M. Dal Cin. XI, 584 pages. 1974.

Vol. 5: Mathematical Analysis of Decision Problems in Ecology. Proceedings 1973. Edited by A. Charnes and W.R. Lynn. VIII, 421 pages. 1975.

Vol. 6: H.T. Banks, Modeling and Control in the Biomedical Sciences. V. 114 pages. 1975.

Vol. 7: M.C. Mackey, Ion Transport through Biological Membranes, An Integrated Theoretical Approach. IX, 240 pages. 1975.

Vol. 8: C. DeLisi, Antigen Antibody Interactions. IV, 142 pages. 1976.

Vol. 9: N. Dubin, A Stochastic Model for Immunological Feedback in Carcinogenesis: Analysis and Approximations. XIII, 163 pages. 1976.

Vol. 10: J.J. Tyson, The Belousov-Zhabotinskii Reaktion. IX, 128 pages. 1976.

Vol. 11: Mathematical Models in Medicine. Workshop 1976. Edited by J. Berger, W. Bühler, R. Repges, and P. Tautu. XII, 281 pages. 1976.

Vol. 12: A.V. Holden, Models of the Stochastic Activity of Neurones. VII, 368 pages. 1976.

Vol. 13: Mathematical Models in Biological Discovery. Edited by D.L. Solomon and C. Walter. VI, 240 pages. 1977.

Vol. 14: L.M. Ricciardi, Diffusion Processes and Related Topics in Biology. VI, 200 pages. 1977.

Vol. 15: Th. Nagylaki, Selection in One- and Two-Locus Systems. VIII, 208 pages. 1977.

Vol. 16: G. Sampath, S.K. Srinivasan, Stochastic Models for Spike Trains of Single Neurons. VIII, 188 pages. 1977.

Vol. 17: T. Maruyama, Stochastic Problems in Population Genetics. VIII, 245 pages. 1977.

Vol. 18: Mathematics and the Life Sciences. Proceedings 1975. Edited by D.E. Matthews. VII, 385 pages. 1977.

Vol. 19: Measuring Selection in Natural Populations. Edited by F.B. Christiansen and T.M. Fenchel. XXXI, 564 pages. 1977.

Vol. 20: J.M. Cushing, Integrodifferential Equations and Delay Models in Population Dynamics. VI, 196 pages. 1977.

Vol. 21: Theoretical Approaches to Complex Systems. Proceedings 1977. Edited by R. Heim and G. Palm. VI, 244 pages. 1978.

Vol. 22: F.M. Scudo and J.R. Ziegler, The Golden Age of Theoretical Ecology: 1923–1940. XII, 490 pages. 1978.

Vol. 23: Geometrical Probability and Biological Structures: Buffon's 200th Anniversary. Proceedings 1977. Edited by R.E. Miles and J. Serra. XII, 338 pages. 1978.

Vol. 24: F.L. Bookstein, The Measurement of Biological Shape and Shape Change. VIII, 191 pages. 1978.

Vol. 25: P. Yodzis, Competition for Space and the Structure of Ecological Communities. VI, 191 pages. 1978.

Vol. 26: M.B Katz, Questions of Uniqueness and Resolution in Reconstruction from Projections. IX, 175 pages. 1978.

Vol. 27: N. MacDonald, Time Lags in Biological Models. VII, 112 pages. 1978.

Vol. 28: P.C. Fife, Mathematical Aspects of Reacting and Diffusing Systems. IV, 185 pages. 1979.

Vol. 29: Kinetic Logic – A Boolean Approach to the Analysis of Complex Regulatory Systems. Proceedings, 1977. Edited by R. Thomas. XIII, 507 pages. 1979.

Vol. 30: M. Eisen, Mathematical Models in Cell Biology and Cancer Chemotherapy. IX, 431 pages. 1979.

Vol. 31: E. Akin, The Geometry of Population Genetics. IV, 205 pages. 1979.

Vol. 32: Systems Theory in Immunology. Proceedings, 1978. Edited by G. Bruni et al. XI, 273 pages. 1979.

Vol. 33: Mathematical Modelling in Biology and Ecology. Proceedings, 1979. Edited by W.M. Getz. VIII, 355 pages. 1980.

Vol. 34: R. Collins, T.J. van der Werff, Mathematical Models of the Dynamics of the Human Eye. VII, 99 pages. 1980.

Vol. 35: U. an der Heiden, Analysis of Neural Networks. X, 159 pages. 1980.

Vol. 36: A. Wörz-Busekros, Algebras in Genetics. VI, 237 pages. 1980.

Vol. 37: T. Ohta, Evolution and Variation of Multigene Families. VIII, 131 pages. 1980.

Vol. 38: Biological Growth and Spread: Mathematical Theories and Applications. Proceedings, 1979. Edited by W. Jäger, H. Rost and P. Tautu. XI, 511 pages. 1980.

Vol. 39: Vito Volterra Symposium on Mathematical Models in Biology. Proceedings, 1979. Edited by C. Barigozzi. VI, 417 pages. 1980.

Vol. 40: Renewable Resource Management. Proceedings, 1980. Edited by T. Vincent and J. Skowronski. XII, 236 pages. 1981.

Vol. 41: Modèles Mathématiques en Biologie. Proceedings, 1978. Edited by C. Chevalet and A. Micali. XIV, 219 pages. 1981.

Vol. 42: G.W. Swan, Optimization of Human Cancer Radiotherapy. VIII, 282 pages. 1981.

Vol. 43: Mathematical Modeling on the Hearing Process. Proceedings, 1980. Edited by M.H. Holmes and L.A. Rubenfeld. V, 104 pages. 1981.

Vol. 44: Recognition of Pattern and Form. Proceedings, 1979. Edited by D.G. Albrecht. III, 226 pages. 1982.

Vol. 45: Competition and Cooperation in Neutral Nets. Proceedings, 1982. Edited by S. Amari and M.A. Arbib. XIV, 441 pages. 1982.

Vol. 46: E. Walter, Identifiability of State Space Models with applications to transformation systems. VIII, 202 pages. 1982.

Vol. 47: E. Frehland, Stochastic Transport Processes in Discrete Biological Systems. VIII, 169 pages. 1982.

Vol. 48: Tracer Kinetics and Physiologic Modeling. Proceedings, 1983. Edited by R.M. Lambrecht and A. Rescigno. VIII, 509 pages. 1983.

Vol. 49: Rhythms in Biology and Other Fields of Application. Proceedings, 1981. Edited by M. Cosnard, J. Demongeot and A. Le Breton. VII, 400 pages. 1983.

Vol. 50: D.H. Anderson, Compartmental Modeling and Tracer Kinetics. VII, 302 pages. 1983.

Vol. 51: Oscillations in Mathematical Biology. Proceedings, 1982. Edited by J.P.E. Hodgson. VI, 196 pages. 1983.

Vol. 52: Population Biology. Proceedings, 1982. Edited by H.I. Freedman and C. Strobeck. XVII, 440 pages. 1983.

Vol. 53: Evolutionary Dynamics of Genetic Diversity. Proceedings, 1983. Edited by G.S. Mani. VII, 312 pages. 1984.

Vol. 54: Mathematical Ecology. Proceedings, 1982. Edited by S.A. Levin and T.G. Hallam. XII, 513 pages. 1984.

Vol. 55: Modelling of Patterns in Space and Time. Proceedings, 1983. Edited by W. Jäger and J.D. Murray. VIII, 405 pages. 1984.